# *Recollections of a Happy Life*

*Being the Autobiography of*

MARIANNE NORTH

---

# *Recollections of a Happy Life*

*Being the Autobiography of*

MARIANNE NORTH

*Volume One*

---

*Edited with an*
*Introduction by*

SUSAN MORGAN

University Press of Virginia
Charlottesville and London

THE UNIVERSITY PRESS OF VIRGINIA


*First published 1993*

Library of Congress Cataloging-in-Publication Data
North, Marianne. 1830–1890.
Recollections of a happy life : being the autobiography of
Marianne North / edited and with an introduction by Susan Morgan.
p. cm.
Originally published: London : Macmillan. 1892; this edition: New York, 1894.
Includes bibliographical references (v. 1, p. ).
ISBN 0-8139-1469-8 (v. 1).—ISBN 0-8139-1470-1 (pbk. : v. 1)
1. North, Marianne, 1830–1890. 2. Botanical artists—England—Biography. 3. Travelers—England—Biography. 4. Women travelers—England—Biography. I. Morgan, Susan, 1943– . II. Title.
QK98.183.N67A3 1993
581'.092—dc20
[B] 93-12326
CIP

Printed in the United States of America

*For Sherry Morgan, sister and friend*

# *Contents*

# *Acknowledgments*

I AM grateful to many people for making possible this new edition: to Barry Chabot, Chair of the Department of English at Miami University, who has continually supported this project through his willingness both to discuss it and to find funds to pay for research; to W. Hardy Eshbaugh of the Department of Botany at Miami, for his welcoming graciousness in helping me along in the world of botany, both past and present; to Sylvia Fitzgerald, Chief Librarian and Archivist at the Royal Botanic Gardens, Kew, for her generous aid in making available to me the Marianne North materials at Kew; to Laura Ponsonby, Chief Education Officer at the Royal Botanic Gardens, whose extensive knowledge turned a chilly walk on a dark November day into brightness and color as she told me about Marianne North; to G. T. Prance, Director of the Royal Botanic Gardens, for his generosity in providing me with time and information; and finally to Nancy Essig, again the perfect editor in her wit, her insights, and her patience.

# *Introduction*

IN SPITE of her lack of reputation in the United States, Marianne North has been publicly enshrined in England for over a century as one of the significant artifacts of nineteenth-century British culture. She was a confirmed spinster, a woman whose life had nothing to do with those conventional female patterns—in life and in novels—of heterosexual romance and domesticity. As North herself commented, "I prefer vegetables."[1] North spent five decades traveling. The last fourteen years of that time she went to all the continents outside Europe painting pictures everywhere she went. Her impartial desire to cover the territory led her to Africa because she had no paintings of plants from that continent. She set out for Australia because Charles Darwin told her it shouldn't be missed. North did not stop traveling until she was almost fifty-five and too ill to brave the tropics any more.

North's status as a Victorian cultural icon began during her own lifetime. Her fame rested primarily on her paintings, which were explicitly described—and admired—as copies, not of other paintings but of exotic natural objects in faraway places. North was a botanic artist, a plant illustrator, a woman who "put a garland about the earth."[2] She enjoyed the role and did what she could to arrange for her temporal fame to acquire a kind of permanence. In 1879, in the *Pall Mall Gazette*'s admiring review of North's private exhibition in London of her paintings, the reviewer suggested that North find a permanent place to house her collection. Whether that suggestion first gave her the idea of donating virtually all her paintings to the Royal Botanic Gardens at Kew, or whether the plan had already occurred to her, is unclear. On August 11, 1879, at Shrewsbury station, having, an admiring biographer tells us, "missed her train and, characteristically, not wishing to be idle even for an hour or two," North wrote to the director of the Gardens, Sir Joseph Hooker, to offer her paintings to Kew. She also offered to pay for building a small museum in which to house them.[3]

North had James Fergusson, a prominent architectural historian, design the building for her. But she did a great deal more than choose an architect and pay his bill. She also carefully supervised the gallery's construction and interior decoration, herself painting the interior frieze and decorations around the doors. Between 1882 and 1885 North took charge of the hanging of 832 paintings, primarily but not entirely of plants, which she had done during her travels. She had the lower walls lined with boards made of the 246 different types of wood which she had collected on her travels. And finally, she paid for 2,000 copies of a catalog compiled at her request by the distinguished and well-known Kew botanist W. Botting Hemsley. With the opening of the North Gallery at the Royal Botanic Gardens, a collection of her work arranged geographically and designated in the catalog as paintings of "plants and their homes," we could say that Marianne North brought the whole world back to England and put it on display.[4]

When the North Gallery opened to the public on July 9, 1882, Sir Joseph Hooker, as director of Kew Gardens and himself a famous botanist, wrote the first official guide. He stressed that North's paintings represent objects already disappearing in the march of progress. "Such scenes can never be renewed . . . except by means of such records as this lady has presented to us." A grateful posterity particularly would praise "her fortitude as a traveler, her talent and industry as an artist, and her liberality and public spirit."[5]

The enshrinement of North carries on into the 1990s. In 1990 Laura Ponsonby offered a fascinating narrative of North's life and artistic achievement, including superb color reproductions of many of the paintings. *Marianne North at Kew Gardens* is a publication of the Royal Botanic Gardens at Kew, on the occasion of the centennial of North's death. Dea Birkett, in a 1992 *New York Times* article on the newly restored North Gallery, pointed out that it is still "the only permanent solo exhibition by a female artist in Britain." Birkett went on to suggest that North's vibrant style, in its "strong female imagery, suggests a 19th-century Georgia O'Keeffe" and to call the gallery "our only record of a lost botanical world."[6] In the "Preface" to Ponsonby's *Marianne North at Kew Gardens,* the 1990 director of Kew offered the explicit purpose of "this beautiful book." The question he asked is, "Who was this extraordinary woman?" And the answer, Director G. T. Prance assured us, will "elevate Marianne North to her rightful position among the great plant explorers of the Victorian age."[7]

WHO MARIANNE NORTH was had everything to do with whom she came from. The North family was long distinguished among English country gentry. In a time and place where the word *breeding* meant almost the same when said of people as it did when said of horses—who sired whom; genetics rather than environment—Marianne North was exceedingly well-bred. Among other notables her ancestors included the famous seventeenth-century Roger North, a lawyer, historian, and musician. He was also the author of an autobiography and of biographies of his three distinguished brothers: Sir Francis North, successful politician who became attorney general of England; John North, distinguished Cambridge professor, Master of Trinity College, and friend of Sir Isaac Newton; and Dudley North, a successful businessman who spent years being a merchant in the Middle East and returned to become a member of Parliament.

Marianne North was born on October 24, 1830. The North family home, built two centuries before by Roger North, was Rougham in Norfolk. Marianne's father, Frederick North, also built a mansion in Hastings, where he was a member of Parliament. Moreover, Marianne's mother, who died when Marianne was twenty-four, was already a widow with a daughter when she married Frederick North. She had inherited from her first husband another impressive property, called Gawthorpe Hall, in Lancashire. In other words, Marianne's extremely privileged childhood and those of her older half sister, Janet, her older brother, Charles, and her younger sister, Catherine, were spent on three large estates, in Norfolk, Hastings, and Lancashire. The Norths moved from one to the next depending on the season.

Nor did the family limit themselves to only the three country homes. Like many other of the wealthy English gentry, they frequently spent part of the year in a town house in London. Also like many of the gentry, they traveled with impressive frequency to various countries in Europe. "Extended visits coincided with the periods when Mr. North had failed to get into Parliament."[8] What was unusual about the Norths' life in comparison with that of most of their peers is that they chose all the options: living on not one but three country estates and spending the season in London and making frequent trips to the Continent. When Marianne was almost seventeen, the Norths went on a European visit which lasted almost three years, only the first eight months of which were spent settled in one place. Clearly, the family were a meandering lot. It is easy to see where Marianne North developed her taste for keeping on the move.

Marianne North was well connected in more than being of impressive descent, wellborn, and rich. Through her family she knew, and often knew well, many important and powerful people. Her mother's father had been a member of Parliament, as was Marianne's father. Frederick North knew a great many famous people. Many of them, such as the novelist Elizabeth Gaskell and the scientist Professor John Tyndall, he met through his travels on the Continent after his wife died in 1855. But many of the Norths' extensive acquaintances and friends were originally family connections or people met through Frederick North's pursuit of his many intellectual and public interests. Frederick North, and thus his family, led an intensely social life both at home and abroad.

Upon the death of her mother, Marianne, as the elder daughter, became the mistress of the household. Frederick North lived with his two daughters in London during the winters. During the summers, when Parliament was not in session, he usually went traveling in Europe with them. All three habitually took their sketchbooks and also kept little travel journals of their days. Marianne North had been collecting and painting flowers and plants since she was a child at Rougham. In the first period after her mother's death she became even more serious about plant illustration and also developed her scientific knowledge of some of the characteristics and varieties of plants.

One of the people the Norths met while rambling in the mountains of Europe was the writer John Addington Symonds, whom Catherine North married in 1864. Catherine's marriage, apparently often rocky but which produced three fine children, was to a man whose place now in cultural history is as one of the famous bisexual writers of the Victorian age. As Catherine's daughter gracefully put it, her mother was "a woman more full perhaps of the deepest spirit of Plato than most women of her times."[9] John Addington Symonds was a poet and scholar who published many works on Italian Renaissance writers as well as many biographies and his own poems. He also expressly struggled with serious health problems and with the complexities of love both for his wife and for young men.[10]

With Catherine's departure the nine years of life shared by the sisters and their father were over. Marianne North was now thirty-four, and the North household had dwindled to two. Over the next five years Marianne and her father took many trips, including a particularly memorable one to Turkey, Egypt, and Syria, painting and collecting plants wherever they went. Then in 1869 Frederick North became ill while at home in England. This did not deter him from starting off on another long European vacation. But in October of that year he got much worse in Austria, and his daughter brought him home to die. Marianne North had just turned thirty-nine. The traveling North family had dwindled to one. Yet the one remaining North would carry on the same meandering family tradition alone, keeping a flat in London and traveling on by herself for the next fourteen years.

One reason North not only never married but also appears never to have even indicated—in personal letters or the three volumes of her *Recollections*—an interest in having a romantic relationship with a man may have been that for this North at least there was hardly anything better than being a North. Another way to put this may be to say that unlike many Victorian women, including rich ones, Marianne North had no practical incentives to marry. Not only did she inherit plenty of money, she was free from entangling family obligations, free to follow her own inclination, which was to go to interesting places and meet interesting people and paint. Moreover, as the elder daughter of a widowed father, she spent approximately fifteen years enjoying the position of being the female head of an impressive household. Since she grew up well before 1882, when the Married Women's Property Act was passed, marriage for North might well have entailed giving up a fair amount of her property to a husband who would have had the financial control of her wealth.[11] It could have been a curtailment of independence, a relinquishment of financial and personal power, and almost certainly a loss of exciting people and events.

Moreover, Marianne North actually did live for thirty-nine years with a man she respected and loved, the man of whom she said, "He was from first to last the one idol and friend of my life," (see below, p. 5) the man who said of her, "She is the main link that binds me to life."[12] Such remarks are tantalizingly provocative to our presumably unrepressed but often rather rigidly post-Freudian minds. But whatever these remarks, and others like them, "really" reveal, it is certain that Marianne North frequently expressed a sense of being deeply happy to live with her father. She recalls in the *Recollections* that her mother, while dying in the early weeks of 1855, "made me promise never to leave my father" (p. 30). It was a heavy weight to put on a young woman of twenty-four. But there is no evidence that Marianne North felt the burden. She represents her father as the one true companion who liked to live as she did, with his death, fifteen years after her mother's, leaving "me indeed alone" (p. 38). Almost ten years later she was still exorcising her sense of being alone by keeping her father present in her thoughts, writing to Dr. Arthur Burnell that "sometimes as I sit and paint, he seems to come and watch me, and the very thought of him keeps me from harm."[13]

We must be wary of assuming that North's life would have been better, that she somehow would have been more fulfilled, if she had married or not been bound for so many years to be the companion of her father. Unlike many privileged Victorian women, Marianne North actually seems to have done what she wanted with her life. Whatever we may want to make of living with her father until she was thirty-nine, she wrote of the situation as emotionally and intellectually liberating and frequently remarked on having escaped her sister's fate of wife and mother. North's written comments about herself are pervaded with a sense of how much she enjoyed her life. Edward Lear, in a fairly typical contemporary account, described North as "a great draughtswoman and botanist, and . . . altogetheraciously clever and delightful."[14]

Marianne North's later career was made possible by her being born a North of the Norths, by her family's unceasing habit of packing up every three or four months and moving on, and by the specific tastes of her father. Frederick North occupied his time in four major ways: first, with his up-and-down career as a member of Parliament for Hastings; second, with the social charms of sustaining a large number of acquaintances, friends, and family; third, with traveling, an activity which often blended with his sociable interests; and finally, with finding, growing, collecting, and painting plants, an interest which blended with his love of travel. In the broadest sense Frederick North enjoyed seeing the world. And he used his money, his position, and his connections to do exactly that.

Luckily for Marianne, North always traveled with his daughter, not only on his literal journeys but on his intellectual trips as well. Frederick North introduced his daughter to the practical study of plants. They learned about plants, hunted for rare varieties of them, painted them, and grew them. Frederick North built three glass greenhouses on his estate at Hastings, one for vines and cuttings, one for temperate plants, and one for orchids. North described living "in those houses all the spring, my father smoking and reading in the temperate regions . . . while I washed and doctored all the sick plants, and potted off the young seedlings" (p. 35).

It is hardly an exaggeration to say that Marianne North's knowledge and her intellectual and artistic achievements were made possible by the personal tastes, the emotional and financial support, and the social connections of her father. It is a situation common enough with sons, much less so with daughters.[15] Frederick North introduced his daughter to a whole circle of scientific men. Sir Edward Sabine, president of the Royal Society, was a North houseguest, as had been Sir Davie Gilbert, another Royal Society president a few decades before Sabine. The Norths took holidays with Francis Galton, a distinguished author, one of the originators of eugenics, and a dear old friend. Sir William Hooker was not only the most famous botanist in England but also virtually the creator, and the first director, of the Royal Botanic Gardens. His son Joseph became the second director of Kew and was Marianne's friend for many years. During the time the Norths were living in London after Mrs. North died, they would often ride over to Kew to admire the gardens and to take home rare plants to paint. Their host was, of course, Frederick North's friend Sir William. Sir William's gift one day to Marianne, then in her twenties, of a hanging bunch of *Amherstia nobilis,* the first ever to bloom in England, made her "long more and more to see the tropics" (p. 31).

WHILE the unusual circumstances of North's background made her achievements possible, they did not make those achievements happen. Marianne was more than her father's daughter. The first—and perhaps the most—significant difference between North's life with her father and her life for the next fifteen years alone after his death is simply where she went. When Marianne was traveling with Frederick North, including the years with her sister Catherine and the earlier years when the whole family moved around, their range from England was fairly typical of the travel patterns of the Victorian upper class. Frederick North must have set the itineraries for his little group. He was a cultural product of the early nineteenth century, before the multiplying discoveries in science, insatiable colonial appetites, and increased ease of travel had expanded the British worldview. For most Englishmen raised in the early part of the century, the world—which meant, of course, the inhabitable, civilized, and culturally interesting world—was directly east. It included the countries of the Continent, particularly France, what would become Germany, Italy, Switzerland, and Spain. And it was bounded on the far reaches of the cultural universe by what was thought of then as the exotic East—and is now called the Middle East—past Greece to the lurid Turkey of Byron's poetry and the spiritual Egypt of the Christian Bible.

At her father's death Marianne North stepped into the new Victorian age. Her traveling range suddenly exploded. She took one decorous journey in the old mode, to Italy in the spring of 1870, immediately following her loss. But never again. Instead of Europe and the Near East, North's destinations by herself were the entire rest of the world. Her first trip in the new mode began in the summer of 1871. She headed west for the cultural wilderness of the Americas, beginning in the north with the United States and Canada and, as winter arrived, continuing on south to Jamaica. On Christmas Eve, 1871, Marianne North, full of joy, had reached the longed-for tropics at last.

A second change from traveling with her father was that after his death North traveled alone. This is not to say that she was always or even most of the time by herself. She often had companions for parts of her trips, and she almost always stayed with old or new friends. North seems to have perfected the art of being a houseguest, perhaps because she tended to spend the day off painting, leaving early and returning late. But with the exception of that first trip to the United States, which she began with Mrs. Skinner (though the two gladly parted by the time they got to Canada), North did not really take trips with another person. She preferred the control, the freedom, of life on her own. How she used that freedom, where she chose to go, was no longer to the Continent but to all the continents, with particular emphasis on the tropics.

When North's travel style and range changed, so did her cultural identity. North went from living what may be described as just an unusually ambulatory version of a familiar life-style for an upper-class Englishwoman to living what may only be described as the distinctly unusual life of the Victorian woman wanderer. This does not at all imply that North was an individual rebel who suddenly threw off early Victorian cultural shackles when her father died, though her changed travel style did reflect the interests of her own generation rather than his. That independence of spirit was a family quality is suggested by an observation of a niece of Marianne's that her mother and Marianne, even as young women touring with their father, "were some of the first to discard the crinoline in travel."[16]

When people in the nineteenth and the twentieth century conjure up that convenient icon of the Victorian age, the British woman traveler—adventurous, unflappable, full skirted, and well-bred—the epitome of the type is surely Marianne North. Isabella Bird Bishop traveled as widely, wrote more books, and probably gained a more extensive reputation. But in terms of birth, background, family connections, that daughter of a clergyman was a Janey-come-lately. Marianne North was a true lady. Her awareness of the importance of her family background in defining her identity may be indicated by the fact that the first sentence of the *Recollections of a Happy Life* establishes North's famous "progenitors." Moreover, North was ubiquitous. Mary Kingsley may have described more daring adventures, but they were so regional, so confined. She hardly left West Africa. North went everywhere and in part because of family connections, proceeded to meet anyone who was "anyone" everywhere she went.

North knew a great many famous Victorians, at least socially, many of whom she met on her travels. She was on friendly terms with other distinguished women, such as Amelia Edwards and Isabella Bird, both famous travel writers; Margaret Brooke, the Ranee of Sarawak; and Julia Margaret Cameron, an early photographer. At fifty-five, with what had become serious health problems at last North settled in England, buying a country home at Alderley, where she made a wonderful garden. She saw through to its finish the enormous project of creating the North Gallery at Kew for her paintings and she also pressed on to finish a draft of what ended up being an immensely long manuscript of her travel memoirs. This huge manuscript, unpublished when she died, was cut down by her sister, Mrs. John Addington Symonds, and published in three volumes in the early 1890s. The manuscript is still extant, but my comments here are based on the posthumously edited and published version.

North occasionally wrote to friends of writing her memoirs, but no one is quite sure when she began.[17] As early as January 17, 1880, before her trip to Australia, she wrote to Dr. Burnell that "I am passing a very pleasant winter at home by the fire, . . . when it is dark and yellow fog, I scribble—I am writing 'recollections of a happy life' and putting all my journals and odds and ends of letters together."[18] It is likely that some of her friends and family who had saved her letters over the years sent them back to her to help her reconstruct her experiences. This is certainly true of Amelia Edwards, since several passages from the many letters Marianne wrote to her friend are reproduced in her manuscript. But as to "what the journals were, and what has become of them," that "remains a mystery."[19]

North's major period of writing her travel memoirs was from 1886 to 1888, after she had given up traveling and settled into her pleasant country life at Alderley. Then came what would turn out to be the disappointing process of trying to get the manuscript published. North consulted her friend Sir Joseph Hooker at Kew about who to send it to and what to expect, while Lady Hooker arranged for the manuscript to be copied. North sent the copy to publisher John Murray in the fall of 1888, and in November he wrote back to suggest some of the drawbacks he felt the "recollections" had. These included "their very great extent & 2ndly their very peculiar character."[20] We can only wonder what is implied by that second objection, but it was certainly true that the manuscript was far too long to publish in a single book, and it also was in need of editing. We can get some idea of the manuscript's length from the fact that having been much shortened after North's death, it came to three substantial books when it was finally published.

But by the fall of 1888 North was already ill, far too ill to face major editing of her manuscript. Afraid she was dying, she sent the original of the manuscript to John Murray. That original is now part of the North family papers at Rougham. The copy, which Murray also returned, was then sent to Macmillan. But this second publishing company wrote back in March 1889 with another discouraging response. The editors definitely thought the memoir was worth publishing because of "the unusual interest of the work," but they also complained of its "immense bulk."[21] Without anyone to take on the job of cutting this extensive work, the subject of publication was closed for Marianne North's lifetime. She died August 30, 1890.

That December, Marianne's younger sister, Catherine Symonds, turned to the work of editing the memoirs for publication. She was not terribly in tune with Marianne's wandering habits and felt that "it was impossible not to wish sometimes, as the years went on, that she might be content to live this pleasant life among her friends, and leave the ends of the earth unvisited—a remnant of them, at least" (*Further Recollections,* p. 315). Still, Catherine worked on the memoirs most of the spring and summer of 1891, and she enjoyed them. "Pop's [Marianne's nickname] diaries are very interesting—I go on & on at them until I know the places by heart. It is horrid chopping them up. . . . Macmillan is going to take two moderate duodecimo volumes at half profits."[22] Certainly, Catherine made many changes, such as deleting Marianne's unkind remarks about her family, including comments about the dullness of the married state as she observed it in her sister's life. Although no one has yet taken on the scholarly problem of a complete comparison of the manuscript and the published version, it seems clear that Catherine edited the memoirs not only to be much shorter but also to be tamer and more conventional.

In 1892 Macmillan published the two volumes under the title *Recollections of a Happy Life: Being the Autobiography of Marianne North.* They appeared in London in 1892 and in London and New York in 1893 and 1894. The first edition sold an impressive 4,000 copies.[23] All the reviews were laudatory, in the kind of language that made clear the extent to which North's writing, particularly this edited and published version, supported rather than challenged or criticized the cultural mores of her time. The reviewer in the *Athenaeum* assured his readers that "Nature revealed itself to her as Nature only does reveal itself to lofty souls dowered with gentleness, courage, reverence, and love." W. Botting Hemsley's review in *Nature: A Weekly Illustrated Journal of Science* was a little calmer, stressing the "entertaining conversational powers" of the book and its important companion function as a commentary on the paintings in the North Gallery. Macmillan's New York branch also published it, and the American reviews again were blandly glowing. The one in the *Dial* called the *Recollections* "a work of the same nature as Charles Darwin's 'Naturalist's Voyage Round the World.'"[24]

The one negative note came in an otherwise typically sweet review in an American journal, the *Nation,* and the sticking point was race. The reviewer commented that North was "very decidedly of the opinion that, except in unfavorable climates, the white man must inevitably drive out the colored—a view which it would be difficult to maintain successfully." No such doubt entered the English reviews, at least about the superiority of this white woman. The *Athenaeum* suggested that North's constitutional fearlessness not only gave her "a wonderful power over animals, but it instinctively attracted something like worship from rough or semi-civilized men and women."[25] The image is of Marianne as the great white lady, gracefully traveling the globe, taming the beasts, two legged as well as four legged.

Macmillan was encouraged enough by the sales and reviews to publish the rest of the manuscript, which had actually been the first part of North's memoirs, consisting of accounts of journeys she had taken with her father around Europe and the mideast, up through 1870. These early chapters, under the title *Further Recollections of a Happy Life: Selected from the Journals of Marianne North Chiefly between the Years 1859 and 1869,* appeared in 1893 in London and New York, and again in 1894 in New York alone. The reviews were blandly delighted, summed up by the tone of the one in the *Athenaeum,* which found that "a more charming volume of travel it would be impossible to name."[26] There were no further reprints of the three volumes.

In 1980, ninety years after North's death, Webb and Bower in collaboration with the Royal Botanic Gardens, Kew, published a book called *A Vision of Eden: The Life and Work of Marianne North.* It included reproductions of many of North's paintings with an accompanying text which provided backgrounds of and further information about the paintings. This text consisted of brief selections from North's three volumes of travel writings (abridged by Graham Bateman). And in 1990, on the centennial of North's death, the Royal Botanic Gardens, Kew, again through Webb and Bower, published *Marianne North at Kew Gardens.* Rather than picking a few selections from North's own books, Laura Ponsonby, education officer at the Royal Botanical Gardens, compiled a beautiful collection of North's paintings and, drawing from many sources, wrote a careful accompanying account of North's life and travels.

This new University Press of Virginia reissue of the full text of the first published volume of *Recollections of a Happy Life* beings readers again to the question posed by the director of Kew Gardens: "Who was this extraordinary woman?" (I leave aside the problematic matter of whether "extraordinary" implies that there is some sort of generic category of "ordinary" women.) Professor Prance's telling answer, that North rightfully belongs "among the great plant explorers of the Victorian age," points to what is unusual and defining about North as a Victorian woman. In an age when many of the great discoverers in the physical sciences—Charles Darwin, Alfred Russel Wallace—were amateurs, when many scientific fields were themselves just being developed, Marianne North was a kind of amateur scientist. Through the acceptably feminine aesthetic genres of travel writing and painting, North lay claim to the masculine world of objectivity and discovery. She did accurate botanic paintings in all the places she traveled. She also collected thousands of specimens to take back to Kew and discovered four species of plants, the *Nepenthes northiana, Crinum northianum, Areca northiana,* and *Kniphofia Northiana.* Sir Joseph Hooker was able to distinguish a new genus of tree in part from her drawing of it while in the Seychelles, and he named it *Northea seychellana* in her honor. Marianne North was that anomaly in Victorian culture, an Englishwoman respected by eminent scientific men. Her reputation rests on her achievements in four major areas: her nonconformist life as a wandering gentlewoman, her travel books, her plant discoveries, and her gallery of paintings.

STARTING a new trip, North wrote that "I am such an old vagabond that I own to being delighted to be perfectly free again."[27] Indeed, North's writings, both her book and her letters, are full of positively gleeful remarks expressing her joy in what she read as her own freedom, her pleasure in the relative distinctiveness of her situation as a Victorian woman, her love for her work as a naturalist, and her contempt for the ties of domesticity binding and deadening the lives of other women she knew. She was certainly indifferent to dress and fairly indifferent to even basic physical comfort, traveling for days on horseback during monsoon season, often eating anything and sleeping anywhere, sometimes neither eating nor sleeping at all.

North consciously dispensed with many of the forms and basic shibboleths of Victorian "civilization," particularly those of her own class. She claimed to W. B. Hemsley that "I rather like to keep my own individuality, not having any respect for any of the North name except my dear old father, some of the time of Old Roger," At her niece's wedding she abhorred the other guests as "strangers of the powdered footman class, with whom I have no two ideas in common."[28] While a houseguest at Government house in Bombay, she "could not stand all the red servants and magnificence and left the next day." The *Recollections* and North's letters abound with comic and contemptuous sketches of the British upper class abroad, "the great people in the usual state of amiable limpness," lolling on their verandas and never looking around at the beautiful world.[29] As she wrote to Dr. Burnell, "When summer comes, and country house dressed up parties, are put as counter temptations to wandering away quietly with my easel and old portmanteau to unseen wonders the other side of the world, I think both you and I can guess which will carry the day." And on a more serious level North was an atheist, seeing herself as culturally advanced, a "Bohemian . . . with thoughts on the most serious things which would perfectly dumbfound most of one's best friends."[30]

For North the eminent Victorians were not those who were wellborn or even those in positions of public power but those who really did something, who worked hard. For the self-proclaimed "heathen" with the "belief in the next to nothing," rejecting the next life meant attending to this one.[31] The great adventure of life was a matter of looking around at the beautiful world. And that looking around, that aesthetic contemplation, was not a static event. It was not to be accomplished with a passive or spontaneous gaze. *Recollections of a Happy Life* is a narrative of unceasing activity, of self-education and self-discipline, of traveling on an ox for eight hours to get on a twenty-foot boat for another day and a night in order to paint for sixteen hours a day at almost 100 percent humidity. Margaret Brooke described Marianne North in Sarawak as "hurtlingly energetic" and "ready for any emergency."[32] Looking at the beautiful world was a matter of rigorous effort, a matter of working to develop a trained eye. It was a matter of science.

North's heroes, the images she patterned herself by, were the natural scientists, the great Darwin, Alfred Russel Wallace, and on a lesser scale the botanists she knew well: the Hookers of Kew, Hemsley, and even an amateur like her father. North was not a botanist. As a woman she was barred from the training necessary for that career. But nineteenth-century botany in England included an impressive number of inspired and to some extend self-educated amateurs.

There is no generic category of Victorian science. Different sciences—biology, geology, botany, entomology—developed in different ways and in different relations to what was going on in various other European countries during the century. The science of botany in nineteenth-century England was in some important senses an amateur activity. This does not in any sense mean that nineteenth-century botany, including in England, was an amateur science. But the development of botany as a science was to a great extent a national affair, with different countries dominating at different periods. During the eighteenth century in Europe the idea dawned that living organisms "had not reproduced themselves unchanged since some moment of divine creation, but had undergone a development in time." The problem of classifying plants in the light of their historical development was the major focus of most botanists, and much of the work, both theoretical and practical, was done in France, at the Jardin du Roi. Founded in the seventeenth century and funded by the government, the Jardin (renamed the Museum d'Histoire Naturelle in 1793) was the "first, and for over a hundred and fifty years, the only national biological research institute in Europe."[33] The Jardin's facilities and its permanent staff of salaried scientists gave France the edge in many fields of natural history.

But in terms of quantitative research the nineteenth century was led by the Germans largely because of the explosion of new universities being created by the separate states jostling for power in what had not yet become a united Germany. In botany the dominant question was plant anatomy. How do plants work? Specific problems, often explored in university laboratories, focused on such matters as cell theory and plant reproduction. With M. J. Schleiden's 1842 *Principles of Scientific Botany* (translated into English in 1851), Wilhelm Hofmeister's 1851 *Comparative Researches* on plant reproduction, and Hugo von Mohl's and Carl. W. Nageli's work on cell theory, modern botany came into its own as a specialized science.

This is not to say that most of the great nineteenth-century scientists were German. Among many others there was Robert Brown, a Scot in London, who described the nucleus of a cell; R.-J.-H. Dutrochet in France, who discovered osmosis; and Count Leszozyc-Suminski, a Pole who made the amazing discovery of the female organs in ferns. And, of course, there was Charles Darwin. But the topics and directions of botanical research were being set primarily in Germany. Interested young botanists from other European countries, including Darwin's son Francis went to Germany to learn the field.

Britain did have its centers of botanical research, primarily in Dublin, Edinburgh, and Kew. The Royal Botanical Gardens at Kew were comparable to the French Jardin du Roi. When the Dowager Princess of

Wales died in 1772, her son George III inherited her property, including the gardens at Kew. King George then made a decision which would shape the direction of British botany, and British colonial life, for the next two hundred years. He chose Sir Joseph Banks, a leading gentleman scientist who had traveled with Captain Cook, to direct the Royal Botanic Gardens. It was Banks, with his particular understanding of scientific activity as literally exploration and discovery, who envisioned Kew as a place to display plants from all over the world. Banks sent trained collectors across the globe to do the gathering. George III and Banks both died in 1820. Nothing much happened until 1840, when ownership of Kew was transferred from the crown to the government. Then in 1841 Bank's legacy was assured. William Jackson Hooker, professor of botany at Glasgow University, was appointed the first director at this new national research center. Just four years after Victoria had become queen, a new and major era of British botany had begun.

Of the three major botanical research centers in Britain, major because they were government funded, Kew was arguably the most important. Certainly botanists in the labs at Kew studied plant anatomy. But a notable part of Kew's function was to display plants from all over the world as well as to map their local habitats. There was a strong emphasis on what I call geographic botany: plant collection and classification, the effort to determine in some complete way which plants exist in the world and where they grow.

A similar emphasis was occurring in American botany as well, but with some key differences. The strong sense in America of living in the New World, in a land which itself needed exploring, meant that for many American botanists the commitment to geographical botany was a commitment to studying indigenous plants. American botanists, also often amateurs, also included quite a few women. The pioneering women in American botany in the nineteenth century were for the most part backyard botanists, collecting, studying, and classifying the plants in their particular area.[34] In the second half of the century Kate Brandegee collected plants in California, Alice Eastwood collected flowers in California and Colorado, Ellen Quillin studied the wild flowers of Texas, and Kate Furbish aimed to collect, classify, and paint the flora of Maine. Those who did leave the continent—such as Annie

Montague Alexander who went to Hawaii, Ynes Mexia who collected in South America, and Elizabeth Britton who collected in the West Indies—all stayed in the Americas. Nineteenth-century American botany, while sharing with British botany an emphasis on plant collection and classification, tended to be regional rather than global, while British botany was both. Amateur British botanists certainly collected and classified plants in their own backyards, but they also, in significant numbers, collected plants throughout the rest of the world.

The difference in emphasis between English and German botany in the nineteenth century also reflects differences in the countries' historical status as nations. While Britain was actively engaged in sustaining, and even extending, the empire on which the sun never set, Germany's imperialistic designs were limited, partly by its internal struggles and focus on the drive toward nationalism.[35] There was no united Germany until 1871. The interesting result was that there was much more government money for academic botany in what we call Germany because there were more governments, each independent state having the option to fund its own university research. There was only one British government, and most of its attention was taken up by the goal of maintaining and extending its colonies. The international geographic emphasis of nineteenth-century English botany is inseparable from English imperialism.

THE COMPLICITY of Kew with the imperialist enterprise is no mere metaphor. Since Britain's imperialist designs, economically speaking, were directed toward attaining control of lands which could supply the raw materials for the industrial revolution going on in the little isle and since Britain's power in the world resulted in part from its ability to manufacture finished products from raw materials, it is clear that the science most closely linked to those designs and that power was botany. For the subject of botany was precisely those raw materials. And, of course, the center of British botany was Kew.

Sir William Hooker had been a professor at Glasgow University because in the first half of the nineteenth century the two great British universities, Oxford and Cambridge, did not consistently offer scientific courses of study, and they would not begin to do so until the 1850s. As a result, Hooker's social milieu was composed of people of professional status rather than of members of the traditional British aristocracy (although to a great extent the two groups overlapped). Sir William's friendship with Frederick North reflects both men's commitment, a commitment North's daughter would inherit, to a kind of aristocracy which develops when class privilege is put to the service of hard work and personal achievement. Hooker's appointment as first director of the new Kew represented the transformation of Kew "from royal garden to state institution." And a large part of what the government explicitly wanted was "to obtain authentic and official information on points connected with the founding of new colonies."[36] The economics of colonial expansion was a key reason for government funding of Kew.

Sir William Hooker reestablished an elaborate system of plant collectors for Kew, a practice first set up by Banks. Hooker's collections, dried, mounted, and classified, formed the basis of what would become the world's largest herbarium. There were satellite gardens in most English colonies, to a great extent controlled by Kew because the director had the right to nominate, and de facto appoint, the directors of those gardens. In other words, one of the main roads of advancement for botanists training at Kew, in Sir William's time and even more so during his son's time as director, was to be sent to oversee colonial botanical gardens. One reason Marianne North's travel memoirs include frequent accounts of her visits to colonial gardens was that part of her social obligation was to bring greetings from Kew to many of the director's colleagues.

If British botany was a fairly small community, for the most part a men's club with the main clubhouse being Kew, it is important to recognize what a truly stellar group made up this particular scientific group. Joseph Hooker, who was director of Kew for thirty years, "was at the center of the world of botany, his reputation unequalled by any other botanist." Kew's control, and thus its director's influence, extended over the entire British empire, supplying by the end of the century seven hundred Kew-trained botanists and gardeners. Hooker's leadership in the British scientific community extended well beyond botany. A close friend of his was Charles Darwin. It was Hooker, along with Charles Lyell and Thomas Huxley (the three called "the greatest botanist, the greatest geologist, and the greatest zoologist, respectively, in Britain then"), who led the defense of Darwin's *Origin of the Species* after its publication in 1859.[37]

The imperialistic aspect of British botany should not be taken to mean that the British simply gathered up raw materials from the countries they were indigenous to and brought them back to English factories. The process was much more active, more creative, and more interventionist than that. Kew's work of plant collection and classification also extended to plant distribution. Kew was, effectively, the depot through which were funneled enormous numbers of species of plants in the nineteenth century, to be classified, to have their uses determined and described, and often then to be sent on to satellite gardens all over the world, but particularly in the tropics and subtropics. I think of rubber, the tree the British took from Brazil in 1876 and cultivated in the huge plantations of Southeast Asia. W. T. Thiselton-Dyer, the third director of Kew (and Joseph Hooker's son-in-law), speaking explicitly on the subject of Kew's achievements, invoked the British saying that the rubber business produced "wealth beyond the dreams of avarice."[38]

Perhaps the most subtle and far-reaching example of Kew's key role in the work of British imperialism, traced in detail by Lucille Brockway, began in 1859–60 and was overseen by Sir William himself.[39] Partly as a response to the poor health, and thus lack of fighting fitness, of British soldiers during the 1857 mutiny in India, the government funded a project whereby British botanists shipped cinchona trees from the Andes and Peru to Kew and on to the satellite gardens in India. There it was successfully cultivated as a plantation crop. Indigenous to South America, cinchona became one of the most important crops in Asia. The bark of the cinchona contains the alkaloid which produces quinine, the wonder drug for treating malaria, the first successful use of a chemical compound to combat an infectious disease. Once the English could produce quinine in quantity, they not only could go anywhere in the world; more importantly, they could remain there. Before the cultivation of cinchona, the English, and more particularly not only English men but English women and English children, wore away and died of fever in the unfamiliar tropical climates. It was mass-produced quinine which enabled the English to take their families along, and thereby to create lasting settlements in the countries of their empire. It was mass-produced quinine which opened up for Marianne North, along with many other Victorian men and women, the sheer physical possibility of touring the world.

BUILT into the geographic side of British botany is a version of the scientist as explorer, probably of independent means in order to meet the expenses of travel, and acceptably an amateur. This particular definition of scientific activity, intertwined with the business of imperialism, is precisely the definition that opened the possibility of North being a botanist. The traditional ways for British women to enter botany were through drawing and writing about gardening, as did Jane Webb Loudon, the wife of a well-known writer of books on botany. There were certainly women botanists who focused on the plants of Britain, writers such as Elizabeth Twining and Anne Pratt. There was even an early nature photographer, Anna Atkins. What makes Marianne North's career unusual is that the plants she collected and illustrated were not British, not even European, but from the far corners of the world.

When the subject matter to be studied was plants outside Britain, the conventions for a successful naturalist were clearly marked. The amateur botanist must travel afar, must observe in an objective and detached fashion, must represent the objects of that observation with accuracy and precision, and must gather up whatever can be gathered and sent home. It is no accident that the North Gallery at Kew displays paintings of the plants of every continent; that when North first set up the display she had not been to Africa and it was Darwin who encouraged her to get specimens—both actual and on canvas—from that continent for her collection to be complete; and finally that the North Gallery is lined with pieces of wood from 240 countries around the world. The North Gallery is an amazing experience. It is an extraordinary monument to one woman's talent and almost superhuman achievement. It is an equally extraordinary monument to a nation's almost worldwide crimes of conquest and occupation.

A similar point must be made about North's *Recollections of a Happy Life*. The autobiographic portrait sketched in these memoirs follows the conventions of the serious naturalist: independent, physically intrepid, with great stamina, indifferent to creature comforts, the objective collector and detached observer with the well-trained eye. It is also, in terms of the cultural significance of such attributes, the portrait of a man—or, more precisely, of the male-identified woman. Here is North writing to Dr. Burnell on marriage: "It is a terrible experiment matrimony for a man especially; as a woman is something like your cat and gets to like the person who feeds her and the house she lives in—but men if they have brains have a romantic idea of companionship in their wife and they discover they have no two ideas in common."[40] Where would we locate the subject gaze of this speaker: with the cat or the "they" who have brains? The North of the *Recollections* represents herself as one of the boys.

North was extraordinarily well placed, extraordinarily talented, and her achievements were extraordinary as well. But was she also intellectually liberated? Her expressed thoughts, on religion, on natural selection and from there to cultural and racial selection, reflected the thinking of many of the most visible scientists of her day. If in some sense Victorian intellectual life was a debate between science and religion, North lined up with the scientists. Like many a supposed free spirit, North was as trapped in the ideology of her culture as the Victorian women who appeared to her so trapped in their domestic and social obligations. The *Recollections* embodies Blake's great insight in *Milton* about the interdependence of conformity and rebellion: "Satan! My Spectre! I know my power thee to annihilate / and be a greater in thy place, and be thy Tabernacle."

Far from being a discourse outside the semantic field of the society from which North depicted herself as having escaped, the *Recollections* and the available letters replicate the imperialistic values of the upper-class botanical community of North's culture. Those values may be glossed individually as masculinity and collectively as science. North traveled very much by the grace of, and in the service of, the men in power in her country. This point may well explain the notable lack of criticism, indeed, the explicitly voiced approval, of her travels on the part of those men. No one seems to have expressed the view that North was behaving too independently for a woman—perhaps because she wasn't. Perhaps because North's individual liberation was located within the enabling context of her particular upbringing and British botanical practices, her books argue for the Victorian colonial enterprise.

A GENRE of writing in which feminist issues continually intersect with issues of class, race, and nationality is Victorian women's travel writing. This is a problematic genre on many familiar levels, such as the question of the canon and women's writings, the intersection of Victorian science and discourses, the relation between place and the female body, the complicity of Victorian literature in the imperialist enterprise, and the possibility of particular works breaking out of that complicity. But before any of these questions can be taken up in critical studies of North's *Recollections* and *Further Recollections,* there is a basic critical complication which, in approaching a work in terms of its relations to a political context, we might all keep in view.

It is easy to point out, in sophisticated and multiple detail, the imperialistic colonial function of such Victorian travel writings as Marianne North's *Recollections*. This is a necessary and illuminating act. But for a critic to perform just this act, particularly but not only about material which is at any historic distance from the moment at which the critic writes, may carry an implicit exoneration of the critic's own writing from the ideologies of his or her own culture. When analyses of the ways in which specific Victorian discourses participated in the constraining patterns of their culture do not go past being exposés, precisely then does energy for change get pressed into service as energy for stasis. Such analyses express what Fredric Jameson called "a deep existential commitment to a rhythm of modernist innovation" and what Rodgers and Hammerstein called "everything's up-to-date in Kansas City."[41] The linear notion of history and the love affair with progress are among the favorite tools in the workshop of twentieth-century American imperialism, the very notions that Walter Benjamin critiqued as "empathy with the victor."[42]

I suggest that a basic premise for approaching Victorian literature might be that we in the United States have not "come a long way, baby." Whenever by the brilliance of our critiques we imply our own cultural advance or our personal emancipations from our culture, we slip back a little more. As Diana Warwick reminded us back in 1885 in

George Meredith's *Diana of the Crossways,* "The moment we begin to speak, the guilty creature is running for cover. . . . I am sensible of evasion when I open my lips."[43]

The ideological coercions contained for a critic in implying, however unconsciously, that uncovering the oppressions contained in a work from a prior age is a sign of one's own cultural or personal progressiveness are particularly seductive when the material being discussed is Marianne North's 1892 first volume of *Recollections*. For the unavoidable fact is that the independent, free thinking, free-spirited wanderer narrating these travel memoirs frequently expressed views which even in Victorian England were conservative and reactionary, particularly on the subject of race.

Here is North, near the end of 1872 in Brazil, where British colonists still had slaves, although the practice was in the midst of being legally, but slowly, phased out. The narrative offers slavery as an admirable, even an enjoyable, institution, certainly better than freedom for the black peoples of Brazil. In language that echoes the claims of apologists for slavery in pre–Civil War America, the narrator assures the reader that "if they have abundant food, gay clothing, and little work, they are very tolerably happy." In March 1873, by which time children eleven years old and under could no longer be gathered up to be enslaved in Brazil, North saw a room full of what she cheerfully described as "remarkably clean little black boys," who had been gathered up by a slave dealer because they were twelve. She reassures her readers that "these boys looked very happy, and as if they enjoyed the process of being fatted up" (1:156).

I remind you all that it had been 1772 when Lord Chief Justice Mansfield had ruled in the James Somerset case that slaves could not be forcibly returned from England to the Caribbean, a ruling that was taken to mean the legal end of slavery in England. Certainly, slavery continued to be legal outside England in many British colonies up through much of the nineteenth century. But by 1872, a full century after the Somerset case, it was also an indisputable, if hardly universal, aspect of liberal Victorian culture to be opposed to, and often repelled by, literal slavery. Clearly, North's writings belong to the reactionary camp.

In approaching North's work, we cannot distinguish between its imperialism and its female narrator's discursive liberations from some powerful Victorian ideologies about the domestic nature of true womanhood. Does Gayatri Spivak's account of *Jane Eyre* as a text where a white woman's "soul-making" is a colonizing act, achieved at the expense of a dark woman, also apply to *Recollections of a Happy Life*?[44] Do North's perambulations around the world take on a morally darker hue because she travels, often quite literally, on the backs of those with darker skins? The answer to both these questions is yes.

North could be an amateur botanist because the Royal Botanic Gardens had impressive garden outposts in many tropical countries whose functions were in large part to cultivate plants in order to service and further British imperial policies. North could be an independent woman traveler because upper-class English people lived in elaborate bungalows all around the world governing local laborers and gathering local goods, and they were delighted to put her up as part of their function of disseminating British culture and supporting the expansion of British knowledge.

But both these facts about the privileges implicit in her situation do not speak to the related issue of North's personal attitudes, her consciousness of her situation. Perhaps it is too much to ask that a Victorian gentlewoman of North's breeding, particular family background, and talents see beyond her class. Yet in many ways North did. Her contempt for shows of magnificence, for laziness, for the blindness which cannot see the worth of people outside one's own class or the beauty of places outside one's own country defines North's achievements as much more than just products of her privileges and again and again light up her prose in the *Recollections*.

And yet the ways in which North did not go beyond class and community do limit her achievement and occasionally darken her prose. Finally, she could love her travels and write about what she saw with an appealing charm and cheerfulness, could describe a "happy life," because North was committed to "science," to the presumably objective observation of the beautiful world. Her sympathy and conscience were continually untouched by the terrible sights of racial injustice or of suffering lives she encountered everywhere she went. Are the natives of India poor? They are stupid and lazy and untrustworthy. Are Brazilian black children enslaved? They are cleaner that way. Should the Europeans, and particularly the British, direct the rest of the world? Of course.

BRINGING home the world through her plant collections and her paintings and her writings, Marianne North served an empire which wished to consume—and almost succeeded in consuming—everyone else. Yet my distance from North cannot be bounded by naming her a colonial and me a "postcolonial." I turn to some other comments North made in late 1872 about slavery in Brazil. "All babies were born free, the consequence of which was that the mothers took no more care of them, as they said they were now worth nothing! In the 'good old days,' when black babies were saleable articles, the masters used to have them properly cared for; and the mothers didn't see why they should be bothered with them now" (1:148).

These remarks offer a familiar Victorian imperialist vision of the developmental primitivism of the Other. North's rhetoric here supports the belief that evolution provided some kind of "objective" evidence for the frequent British conviction that they had a right to take over other people's countries, and even other people's very lives, because certain races had evolved further than others. In 1878 she wrote to Dr. Burnell, worried about his health and urging him to stop his relief work among the poor in India, that "I would have you do good to those who are nearer yourself—& farther from monkeys."[45] Yet her comments about black mothers in Brazil offer an even more extreme version of Victorian social Darwinism. The passage represents darker foreign peoples not only as subhuman but as even lower evolutionarily than monkeys. After all, North might easily have acknowledged that monkey mothers did care for, and care about, their young without economic inducement.

Yet the shock of such a passage depends in part on its being narrated by a woman's voice, albeit one self-identified as belonging among the scientists. That voice challenges conventional cultural expectations by exhibiting not the slightest sympathetic feeling for the children of another race, even as she defines that other race as inferior on the very grounds of its women's lack of feeling for their children. We are drawn to notice the narrative's lack of sympathy in part because ideological expectations about the female self-representation are yet again confounded, as they are again and again throughout the narrative.

But the passage does have another important dimension. It undermines its own implicit claims for the superiority of the narrator's race. The British "masters" cared for the babies not from superior British humanity, which the voice is showing to be lacking, but precisely because those "babies were saleable articles." Indeed, valuing human life on the basis of economics is the basic principle of the institution of slavery, the very institution which white men introduced to indigenous peoples in the Americas through importing black slaves. Slavery is the major cultural gift from the European foreigners. And it is slavery, the traffic in human life, which the passage presents as the cause of these black women's present inhumanity. By the logic of the passage, it was their British and Portuguese masters who taught them to view babies monetarily, in terms of an infant's cash value or lack of it.

If North's own expressed lack of maternal or nurturing feelings links her with the black mothers who don't bother with their children, her implied critique of the economic basis of human value separates her from the British and Portuguese masters. Insofar as these mothers are dehumanized, what has deprived them of their humanity is not far to seek. They, or more probably their grandmothers or great-great-grandmothers, were stolen from their towns and turned into commodities in order to be brought to Brazil to work the colonial plantations and mines. It might have been relatives of these very women North saw who did the physical labor which made possible the gathering of cinchona in the Andes to send to the satellite gardens in India. If North's narrative voice speaks of distance from these black mothers, it also tells of deep connections. For she, through her own ancestors, her own ties to British botany and British colonialism, has surely been dehumanized as well. Finally, there is nothing politically innocent about a nineteenth-century British woman painting rare tropical flowers and writing about her experiences. But there is nothing politically simple about it either.

*Susan Morgan*
*Miami University*

## Notes

1. Marianne North, *Recollections of a Happy Life: Being the Autobiography of Marianne North,* ed. by her sister, Mrs. John Addington Symonds, 2 vols. (London: Macmillan, 1892), 2:99.

2. Quoted by Dea Birkett, "A Victorian Painter of Exotic Flora," *New York Times,* Nov. 22, 1992, p. 30.

3. Laura Ponsonby, *Marianne North at Kew Gardens* (Exeter: Webb & Bower, 1990), p. 9.

4. The phrase is quoted in W. Botting Hemsley's own review of "The Marianne North Gallery of Paintings of 'Plants and Their Homes,'" *Nature: A Weekly Illustrated Journal of Science* 26 (June 15, 1882): 155.

5. Sir Joseph Hooker, "The North Gallery, Kew," *The Journal of Horticulture [Cottage Gardener] and Home Farmer: A Chronicle of Country Pursuits and Country Life,* 3d ser., 51 (Oct. 19, 1905): 364.

6. Birkett, "A Victorian Painter of Exotic Flora," p. 30.

7. *Marianne North at Kew Gardens,* p. 7.

8. Ibid., p. 11.

9. Margaret Symonds, *Out of the Past* (London: John Murray, 1925), p. 59.

10. See the chapter entitled "The Problem" in Phyllis Grosskurth's *John Addington Symonds: A Biography* (London: Longmans, Green and Co., 1964), pp. 262–94.

11. For a full discussion of the legal situation of Victorian married women, see Mary Lyndon Shanley's *Feminism, Marriage, and the Law in Victorian England, 1850–1895* (Princeton, N.J.: Princeton Univ. Press, 1989).

12. Quoted by Brenda E. Moon, "Marianne North, 1830–1890," in Marianne North, *A Vision of Eden: The Life and Work of Marianne North,* ed. Graham Bateman (Exeter: Webb and Bower, 1980), p. 235.

13. Royal Botanic Gardens, Kew: Marianne North Letters to Dr. Burnell, 12 (Feb. 5, 1878).

14. Ibid., 4 (Aug. 17, 1877).

5. Marcia Myers Bonta, in her fascinating account of *Women in the Field: America's Pioneering Women Naturalists* (College Station: Texas A&M Univ. Press, 1991), points out that among these women "nearly all had had male mentors early in their careers" (p. xiii).

6. Symonds, *Out of the Past,* p. 48.

7. For a thorough discussion of this subject, see Brenda E. Moon's excellent article, "Marianne North's *Recollections of a Happy Life:* How They Came to Be Written and Published," *Journal of the Society for the Bibliography of Natural History* 8, no. 4 (1978): 497–505.

8. Royal Botanic Gardens, Kew: Marianne North Letters to Dr. Burnell, 31 (Jan. 17, [1880]). I am following Brenda Moon's assumption that this letter, dated only Jan. 17, was in fact written on Jan. 17, 1880.

9. Moon, "Marianne North's *Recollections of a Happy Life,*" p. 500.

0. Ibid., p. 501.

1. Quoted by Moon from the Marianne North papers at Rougham Hall, ibid., p. 502.

2. Ibid., p. 503.

3. Ibid.

4. *Athenaeum,* Feb. 27, 1892, p. 270; *Nature,* April 28, 1892, p. 602; *Dial,* May 1892, p. 15.

5. *Nation,* June 2, 1892, p. 418; *Athenaeum,* Feb. 27, 1892, p. 269.

6. *Athenaeum,* June 17, 1893, p. 756.

7. Royal Botanic Gardens, Kew: Marianne North Letters to Dr. Burnell, 27 (July 27, 1878).

8. Ibid., Letters to W. B. Hemsley, 2:44 (June 27, 1882), 68 (Nov. 5, 1882).

9. Ibid., Letters to the Shaen Family, 1875–84, 3 (probably early 1876), 2.

0. Ibid., Marianne North Letters to Dr. Burnell, 37 (March 20, 1878 [but must actually be 1879]), 5 (1878).

1. Ibid., 15 (Feb. 26, 1878), 35 (Feb. 21, 1878 [but probably 1879]).

32. Margaret Brooke, Ranee of Sarawak, *Good Morning and Good Night* (1934; rpt. London: Century Publishing, 1984), pp. 171, 173.

33. A. G. Morton, *History of Botanical Science: And Account of the Development of Botany from Ancient Times to the Present Day* (London: Academic Press, 1981), pp. 288, 294.

34. I am indebted throughout the following discussion of American botany to Marcie Myers Bonta's excellent book *Women in the Field.*

35. See the "Introduction" to Edward Said's *Orientalism* (New York: Random House, 1979) for a discussion of different sorts of European imperialism.

36. Lucile H. Brockway, *Science and Colonial Expansion: The Role of the British Royal Gardens* (New York: Academic Press, 1979), p. 80. I am indebted throughout my discussion of nineteenth-century British botany to this compelling and marvelously detailed book.

37. Ibid., pp. 92, 96.

38. Ibid., p. 101.

39. "Kew and Cinchona," ibid., pp. 103–40.

40. Royal Botanic Gardens, Kew: Marianne North Letters to Dr. Burnell, 9 (Jan. 20, 1878).

41. Fredric Jameson, "Third World Literature in the Era of Multinational Capitalism," *Pretexts: Studies in Writing and Culture* 3, nos. 1–2 (1991): 82–104.

42. Quoted by Patrick Brantlinger, *Crusoe's Footprints: Cultural Studies in Britain and America* (New York: Routledge, 1990), p. 193.

43. George Meredith, *Diana of the Crossways,* in *The Works of George Meredith* (New York: Charles Scribner's Sons, 1910), 16:428.

44. Gayatri Spivak, "Three Women's Texts and a Critique of Imperialism," in *"Race," Writing, and Difference,* ed. Henry Louis Gates, Jr. (Chicago: Univ. of Chicago Press, 1986), pp. 262–80.

45. Royal Botanic Gardens, Kew: Marianne North Letters to Dr. Burnell, 7 (Jan. 12, 1878).

# RECOLLECTIONS OF A HAPPY LIFE

VOL. I

*[Frontispiece: Photograph of Marianne North by Julia Margaret Cameron, from London 1892 edition of* Recollections, *vol. 1, facing p. 315, substituted in this facsimile edition for bust-length photograph used as the frontispiece]*

# RECOLLECTIONS

OF

# A HAPPY LIFE

BEING THE AUTOBIOGRAPHY OF

MARIANNE NORTH

EDITED BY HER SISTER

MRS. JOHN ADDINGTON SYMONDS

IN TWO VOLUMES

VOL. I.

New York

MACMILLAN AND CO.

AND LONDON

1894

# PREFACE

This story of my sister's life was the work of her first two years at Alderley. She put it into the hands of her friend, Sir Joseph Hooker, who undertook to negotiate with Messrs. Macmillan for its publication. They agreed to take it upon condition of certain retrenchments, which she was then far too ill to make, and the manuscript was put by till after her death.

As it now stands, her earlier journeys in Europe, Egypt, and Syria have been cut out entirely and the first chapters compressed. The later and most interesting journeys remain, except for minor points, which had to be revised, just as she left them.

By the advice of her oldest friend (and my own), Mr. Francis Galton, the spelling of Indian names and places has been altered to the system now adopted by the Government of India. These chapters have been kindly revised for me by our friend and neighbour here, Major-General M. R. Haig. . . . Mr. Galton had the still more puzzling names of Java verified and corrected for the book at the Royal Geographical Society.

The proofs of the diaries throughout have been most kindly read and revised by Mr. Botting Hemsley, of the

Herbarium, Royal Gardens, Kew, without whose generous help the book must have contained many botanical errors. My sister was no botanist in the technical sense of the term: her feeling for plants in their beautiful living personality was more like that which we all have for human friends. She could never bear to see flowers uselessly gathered—their harmless lives destroyed.

Of the portraits (reproduced by Obernetter), the vignette by Williams, at the age of thirty-four, is the best likeness. The frontispiece to the second volume, which represents her standing on the doorstep of her house at Alderley, was done by a neighbour, Mrs. Bryan Hodgson, on her first arrival there, and sent out to her old friends as a standing invitation to visit her in her new home. It has her signature, and tells its own story: the last line of the autobiography, the close of her happy life.

J. C. S.

DAVOS.

# CONTENTS

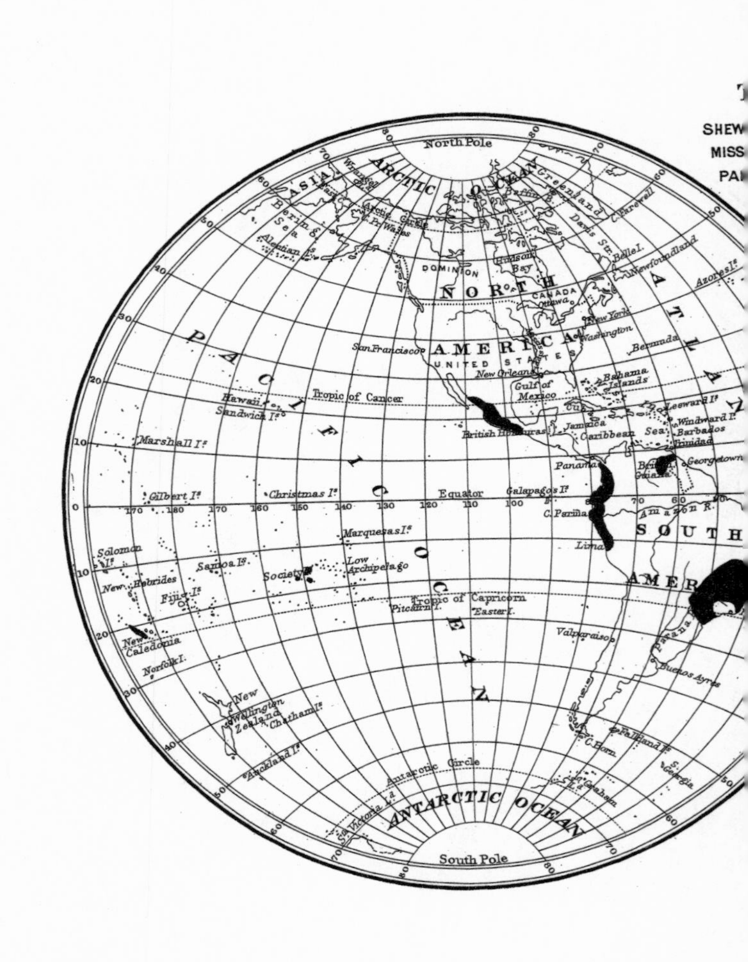

North Pole
ARCTIC OCEAN
ASIA
Bering Sea
Aleutian Is.
Arctic Circle
C. P. Wales
Greenland
Baffin B.
Davis Str.
C. Farewell
Hudson Bay
DOMINION
NORTH
CANADA
Ottawa
Belle I.
Newfoundland
Azores Is.
New York
Washington
San Francisco
AMERICA
UNITED STATES
New Orleans
Bermuda
Bahama Islands
Gulf of Mexico
PACIFIC OCEAN
Tropic of Cancer
Hawaii
Sandwich Is.
Jamaica
Caribbean Sea
Leeward Is.
Windward Is.
Barbados
Trinidad
British Honduras
Marshall Is.
Panama
Georgetown
Gilbert Is.
Christmas Is.
Equator
Galapagos Is.
C. Parina
Amazon R.
SOUTH
AMER
Lima
Marquesas Is.
Solomon Is.
Samoa Is.
Society
Low Archipelago
New Hebrides
Fiji Is.
Tropic of Capricorn
Pitcairn I.
Easter I.
Valparaiso
New Caledonia
Norfolk I.
Buenos Ayres
New Zealand
Wellington
Chatham Is.
Falkland Is.
C. Horn
S. Georgia
Auckland Is.
Antarctic Circle
ANTARCTIC OCEAN
Victoria L.
South Pole

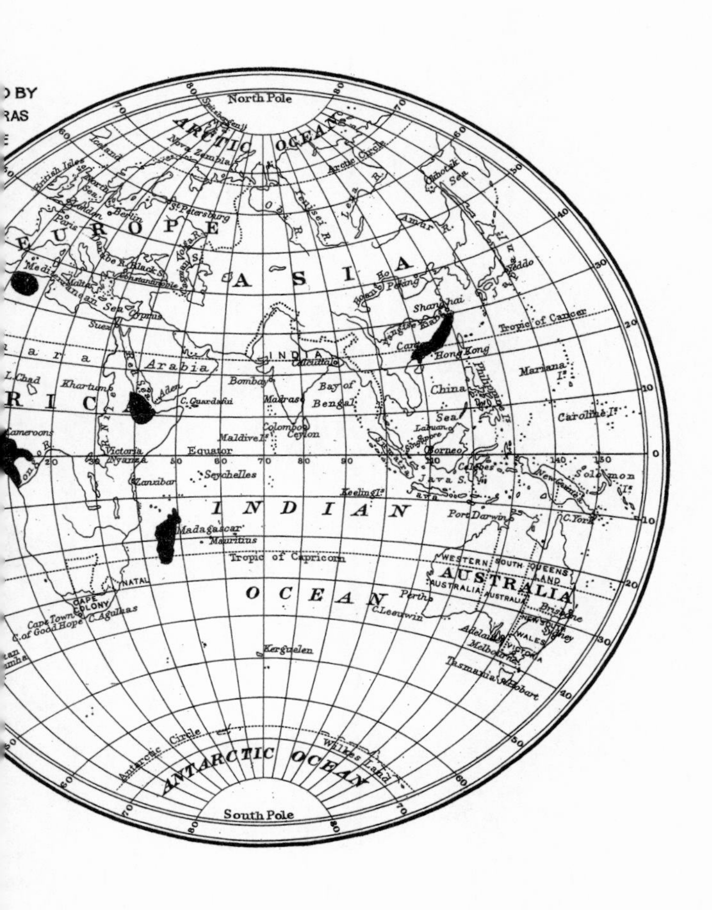

North Pole
ARCTIC OCEAN
Iceland
British Isles
North Sea
London
Paris
Berlin
St. Petersburg
Nova Zembla
Arctic Circle
EUROPE
Ob R.
Yenisei R.
Amur R.
Okhotsk Sea
ASIA
Hoang Ho
Peking
Yeddo
Shanghai
Black S.
Constantinople
Malta
Mediterranean Sea
Cyprus
Suez
Tropic of Cancer
Canton
Hong Kong
Arabia
Aden
INDIA
Calcutta
L. Chad
Khartum
Bombay
Bay of Bengal
Madras
C. Guardafui
China Sea
Philippine Is.
Mariana Is.
Caroline Is.
Colombo
Ceylon
Maldive Is.
Labuan
Singapore
Borneo
Celebes
Cameroons
Victoria Nyanza
Equator
Java S.
Zanzibar
Seychelles
Keeling Is.
New Guinea
Solomon Is.
INDIAN OCEAN
Port Darwin
C. York
Madagascar
Mauritius
Tropic of Capricorn
WESTERN AUSTRALIA
SOUTH AUSTRALIA
QUEENSLAND
AUSTRALIA
NATAL
CAPE COLONY
Cape Town
C. of Good Hope
C. Agulhas
Perth
C. Leeuwin
Brisbane
NEW SOUTH WALES
Sydney
Adelaide
VICTORIA
Melbourne
Kerguelen
Tasmania
Hobart
Antarctic Circle
ANTARCTIC OCEAN
Wilkes Land
South Pole

# CHAPTER I

## EARLY DAYS AND HOME LIFE

It began at Hastings in 1830, but as I have no recollections of that time, the gap of unreason shall be filled with a short account of my progenitors. My fourth great grandfather was Roger, the youngest son of Dudley, fourth Lord North of Kirtling, and Anne, daughter of Sir Charles Montagu.[1] He had been Attorney-General under James II., and wrote the lives of his three brothers—the Lord Keeper Guilford; Sir Dudley, Commissioner of the Treasury to King Charles the Second; and Doctor John North, Master of Trinity College, Cambridge.[2] The portraits of these famous brothers and of their grandfather, the third Lord North, were among the first things which impressed me with childish awe, in our dining-room at home.

For Roger I had an especial respect, as the brown curly wig was said to be all his own, and not stuck on with pins driven into his head as my doll's wig was, and I thought he used to look down on me individually with a calm expression of approval. I liked to make my father tell me stories about him, and how the great Lord Clarendon had written in his journal "that he and one other were the only two honest lawyers he knew," that he was one of those who found time for everything, for music and painting as well as law, and when he tired of the latter work and the political squabbles of the time, he retired to the old hall at Rougham, in Norfolk, and employed Vater Schmidt to build him the best organ which could be made, on which he used to play the works of Corelli and Purcell.[3] He wrote histories of music, building, and architecture, and covered his walls with pictures, including duplicates of many of the lovely but soulless beauties of his friend, Sir Peter Lely. He had also built a library attached to the church, in which he deposited many curious and valuable books, including the Oriental manuscripts collected by his niece Dudleya North, whose great knowledge of languages would have been remarkable in any age.[4]

Roger lived on at Rougham ("out of the way," he called it truly) to a good old age, and as he wrote on his own epitaph, "freely communicated to all, without fee or reward, that great knowledge of the laws whereby he had formerly acquired the moderate fortune he died possessed of."

His son Roger had not wit enough even to add the date of his father's death to his epitaph. He had a vile temper, and flogged his son Fountain to such a degree that the boy ran away to sea, and stayed there till his father's death left him Squire of Rougham—a place he hated from old associations—so he never went near it again, and ordered the house to be blown up with gunpowder, as it was too solidly built to be pulled down easily. All its contents were dispersed and sold by public auction, and even now rare old books and pictures may be found at sales and markets in the neighbourhood which once belonged to this collection, while the beauties of Lely still simper out of their frames at Rainham or Narford. The sailor-squire cared only for the sea, and in his old age settled himself as close to it as he could in the first lodging-house ever let at Hastings, dividing his time between it and a house he built at Hampstead, with a flat roof, bulwarks, and portholes, like a man of war's deck, on which he used to pace up and down, firing off cannon from it on all great occasions and birthdays. His wife was a farmer's daughter with no education, but she was a most indefatigable worker of worsted-work, and through her (my father used to say) a certain amount of common sense was reintroduced into the family. The old Squire had two beautiful sisters, great riders, who often went over from Rougham to Houghton to watch Mr. Boydell at his work of engraving the famous collection of pictures there, which were afterwards sold to the Empress Catherine of Russia, and lost in the deep sea on their way out to her.[5] He married one of these Miss Norths, while Kent the architect (also employed at Houghton) married the other.[6]

My grandfather Frederick Francis also lived all his life at Hastings, and never went near Rougham. He married the Rector's daughter, and did nothing to distinguish himself but have the gout, which gave him an excuse for a bad temper; and my poor father had a Latin grammar thrown at his head almost before he could speak, "early education" being part of my grandfather's faith. He had five sons and a daughter, of whom my father was the eldest; he was born in 1800, and when a mere child of eight years old was sent to Harrow to fight his way among his elders, and endure many a hard hour of bullying and fagging. But he always spoke with pleasure of those days at school, and his sorrows came more in the holidays at home. Years afterwards, when opposing the election of Mr. Brisco, he used to say, it "vexed him to have to do so, as he could not help remembering how he (a big boy at Harrow) had interceded with the others to put little North on the top of the victims who were to be folded up in a press bed, he was so very small" (a mode of torture very fashionable amongst school bullies then).

My father stayed at Harrow till he was Captain of the school in Dr. Butler's house, and the old Dean used to say jokingly in his latter years that he would never have been able to get married, if my father had not kept such good order in the school and given him time to go a-courting.[7] His daughter was one of my first friends, and is my best friend still. From Harrow my father went to St. John's, Cambridge, and in due time took his degree of Senior Op., spending his vacations with an old farmer at Rougham in preference to his ungenial home, and getting a liking for the old place, its noble trees and poor neglected people—a liking which increased with years.

After leaving college he went to Switzerland, put himself to board with a Geneva family to learn French, walked round Mont Blanc, picked up some crystals, and finally came back to the Temple to study law, but instead of doing so, fell in love with my mother, the beautiful widow of Robert Shuttleworth of Gawthorpe Hall, Lancashire, and eldest daughter of Sir John Marjoribanks, Bart., of Lees, M.P. for Berwickshire; her mother was a Ramsay of Barnton, and remarkably small and precise in her ways, and when the almost gigantic Baronet proposed to her in a box of the Edinburgh theatre, he received a "hush!" for his answer, and, "dinnae speak sae loud, or the folk'll hear," was all the encouragement he obtained. They both left the world before I came into it. My mother herself had lived little with them, but was brought up by her grandfather at Lees, passing her time pleasantly in picking up pebbles, and throwing them back again into the clear running river Tweed, and dreaming over Sir Walter Scott's romances as they came out, having an additional interest in them from personal acquaintance with the author, on whose knee she often sat while he told her stories. Her Aunt Marianne taught her the little she did know; I was called after her in gratitude for the teaching, and the name gained me a legacy from one of her uncles, who on his return from India found no other memento of his favourite sister but myself, a small baby.

My mother's first marriage was soon over; her husband having upset a coach and four he was driving, died himself, and nearly caused the death of his wife and of the delicate child Janet, who was born afterwards. When my father first saw my mother she was in deep widow's weeds, trying to keep the tiny heiress alive in the mildest climate of England, Hastings. In those days it was a very different place from what it is now. There were not half-a-dozen regular lodging-houses, it was (though first of the Cinque Ports) merely a fishing village. There was no St. Leonards at all, the great "White Rock" to the west was afterwards removed bit by bit: it is now only marked by the name written on a portion of the long two miles of continuous houses that join the towns of Hastings and St. Leonards. My father was elected member for the town in 1830 by ten "Freemen," one of them being himself.

My first recollections relate to my father. He was from first to last the one idol and friend of my life, and apart from him I had little pleasure and no secrets. He used to carry me on his shoulders over the hills and far away, down on the beach to see the fishing-boats land, and the heaps of glittering slippery fish counted and sold by Dutch auction; and I well remember the old fishermen, covered with silver scales, calling out, "Make way for Muster North and his little gal!" giving me kind pats with great salt hands as I passed perched high on my father's shoulder through the crowd. People tell me this is impossible, but I have a strong recollection of seeing the great dinner given after the passing of the Reform Bill, for which my father voted, riding or walking home night after night after the heated divisions to his house at Notting Hill, and arriving in the small hours of the morning. When that was over, his health broke down, and he had to give up Parliament for awhile, and had the more leisure to attend to me.

We had much variety in our life, spending the winter at Hastings, the spring in London, and dividing the summers between my half-sister's old hall in Lancashire and a farm-house at Rougham. We saw many pleasant people in all these places, but especially at Hastings. The one who made the strongest impression on me was Lucie Austin, then at school with Miss Shepherd at Bromley Common.[8] She used to spend many of her holidays with us while her parents were abroad, and inspired me with the most profound respect and admiration, as one raised above ordinary mortals. Her grand eyes and deep-toned voice, her entire fearlessness and contempt for what people thought of her, charmed me; then she had a tame snake, and must surely have been something more than a woman to tame a snake! She used to carry her pet about with her, wound round her arm, inside the loose sleeve which was then fashionable, and it would put its slender head out at the wrist-hole and lap milk out of the palm of her hand with its forked tongue. It was as fond of glittering things as Lucie herself, and when she took her many rings off her fingers and placed them on different parts of the table, it would go about collecting them, stringing them on its lithe body, and finally tying itself into a tight knot, so that the rings could not be got off till it pleased to untie itself again. Sometimes Lucie would twist the pretty bronze creature in the great plait of hair she wore round her head, and once she threatened to come down to a dinner-party of rather stiff people thus decorated, and only gave up when my mother entreated her, with tears in her eyes, not to do so. She used to sit for hours together in a rocking-chair, reading Shakespeare to us, and acting and declaiming her favourite parts over and over again, till I knew them by heart myself, and Beatrice and Portia became my personal friends. When my sister Catherine was to be christened, Lucie thought she would like to be christened at the same time,—her mother, who was one of the Unitarian Taylors of Norwich, had of course never thought of such a thing, but when (at my father's suggestion) she wrote to ask her parents' consent, Mr. Austin wrote back that she was welcome to do as she liked in the matter; and I well remember the curious scene of our good old Rector in a highly nervous state, performing the ceremony for the baby in arms and the magnificent lady of eighteen in the ugly old church of St. Clement's, Lord Monteagle, Miss Shepherd, and my mother being the sponsors.

Soon after that Lucie was engaged to be married to Sir Alexander Duff Gordon, a very handsome man, who used to come down for weeks at a time, and draw wonderful devils in our scrap-books, and walk about with Lucie wound up in one plaid, both smoking cigarettes. They specially liked doing this on the roof at Gawthorpe, to the horror of my mother, who thought the neighbours might think it was Janet, who was quite innocent of everything but good works—schools. lending-libraries, church-building were her delight, and she generally sat in one of the great recesses of the long gallery working out her plans like Dorothea in *Middlemarch*.

This room was extremely beautiful, stretching along the whole front of the house, one side and the two ends covered with stone mullioned windows, three of them deep bows or recesses, like small rooms, the ceiling richly moulded with cones and leaves, and long hanging pine-apples in the middle of each pattern; grim old family portraits lined the whole length of the room opposite the windows, and in the middle was a curiously carved chimneypiece, over which was inscribed, "Fear God, honour the King, seek peace and ensue it." At one end was a huge iron-bound chest, the very original (we children thought) of the Mistletoe-bough story, and of which no one living could open the lid, it was so heavy. Near it was a winding corkscrew stone staircase (it made one giddy to go down it), and opening on it were wainscoted rooms and secret closets, and the lady's boudoir, with a sliding panel, from whence she could give her orders to the musicians in the black oak gallery, or look down on the guests who were feasting on the raised floor of the dining-hall below. That was a grand room too, with a great arched chimney under which one could sit in arm-chairs on each side of the burning logs, and look up at the stars above (when not too smoky). A secret chamber, with some chests of old plate, had once been found up that chimney, we were told, and the whole house was full of mystery.

Outside, the house was all windows. I often counted them, but could never find where they were inside; there was a moat also, which kept it nice and damp, though the course of the river had been changed and no longer ran through it. The river Calder was itself spoilt by the numbers of factories which threw in their surplus dyes, and its colour varied from orange to scarlet or purple. The noise, smoke, and general griminess of every body and thing in that country were most unattractive to me, and I was always glad to move from it to clean dull old Norfolk, with its endless turnip-fields and fir-plantations, pigs and partridges, and where I had a most remarkable donkey to ride. That donkey was a genius! He could open every gate in the parish; neither latch nor chain could keep him out. We called him Goblin, after the Fakenham Ghost, and he soon found me inconveniently heavy, and made riding unpleasant by taking me into ditches and under low prickly hedges, when my only chance of avoiding being torn in pieces was to lie flat on his back or roll off; pulling at his mouth was as useless as pulling at the church-tower, and I was not sorry when I was raised to the dignity of riding a pony, on whose back I spent the chief part of my days, following my father about from field to field, tying up the pony while he was busy with his axe, and devouring Cooper's novels under the trees he had planted, till I fancied myself in the virgin forests of America.

Governesses hardly interfered with me in those days. Walter Scott or Shakespeare gave me their versions of history, and Robinson Crusoe and some other old books my ideas of geography. The farm-house we lived in had been originally the laundry of the Hall, and consisted of one large centre room on the ground-floor, with sufficient bedrooms over it and offices outside. This room had nine doors, three of them outside doors, so it was very airy. It had a great open fireplace in which we burnt huge logs of wood, and a steep ladder-like stair more fit for a ship than a house. The garden was full of old-fashioned flowers; it had tiny paths and beds edged with box hedges, leading up to a quaint old pigeon-house covered with ivy, and beyond that was the park full of grand trees, and the church and village. Everything was most unconventional; the Methodists and Wesleyans had their own way, there having been no resident clergyman within the memory of man, and no school of any sort except theirs. The Rector lived at Whitehaven; his son-in-law was his curate, and kept a school somewhere else; he paid a hack parson, Mr. York, to come over every Sunday and "get through" the service somehow. He used to stroll in with his pipe in his mouth, within an hour or so of the proper time, and after he had finished his task, in an almost empty church, came in to dine with us, and have a game of chess with my mother afterwards. He was not a bad man, but uneducated, had been originally a ploughboy, and had won the heart of a farmer's daughter, who first married him and then took him to Cambridge to make a gentleman of him by having him crammed sufficiently to get him ordained; they lived in a cottage three miles off, and he got a poor livelihood by taking "hack duties" as they were called.

All our parish were Dissenters of different sorts, but chiefly "ranters" or Primitive Methodists, a sect whose chief preachers were women.[9] When we came, however, a good many usually came to church in order to have a real good look at the Squire, who was always popular; and my mother started a Sunday School "in order to bring more people to church," a result of which I could never quite see the benefit.[10] She induced the leading Methodist to bring all his school-children into a sort of lean-to, or side-aisle, of his tumble-down old church (Dudleya's original Library), and got one of the farmer's wives, with our governess and butler, to go and teach them, bribing the children to stay through Mr. York's service and sermon by giving them small tracts with pictures in them afterwards. I remember well the first Sunday these children were seated round the chancel outside the altar rails. The parish clerk was rather astonished that my father objected to his seating himself *on* the altar table during the sermon, with a long pole in his hands to touch up the heads and backs of those who went to sleep or did not behave with due solemnity. My mother also started an evening class for young men, for until then few of the villagers could either write or read. They were most eager to improve themselves in this way; far more so than they were after a properly organised schoolmaster, mistress, and house were established in the village.

Our life at Hastings was very different, and our comfortable house was generally full of guests. The Davies Gilberts often came over from Eastbourne; the P.R.S. was a gentle lovable old man; his clever wife a most inveterate talker, full of philanthropic schemes for improving the condition of the labouring classes, a subject not so much thought and written about in the England of that day as it is now. She used to carry models of ploughs, draining-tiles, and other machines in her huge pockets, and the slightest gap in the conversation brought them out, with all her arguments for and against them. Another old lady used to come and stay with us, a Mrs. Stock, who impressed me greatly, as she wore stick-up collars, played splendidly on the piano, and had a mania for phrenology. Whenever she came victims were collected, with their back hair let down, to have their bumps felt and registered, and the drawing-room looked like a hairdresser's shop; under her influence my mother became quite a believer.[12]

My half-sister Janet had been a good deal away for some time with her cousin, Mrs. Davenport, afterwards Lady Hatherton, and one morning wrote the astonishing intelligence that she was engaged to be married to Dr. Kay, the great educationalist. My mother had never seen nor heard of him before, and was perfectly dumfoundered by the news. Dr. Kay came down to see his future mother-in-law. He was twelve years older than Janet and very bald, and as he took my sister Catherine on his knee and petted her to hide his nervousness, she made the deliberate and somewhat embarrassing remark, "Dr. Kay, why does your head come through your hair?" Terrible innocent!

Catherine was about four years old then, I twelve, and Charley two years older. The wedding was all fun to us, but I remember thinking that a medicine-chest was an uncomfortable kind of thing to be stuck between bride and bridegroom in the yellow chariot as they rolled away.

The next great event was a journey to Scotland, where my father had to look after the property of his ward, Sir John Majoribanks. He and my brother joined us from Eton, and we all went down by sea from London to Edinburgh, coasting under the cliffs of Scarborough and Tantallon, then close under the Bass Rock. From thence we went by coach to Lees on the Tweed, where my mother's girlhood had been passed. It is not an old place, but very lovely, with the clear river running round the Lees or meadow on which the house is built, and the Eildon Hills in the distance. The wild cherries were white with blossom then, and reflected in the water, through which one could see the salmon glittering as they glided along deep below the surface. Men used to sit on raised platforms on the bank, watching for the big ones to come up from the sea, and gave notice to the fishermen to look out for them. One story I heard there which impressed me much. The factor had a large tom-cat, which used to sit and watch him fishing, and got so excited when any big fish was hooked that it would rush into the water and help to land it! A curious instance of the love of sport overpowering its cat-like dread of wetting its feet! Poor puss at last fell a victim to its own vices, and died from a fish-hook which stuck in its throat when devouring a stolen fish too eagerly.

When we went south again, we caught the railway somewhere beyond York. There were only bits of railways in those days, and we generally drove a long way to reach them, and then used to sit in our own carriage, which was tied on a truck, surrounded by all our luggage. My great delight was to sit with my father on the rumble or coach-box, biting my lips hard and shutting my eyes when we went through the tunnels and bridges to keep myself from calling out with fright, though I knew I was safe with my father's arm round me.

Our journeys from Hastings to Norfolk every year were a long week's work, and we were treated like old friends at all the inns on the road. My father often drove himself, with me on the box beside him. We also rode some hours each day. I knew every big tree, pretty garden, or old farm-house, with the wooden patterns let into the walls, and yews and box-trees cut into cocks and hens, and I sadly missed them when the days of "improvement and restoration" came. Ely Cathedral and Cambridge also made up for the monotony of the low fen-country, and we always stopped at the latter place to visit old haunts and take a load of books from the library—such books as we could not get elsewhere. Amongst others, Mrs. Hussey's two large volumes on British fungi were my great delight one summer, and started me collecting and painting all varieties I could find at Rougham, and for about a year they were my chief hobby.[13] One, I remember, had a most horrible smell;[1] it came up first like a large turkey's egg, and in that state was inoffensive; and as I was very anxious to see the change, I put it under a tumbler in my bedroom window one night, and the next morning was awakened by a great crash. Behold the tumbler was broken into bits, and the fungus standing up about five inches high with a honey-combed cap, having hatched itself free of its restraining shell,

[1] *Phallus impudicus.*

and smelling most vilely. Good and bad smells are merely a matter of taste, for it soon attracted crowds of a particular kind of fly, which seemed thoroughly to enjoy themselves on it.

At last some one told my mother that I was very uneducated (which was perfectly true), so I was sent to school at Norwich with Madame de Wahl, one of the three sisters of Lady Eastlake who had committed the folly of marrying Russian nobles while students at Heidelberg. She had lived to repent, and escaped after much trouble, bringing home to England a son and a daughter, whom she had to educate and bring up by her own earnings. She was very handsome; it was impossible not to love her, but school-life was hateful to me. The teaching was such purely mechanical routine, and the girls with one exception were uninteresting. The only bright days were when my father used to ride over for the assizes or some other business and take me with him; one day he took me to see Bishop Stanley in his old house by the Cathedral, and made him promise to compel the next Vicar of Rougham to live there, which he did. That bishop was a beautiful old man, and more energetic than Norfolk sporting parsons cared for; they said he was undignified, and ought to sit still in his carved stall at the Cathedral, instead of starting off on his pony, no one knew where, on Sunday mornings, and pouncing down on some wretched preacher and empty church, in which he was not found out till his fine voice gave out the blessing at the end.

At last the happy time came, and I left school. My months there had not been many, but they were very long ones to me, and soon after it was decided to let Hastings Lodge and to go abroad for three years.

In August 1847 we went to Heidelberg, where we settled for eight months in the two upper storeys of a large ugly house outside the town gates, on the Mannheim road.[14] We were a large party—my father and mother, three English maids, a German cook, my sister, myself, and an old English governess we used to call "Pietra Dura." Her name was Miss Stone, and she knew the peerage by heart. My mother believed in her; we hated her. Music was then my mania, my master's violoncello generally lived in our schoolroom, and it was a real delight accompanying it both with voice and piano. There were few English then in Heidelberg; our friends were all German. The winter was cold and bright; continuous frost and sunshine, with neither fogs nor thaws. The students in all their gay finery of caps and bands and tassels drove jingling sledges up and down the High Street, and skated on the river by torchlight, pushing gaily dressed ladies in perambulators before them. They also gave excellent concerts every week; and the walks over the beautiful Berg-strasse, covered with crisp frost or snow, were most enticing. On Christmas Day we joined Professor von Mohl's family party round the blazing tree; and as we walked to his house down the long High Street every window was illuminated with the same trees, and family parties round them. No shutters were closed on that night, so that those in the street could also get some reflected warmth from the Christmas tree.

My father often took me expeditions, starting by rail, and then plunging into the forests, over hills and valleys, where we met pretty roe-deer, hares, or foxes, and gathered great bunches of lilies of the valley; all was apparently so calm and peaceful, though at that moment great revolutions were hatching all over Europe.[15] Shortly after, Louis Philippe fled from France, which was soon in the hands of a Provisional Government. Revolutionary ideas are infectious, and soon crossed the Rhine to our students, who strutted about with cocks' feathers added to their gay caps, and dressed their big dogs' necks with the colours of united Germany, "Roth-schwartz-gelb." The first great meeting to promote that end was held in the court of the castle of Heidelberg on the 26th of March 1848. It was crowded with many thousands of people, who listened very quietly to hours of dull speechifying, a good deal of pistol-shooting, and a little national music.

Before April was over we left Heidelberg, and steamed up the Neckar to Heilbronn. From thence we drove in a lumbering kind of omnibus, bag and baggage, through Ulm to Augsburg and Munich, where we took a flat belonging to a Bavarian grandee, over the Prussian Embassy, and next to the English one, which pleased my mother, but my father called it splendid discomfort; and indeed, though the reception-rooms were fine, the bedrooms and all usual comforts of life were as deficient as they could well be. The whole of modern Munich had the same mushroom character, the fancy of a poetical old king, who just before our arrival had outfooled himself under the reign of Lola Montez, and had been forced to abdicate: the shop-windows were full of portraits of him tearing his hair at the sudden departure of the "Despotinn" and his crown.[16]

I soon fell a victim to typhoid fever; but I had time before it came to hear Don Giovanni for the first time, and to see the Sleeping Faun and some other masterpieces in the wonderful Glyptothek. When my fever abated I was taken to the Lake of Starnberg—just the place to recover in—a still clear lake on which one could be rowed without fatigue or hurry. Lovely Alps in the far distance, quite too far to think of going to, no particular expeditions to tempt one, a general prettiness and freshness without anything the least grand or exciting. On the banks of the lake was a small hunting-box of the young king's, and as he had been made Lord High Admiral of the new united German fleet, he had a gunboat or yacht launched on the lake, and came down once to try it, but was very sea-sick, and never came again; such was the story.

My brother came out for his vacation, and did an amazing amount of fishing, and then we all packed ourselves into a huge three-bodied vehicle, and drove to Salzburg, taking about three days on the road.

There was a great charm then about the old inn of the

Schiff, with the splashing fountain, backed by the Cathedral, in front, and Mozart's minuet perpetually chiming overhead (sadly out of tune), as well as the sunset guns and military bands, which seemed also to belong to it.

We did not stay long there, but settled ourselves in an old manor-house with four turrets and about seventy windows, two miles east of the town; for all this my father paid £16 for the summer, so could afford also to hire a nice open carriage with a pair of strong little horses, which took us to Berchtesgaden, König See, Hallein, and other expeditions, whenever we were in the mood. Nearly every afternoon we had a thunderstorm; often they were more violent than our old roof could stand. The water used to rush down the winding stairs like a river from the upper storey. But the sky effects at those times were superb, and from every side of the house one could see magnificent and varied views.

My brother got as much fishing as he could manage, and an Austrian gentleman one day took him to see some native sport in the forest; they tied an owl to a post with the sun in its eyes, when it blinked so hard that all the birds of the air came to look at it, then the sportsmen shot at them from a sort of wigwam of branches they had made to hide in. They brought home two hawks, a jay, and some other game more curious than eatable.

Our next move was to Ischl, thence to Lake Gmunden, and on by a horse-car to Linz, and down the Danube to Vienna, meaning to pass the winter there. We began house-hunting at once, and had nearly decided to take one on the ramparts; but the good landlord advised us to wait till the next day before agreeing about it. He was an honest man, and knew, what we strangers did not know, that disturbances were already beginning in the city. My father and mother were in the Cathedral when the mob broke in to fetch out the seats and other movables to make a barricade outside; the firing began, the tocsin sounded its great muffled bell to call in the people from the suburbs, and armed men pointed their guns at a door behind the pulpit.[17] As no one came out of it, no one was shot at, but when they began to bring in wounded from outside, my father got my mother out by a back door, and home through side streets. That night Latour, the Minister of War, was hanged on a lamp-post and shot at by the students, who burned down the arsenal, and dressed themselves up in old breastplates and helmets, strutting about in them like Bombastes Furioso. My father asked one who was pointing a gun over the gate nearest to our hotel if we could leave the town, and was told we should be shot down if we did. So he went at once and secured two good porters with wheelbarrows, on which our travelling luggage was placed. We all marched out with it, and arrived safely at the railway station in the Vorstadt; from thence by slow degrees to Baden, proposing to stay there till the row was over, and never dreaming that poor Vienna would be shut up for a month. After we left, indeed an hour after we passed through that gate, it was barred and bolted.

Baden was crammed already with refugees, and not a room to be had, so the next train took us on to Neustadt, where my mother was so completely knocked up that she declared she would sit up in the dining-room of the first inn we tried for the night; while the rest of our party wandered forth again from inn to inn, under the bright moonlight, hearing the watchman call ten, eleven, and twelve, when at last we found comfortable beds and a kind host in a little inn outside the town. The "Angel" was its name, and its nature, we thought, as we rested our weary feet and pitied my poor father, who returned to assist my mother to "sit up" on a hard bench among the smoking and beer-drinking Austrians till morning in a dirty salon. He was not well, and few men could have managed as he had done to get eight women and all their luggage safely out of a fortified city in the hands of revolutionists. The next morning's train took us to the foot of the Semmering Pass, and the mouth of the first great tunnel which pierced the Alps. After a three hours' hunt we found an omnibus and horses to carry the party over the pass. I and my father walked, and enjoyed doing so. The scenery was splendid; an Italian regiment was before us, singing and rejoicing as they got nearer their native land; the savage grandeur of the mountains was softened by these distant voices and melodies into something more kind and gentle; and the whole was flooded with a rosy sunset as we reached Murzuschlag, where we were told no train would leave till three the next morning. Of course the soldiers had taken every lodging. We thought ourselves happy to be allowed to camp in the third-class waiting-room for the night at the station, and to make our supper off the few scraps of cold plum-pudding and other national food we had left in our baskets, which had been filled the week before at Ischl. One of the railway-guards opened a drawer in a press and pulled out some blankets, which he courteously offered my mother, who as courteously refused, on which he took off his boots, lighted his pipe, curled himself up amongst them in the drawer, and smoked himself to sleep. We all did our best to sleep also.

The train did not really start till six the next morning, when we were far too weary to enjoy the scenery of the Mur Valley; but the clean rooms, baths, and breakfast at the "Black Elephant" at Gratz, were among the greatest luxuries I ever met in my life. It was a genuine specimen of a "gasthaus" of the old school, where the landlord treated his guests like friends. All scraps and bones were made into soup and given away to a crowd of poor people who collected in the court every afternoon.

During the month the siege of Vienna lasted Gratz was also in a panic; but all the unquiet spirits went off to assist the confusion of the capital, and the Styrians kept perfectly good-humoured and quiet, though every one was most anxious for the future, and the banker even refused to cash circular notes or cheques on Coutts's letter of credit, which was inconvenient.[18] Our heavy luggage had been sent from Ischl by an "expeditor," and after sticking in Vienna for six weeks, reached us safely at last, when we were already well settled in a large new house near the Castle Hill, an isolated rock in the midst of the wide rich valley of the Mur.

The cathedral was connected by a covered bridge with the theatre; the fiddlers and singers were common to both institutions, and walked over it from one to the other, performing equally well in both. We heard many of the best operas very respectably done for two shillings a stall; but one thing the Gratzers insisted upon—they must have their suppers by ten o'clock; so that if the opera was a long one, it had to begin earlier so as to be over at the required time, or they broke the manager's windows the next morning. We often had to walk in by daylight to suit this rule.

I had a delightful old singing-mistress who had been famous in her youth, and had come out at Prague as the "Queen of the Night" in the *Zauberflöte*, and even then could reach F above the lines with ease. With her I went through all Mozart's operas and masses, singing and transposing all the solos and duets; and between singing and playing I often passed eight hours a day at the piano. I learnt many of Beethoven's sonatas by heart. My dear old mistress used to take me also to sing amongst her daughters and pupils in the organ-gallery of the cathedral, where I heard many great works, including Bach's *Passion* music and Haydn's *Last Words*. Of course Gratz was full of soldiers, mostly raw recruits, Border Men and Croats, who were brought there to be drilled before they were sent off to the wars then going on in Hungary and Italy. These poor creatures came from their homes in picturesque sheepskin coats, with the wool inside and embroidery outside; their well-shaped sandalled feet were forced into regulation boots, while their bodies were squeezed into tight uniforms. They looked very miserable, asking every one to change the paper florins (one of which had to be divided between three of them), though no one had any change to give them. Metal money was nowhere to be seen; the notes were usually torn into quarters, each quarter used separately till it became an illegible shred. All living was amazingly cheap. £7 a week covered the whole of our expenses for nine persons, lessons and operas included. We were the only English people in all Gratz.

About the end of April we turned northwards again. One day of railroad took us to Vienna, but we did not care to lodge a second time within the walls, the experiences of the autumn having given us a horror of being "shut in." We found the beautiful old city much knocked about by the siege.

We went on by slow railway journeys to Brunn and Prague, where we saw Kaiser Ferdinand at his devotions in the chapel of the noble Hradschin Palace, lingered on the old bridge among the statues, pulled some hairs out of the latest tail of Wallenstein's stuffed horse, and stared at the gorgeous Bohemian glass in the shop windows—it was just then the most popular of chimney ornaments; after a while the Venetian reproductions of Salviati made it look vulgar, then the still older potteries of Japan became the fashion, and what next, I wonder!

Our next move was eventful; for after floating peacefully down the beautiful Elbe, through the picturesque rocks and forests of Saxon Switzerland, we reached Dresden, and the very day after our arrival the revolution began. The king fled up to his fortress on the Königstein, taking all his most valuable treasures from the "Green Vaults" with him, and sending for soldiers from Prussia to come and reduce his beloved subjects and their city into order, by fair means or foul; while his pet Wagner, the Composer of the Future, harangued and led the mob against them. Barricades were raised in every street, and all traffic over the bridge leading from our side of the river was stopped by soldiers. My father bribed a boatman to take him across to see what was going on, and thought the Saxons were not so much in earnest over their civil war as the Viennese had been; but when the Prussians arrived and regularly besieged the place, then the fight began only too seriously, and the noise of cannon and guns went on incessantly for some days. The Prussian commander lodged in our house, and we saw from its windows the poor Zwinger and other fine buildings in flames. Most of the ambassadors also crossed the noble river to seek safety with us. But our own representative, Mr. Forbes, sent his sister only, and stayed, like a brave man, at his post, with the English flag flying over his house, ready to shelter any of his country people who wished to claim its protection.

It was a fearful week of anxiety, but at last the soldiers gained the day; the insurgents submitted unconditionally, and we saw the Prussians march over the bridge with green branches on their helmets and bayonets, amid much cheering and music. Of the poor wounded and killed we also saw enough to give a vivid horror of war to our minds ever after; but as far as we could judge, the good-humour and discipline of the Prussians on that occasion were very remarkable. They were soon billeted and quartered in every house in the city, and many were the stories told of the behaviour of the men, especially of the Alexander Guards, who were all of noble birth and education, though privates in rank, and who astonished their hostesses by playing the piano and other accomplishments (darning their socks also between whiles).

We were not sorry to escape and get into a clean roomy apartment the other side of the city. It was hot at Dresden in summer, but I worked hard and enjoyed three months there exceedingly. Ceccarelli, the chief singer of the king's chapel, found out my voice was contralto instead of soprano, so I tried no more high tunes. I learnt with him to know the grand old sacred music of Italy—Marcello, Pergolesi, Stradella, as well as Haydn, Hasse, and Bach.

We had many delightful days there, and of course learnt to know the famous pictures by heart—all that is too well known to need description. But I think what made the most impression on my memory were the visits we used to pay to old Moritz Retsch and his wife. They lived five miles from Dresden, quite in the country, in a small cottage amongst the vines; under those very vines the old artist is said to have found a baby asleep when he himself was a boy of twelve years old, and to have vowed that that baby should become his wife when she grew up. He superintended her education, and never changed his mind. She was an only child, and succeeded to her father's home, vines, and farm, and when we knew her was still a very beautiful old lady. Retsch had won many honours by his genius for illustration, but could never be tempted away from his quiet country home. He had never even been to Berlin, and only went into Dresden when absolutely obliged by his professorial duties; but although he had seen nothing of the world, his variety of fancy knew no bounds, and he worked it from morning till night. His original drawings were done with pencil, shaded with the greatest fineness, and were very unlike the bold outlines which have been engraved, and by which his name has become known to foreigners. He used to get five or ten guineas for one of these small pencil drawings—a great price in Germany at that time. He was exceedingly simple and quaint in his manner and talk, delighted in showing his children (as he called his drawings), and telling marvellous stories about them, while his wife brought out her famous coffee and cakes. In her room I saw for the first time a sofa with real growing ivy trained over it as a canopy. We also knew the painters Dahl and Vogelstein.[19] The latter gave my mother a beautiful study for his large painting of a martyr taking leave of her child through the prison bars, as he said the face resembled hers.

About the end of August we moved on by Berlin to Stettin on the Oder, and down that river in a barge towed by a steamer to Swinemunde. We went on the same day by a small steamer, and in about six hours landed in the Isle of Rügen, and soon settled ourselves in a nice little house at Putbus. The name of that place had become known to English people chiefly through its Prince having been the representative of Prussia at the coronation of our Queen, when he came over in great state to attend it. He was said to be one of the richest nobles of Germany, possessing a great part of Pomerania, which is famous for its corn (even more than for its "plum-pudding" dogs). The Prince and Princess lived in a large palace at Putbus, and were most kind and hospitable to all strangers, allowing every one to ramble as they liked through their lovely gardens and park. Of course the newly arrived English family were asked at once to dine at the Palace at four o'clock.

The chalk heights in Rügen are all sprinkled over with granite boulders, and Hünengräber are scattered in every direction, some of them very large. Fine beech-trees begin also with the chalk; though so much exposed to the north winds, the trees actually hang over the very edge of the Stubenkammer precipice, 400 feet above the sea. A small "gasthaus" had been built there, and about a quarter of a mile behind it was the temple of the goddess Hertha, resembling an enormous raised-pie of granite boulders, covered with earth, bedded in the thickest wood, and fringed over its very sides with beech-trees of large growth. Its walls were at least 50 feet high, their scarp still defined and steep; the edge, on which an easy path is traced, being just broad enough to hold it. This seemed most wonderful, as their date was beyond tradition. On one side the walls opened upon a dismal lake covering five or ten acres, most funereal-looking even in the brightest sunshine. Great beeches shaded it, and gigantic rushes fenced in the whole of its circular edge. Its waters were warm, and its depth was said to be unfathomable ; it had no apparent outlet or inlet. Higher up to the west was one large barrow or Hünengrab, among many others, from which the view was magnificent. The sun could be watched as it set gloriously in the sea towards Holstein, and one could sit and wonder at the hundreds of Hünengräber, and the still more mysterious granite blocks of which they were built. How were they brought to those high chalk cliffs, with which they had no possible geological connection? The glacier theories alone could explain it, and to a woman's mind the problem did not seem so difficult in this cold Baltic region as it did in the tropics, where I again encountered it after many years. A fine chalybeate spring scenting the air with sulphur added its odours to the wonders of the Stubenkammer, bursting out of the chalk cliff half-way down the steep path to its foot. Beech-trees followed it down, and grew close to the edge of the sea, with its hard bed of granite and flint pebbles.

The bathing at Putbus was delicious. Half an hour's drive brought one to a pretty wood, where a circular clearing had been made for the carriages and their drivers to wait in. Five minutes more took one on foot through the thick screen of bushes to a lovely sandy cove, with dressing-huts built on piles round it, and steps descending at different depths to suit people's ages and fancies. The water was always calm and clear as crystal, but only half salt; there was never more than a foot or two of tidal difference in the water's edge. Beautiful jelly-fish floated about, set with stars of all the purest colours, and could be easily caught in the hand, but melted away on dry land. Lovely sea-weed tempted one to collect, but shrank to a formless nothing when captured. There never was a more enjoyable bathing-place, and the old Princess herself used it in the same simple manner we did, except that her "shakebus" (no better than ours) was drawn by six fine horses. There was another small bay set aside for the men bathers. The shores of Rügen are never straight in any direction for half a mile; from every quarter little bays and inlets run up to the centre of the island, and the view from its capital, Bergen (which with its old Runic castle Rugard forms the very heart of it), was most remarkable,—one saw far more sea than land, while the abundance of fine old trees and high cultivation reminded one of England.

The park at Putbus was full of deer, so tame that they would take bread from our hands. The Prince's Kapelmeister introduced me for the first time to the music of Handel—such a great event in my life that I found myself wondering the other day when looking over my father's journals why he did not mention it, till I came to my senses, and remembered that "all music" (to him) was "a horrid noise, which must be submitted to for the sake of others who like it." Herr Müller used to take me to the organ-gallery of the Palace Chapel, and accompany all the noble old songs from the oratorios on the organ, transposing those that were out of my reach in their original keys, as Madame Sehr had done, thus giving me the delight of learning their glorious melodies, which once learnt can never be forgotten. They have filled many a wakeful night and weary day of voyage with pleasant memory. None who really love music can be dull, and it can be thus enjoyed without disturbing others by "a horrid noise." I wish practising could be done equally silently, and fear I must have been a perfect nuisance to all my neighbours in those days.

We stayed till all the other guests were gone, and the little steamer had ceased to run; so the one real carriage at Rügen was placed at my mother's disposal, while the rest of the party and our twenty packages were packed into carts. We drove over the sands to the ferry, and crossed to the old fortified island of Stralsund, with its many towers. The pavements were barely above the level of the sea and moat.

Hagenow was the end of our drive, whence the railway took us to Hamburg in a few hours, where we rested and looked at the busy streets and muddy canals and the reflected gas-lights in the great Alster Dam from our windows at night. Started again by rail to the semi-English Hanover, then full of royal red liveries and soldiers, and by Cologne to Brussels, where we again settled in a lodging for six months, and my practising became more incessant than ever under Herr Kufferath, a pupil of Mendelssohn, to the study of whose works I devoted myself that winter.

Two people I saw and heard at Brussels whom I cannot forget, Madame Sontag and Mrs. Norton.[20] The latter I found one day talking to my mother in our sitting-room, and without knowing anything about her, felt quite awed by her grand beauty and deep bell-toned voice. She was a real queen. Madame Sontag I only heard at a concert, but having once heard I persuaded my father to take me to each of her performances. Her singing had a manner and perfection peculiar to herself; and her history, as we heard it, gave me an extra interest in her. She had been for many years the wife of Count Rossi, the Sardinian Minister at Berlin, who was then pauperised by the Revolution. She had taken up her old profession to support him and her children. Report also told how Jenny Lind had retired from the opera in England to make way for the older woman, who had shown her much kindness when unknown and in want of friends—a story I liked to believe true, whether it were so or not. Another wonderful woman I saw in Paris a few weeks afterwards, whose image does not give me pleasure but a shudder to think of, I mean Rachel as I saw her in *Phedre.* It was fearful, and the acting must have been horribly real to engrave itself so vividly on my memory. I can hear the tremulous thrill of her voice now.

In London that year I had some lessons in flower-painting from a Dutch lady, Miss van Fowinkel, from whom I got the few ideas I possess of arrangement of colour and of grouping, and then we recommenced the happy old life at Rougham, I passing hours and hours of every day on horseback, painting and singing with little fear of interruption. Our neighbours were kind, but we were not what is called a sociable family, and the few balls or parties were to me a penance. I hated the dressing up and stiffness of them, and the perpetual talk of turnips, partridges, or coursing, of my partners. The farmers tried to look like squires, the squires like gamekeepers; the women had the same ideas (though of course there were exceptions).

The next season I saw the opening of the first great Exhibition (1851), with the Chinaman admiring the real live Duke of Wellington, who thought the whole thing very dangerous.[21] I also went to a drawing-room and began studying singing again under Miss Dolby. I never had any other mistress, but learnt to admire her more and more, till both our days for singing were over. I loved her for herself, as well as for her voice, and I believe she liked my singing, as she used to make me take the contralto solos in the concerts she gave her pupils in concerted music, while the other solos were sung by professionals; but I grieve to say I never did well on those occasions, having a most provoking habit of nervousness; when told to stand up and show off, the room seemed to go round and round, and I could not keep myself from shaking all over.

Bartholomew also gave me a few lessons in water-colour flower-painting: the only master I longed for would not teach, *i.e.* old William Hunt, whose work will live for ever, as it is absolutely true to nature.[22] We used to see a good deal of him at Hastings, where he generally passed his winters, living in a small house almost on the beach under the East Cliff, where he made most delicious little pencil-sketches of boats and fishermen. I can see him now, looking up with his funny great smiling head, and long gray hair, above the poor dwarfish figure, and his pretty wife, with her dainty little openwork stockings and shoes, trying to drag him off for a proper walk on the parade with her daughter and niece, where he looked entirely out of character. I remember "That Boy," too, whom Hunt taught to be anything he chose as model, blowing the hot pudding, fighting the wasp, or taking the physic; the apple-blossoms and birds'-nests, with their exquisite mosses and ivy-leaved backgrounds, were found in the hedges and gardens about Hastings. Prout also lived in a Hastings lodging in George Street. He was very delicate, and used to draw even in his bed. These two, as well as other artists, used to spend much of their time in the house of Mr. Maw, a man of great taste, who, when he had retired from business, settled in a house on the West Hill, where he made an exquisite collection of Turner's water-colours. He also fitted up a studio with rare old oak, which became the background to many famous historical pictures. His daughter was my great friend as a child, and his son has since made himself famous by his beautiful book on crocuses. But the neighbour in whom we delighted most was the Rev. Julian Young, son of the actor Charles Young, who had insisted on his going into the Church when his natural talent and inclination were for the profession of his father. He could not tell the smallest tale without acting, bringing his face unconsciously into the likeness of the person he was speaking of. His voice was as elastic as his face, and his singing as good as his acting. He was in our house quite three times a week, it being a convenient halt between his home at Fairlight and the town. The days seemed brighter when he came. He did his duties as a clergyman most conscientiously; but his congregation did not appreciate him, and they were much shocked one Sunday when he kept them all out on the edge of the cliff to see the Baltic Fleet pass up the Channel till long after the usual time for the service to begin, and then talked of it in his sermon, and of the men from the Signal Station on those cliffs who were in that fleet, and prayed that they might return safely again: it was considered a most unorthodox and undignified proceeding.

Our garden was much of a weedery in those days at Hastings, but in spite of neglect many tender shrubs grew high in the mild climate and sheltered situation between two hills; myrtle, sweet bay, and fig-trees flourished there. The latter tree tempted another artist, Edward Lear, the author of the *Books of Nonsense*, to settle himself as a lodger in the cottage of our gardener close by, and finish there his great pictures of the Quarries of Syracuse and Thermopylæ, with our fig-tree in the foreground of the former, a group of ravens in the latter, all of them painted from one old specimen with a broken leg, which was fastened to an apple-tree opposite his windows.[23] He also painted a great view from Windsor for Lord Derby, with some Southdown sheep in the foreground, which my father bought on purpose for him, and kept in the field within sight of his room—a kindness he never forgot, and repaid in friendship to his children and grandchildren. He was most good-natured in letting us watch him at work, and used to wander into our sitting-room through the windows at dusk when his work was over, sit down to the piano, and sing Tennyson's songs for hours, composing as he went on, and picking out the accompaniments by ear, putting the greatest expression and passion into the most sentimental words. He often set me laughing; then he would say I was not worthy of them, and would continue the intense pathos of expression and gravity of face, while he substituted Hey Diddle Diddle, the Cat and the Fiddle, or some other nonsensical words to the same air. I never was able to appreciate modern poetry, and still think it is sense worrited, and often worrit without the sense.

In May 1854 my father became again M.P. for Hastings, being elected without opposition on the death of Mr. Brisco.

On the 17th of January 1855 my mother died. Her end had come gradually; for many weeks we felt it was coming. She did not suffer, but enjoyed nothing, and her life was a dreary one. She made me promise never to leave my father, and did not like any one to move her but him; he was always gentle and ready to help her, and missed her much when she was gone, writing in his diary in his own quaint way: "The leader is cut off from the main trunk of our home, no branches, no summer shoots can take its place, and I feel myself just an old pollard-tree."

My father let Hastings Lodge, and took a flat in Victoria Street. Soon it became more like home than any other to me, and was a great rest after the big house at Hastings with its perpetual visitors; for my father tried to be civil to everybody, and always knew the principal people who wintered there. In our flat we had few servants, and needed no bells, as they were all within call. None but real friends came to see us, as eighty-seven steps were a trial to any friendship, and kept the mere acquaintances away. We had one friend, an old retriever dog, who did not mind them at all. He was called Jill (his brother Jack having tumbled down and broken his crown when young). He and I used to accompany my father to Westminster when he went in the morning to his committee-work, the dog scampering all over St. James's Park like a wild thing till we reached the Abbey, when he was told to go home with me. He would put his tail between his legs, his nose in my hand, and pace solemnly down Victoria Street by my side, never even looking at a dog till he had seen me inside the door; then he would give a sharp bark of joy, and gallop back to wait for his master in Westminster Hall, and would stay starving there till the small hours of the morning. Jill knew all my father's haunts and ways, and if he missed him anywhere would go and look for him at the Athenæum about tea-time, where he was often to be seen sitting like a sphinx on the steps, much patted by the bishops and other great people. All the House of Commons policemen knew him, and used to lead him home when the House was counted out—a mode of proceeding Jill could never understand. His end was sad; he was run over by a cab, and we mourned a real friend.

We rode often to the Chiswick Gardens and got specimen flowers to paint; were also often at Kew, and once when there Sir William Hooker gave me a hanging bunch of the *Amherstia nobilis,* one of the grandest flowers in existence. It was the first that had bloomed in England, and made me long more and more to see the tropics. We often talked of going, if ever my father had a holiday long enough.

London was full of delights for me, though I never went through much of the treadmill routine called "Society"; when my father had a holiday he liked to spend it at home. At Rougham the usual round of farming, shooting, and petty country traffic went on, then back to Hastings with its heavier traffic of big dinners and pleasant music parties, half amateur, half professional, of which Baron de Tessier, a naturalised French refugee, was the centre, his zeal being occasionally greater than his skill. One musical evening at Hastings Lodge about this period is memorable in my mind. The "Toy Symphony" of Romberg was performed with a distinguished cast, Mr. H. Brabazon (an accomplished amateur) at the piano, while Prosper Sainton and Carl Deichmann took the two violins, and Madame Sainton Dolby played the big drum with a will.

Agnes Zimmermann, too, was often there, from the time when we first knew her—a pale over-thoughtful child of eight, to whom little of childish pleasures ever came. She used to play Beethoven's sonatas through by heart, perched on a high music-stool, till she grew to be the thorough musician she is now. She has been one of my life-long friends.

But the society we enjoyed most was a set of rare old friends who came to us every winter, and talked together delightfully. My father was no great talker himself, but enjoyed listening to others, when his deafness allowed him—that was his great trouble; but some voices he always heard, mine especially, however low I spoke. There is no greater mistake than to shout to a deaf person. I am quite deaf in one ear myself, but the nerves of the other are most tender, and some harsh voices give me positive pain.

Next season we went another way down to Norfolk by Harrogate, where an Indian uncle was drinking the waters. How that place smelt of sulphur! Thence to Kendal and the Lakes, all a land of delight to us; and after going the usual tourist round, we settled for a week with our old friends, the Francis Galtons, near Grasmere, in a farm-house on the hillside, to rest and talk in quiet without "moving on."[24] After a month of mountain air we returned to the flats of Norfolk, where soon after my brother engaged himself to the eldest daughter of our neighbour, the Hon. and Rev. Thomas Keppel, Rector of North Creake, and they were married on the 16th of March 1859.

After that a stormy Parliamentary session succeeded; my father forming one of a small party of Liberals who called themselves then the St. Stephen's Club.

We let the house at Hastings for that summer (and the two next also) to Count Poutiatine, the famous Russian Admiral, who ran the blockade of the White Sea so cleverly during the war. He had an English wife and a most polyglot family. The children talked five languages, besides studying all other accomplishments. They had an English tutor, a German governess, a French *bonne*, an Italian valet; they studied Latin, Greek, and Hebrew, besides their own difficult Russian tongue. Our old housemaid stayed on with them, and declared it to be as good as being in the tower of Babel to live at Hastings with these people. When the House was up we three wandered off abroad, starting by way of Jersey for the Pyrenees and Spain, returning in an English ship from Cadiz to the Thames on the 3d of January 1860.

After my brother married, my father gave up the old house at Rougham to him, and each summer, when the Parliamentary session was over, we three, with our three old portmanteaux (their collective weight nicely calculated under the 160 lbs. allowed on Continental railways), used to start forth on some pleasant autumn journey. My father loved the deep romantic valleys round the southern slopes of Mont Blanc and Monte Rosa, and there summer after summer we found ourselves walking over easy passes, with just enough of necessaries to be easily carried on an Alpine porter's back, staying a while at Macugnaga, Gressonay, Courmayeur, or Varallo, till we joined the welcome portmanteaux again at some point on the Italian Lakes. Baveno was the place he liked best; it had no huge hotel in those days, but a pretty primitive old inn, painted pink, on the Lake shore, with a kindly landlord who made old friends welcome.

In the autumn of 1861 we made a longer journey to Trieste, Pola, Fiume, by the Hungarian Lake of Balaton, where grew such grapes as I have never seen elsewhere in Europe, to Pesth and Debreczin. Here we were lucky enough to see the wild humours of a great Hungarian fair, with horse-races, and a superb gipsy band. Then down the Danube and across the Black Sea to Constantinople, Smyrna, Athens, and home by sea to Marseilles.

The winter of 1863-64 was a merry one. We had a succession of nice people staying with us, whom our young cousins in the Croft used to describe with youthful flippancy as "Old Couples without Encumbrances." Sir Edward and Lady Sabine (he was President then of the Royal Society, and she no less wise than he, though as unassuming as a gentle child), the Benthams, F. Galtons, Erskine Mays, Knoxes, and Lady Hawes, with a queer household of polyglot servants.

The great event of the winter was a fancy ball given at Beauport by the Tom Brasseys, most hospitable of youthful hosts.[25] Old Mr. Brassey, the father, was a grand specimen of an Englishman, with all the instincts of a real gentleman, generous, honest, and most simple in all his ways, though he left more than three millions among his three sons. He delighted in telling how he had saved up his first hundred pounds, and then been helped on by Robert Stephenson. Long before we met him we had heard his praises sung at Smyrna, for the way he took care of his men and their families while making the railway through those feverish plains, keeping, at the same time, such discipline among the navvies employed, that the English name was not lowered as in other remote countries by tales of drunkenness and dishonesty.

In the summer of '64 we went to Pfeffers, crossed the Julier Pass to Samaden and Pontresina, and settled ourselves in that paradise of Alpine climbers, the Old Crown Inn. In those primitive days it was nearly all built of wood (as was also the old Chalet Inn at Mürren), and as the majority of its frequenters delighted in getting up in the very smallest hours of the morning, putting on heavily-nailed boots, and shouting at one another from room to room, it could not be called quiet quarters. A merry party of young people were collected there, including Mrs. C. and all the Zigzaggers, whose adventures our friend Miss Tuckett illustrated so capitally, also John Addington Symonds, whom we had met at Mürren the previous summer, and Professor Tyndall.[26] The latter invited my father to join him in a search-party to look for his watch, which had been swept out of his pocket during a wild ride on an avalanche. They found the watch (a gold one) after many hours, on the glacier, safe under a stone, which had sheltered it from the sun's rays. It had quietly run itself down, and when wound up went on as merrily as usual : so did our brave old father on that adventurous walk.

Mrs. Gaskell was also at Pontresina at that time, and had taken a quiet room outside the village to work in peacefully.[27] There she finished a great part of her last story, *Wives and Daughters.* She was very beautiful and gentle, with a sweet-toned voice, and a particularly well-formed hand.

1864.—On the 10th of November my sister was married to

John Addington Symonds, in St. Clement's Church, Hastings, and narrowly escaped being given away by her masterful great-aunt, Lady Waldegrave, instead of her father. That old lady had for many years ruled over our family at Hastings with no gentle sway, for which its younger members did not love her, but she was the last of a generation now long passed away, and her stories of the good old smuggling days and the primitive ways and dissipations of the Hastings fashionables of her youth would have been worth chronicling.

1865.—In July came a general election, when my father lost his seat by only nine votes. As it is an ill wind that blows nobody good, we were able to utilise this period of unwished-for leisure, and carry out a long-cherished plan of a journey to the East. He and I started from Trieste by the Austrian Lloyd boat, which coasts the Adriatic by Spalatro, Ragusa, and Cattaro, spending eleven unwilling days in quarantine in the harbour at Corfu: then going by Beyrout to Damascus.

We spent the winter on a Nile boat, going as far as Assouan; the spring of the following year in Syria, returning to England early in the summer of 1866 by Carinthia and Tyrol.

We did not stay much in London that next season, 1867, but devoted ourselves to the Hastings garden. My father built three glass-houses: one for orchids, another for temperate plants, and another quite cool for vines and cuttings. We lived in those houses all the spring, my father smoking and reading in the temperate regions, where we had a table and chairs, while I washed and doctored all the sick plants, and potted off the young seedlings. It was delightful work, and though the constant change from damp tropical heat to cold English east winds brought on a most irritable rash on my face and hands, I thought the pleasure of seeing the growing wonders compensated amply for the pain. We established water-pipes all through the garden, so that we could give the plants any quantity of irrigation, and we had all sorts of shady nooks under the great bay-trees, every variety of aspect and ground, and could grow most things well. We used to work like slaves, and were often working till it became too dark to see our flowers any longer, having only a young Wiltshire gardener of eighteen to help us, with a Rougham boy under him, and Garibaldi the poodle to look on. That dog was supremely wise; as a puppy he had been the hero of a story which Miss Cobbe has rendered classical. We had gone to a play in London. Baldi was left in the dining-room, with the supper ready laid on the table. He took a pigeon out of the pie and ate it, then looked about till he found the sponge my father used to wipe his pens on, then put it in the pie in the place of the pigeon, where our old servant Elizabeth found it. He was very fond of being washed, and used to collect all the things he considered necessary for the ceremony as soon as he saw his bath put out, and bringing them to it, sit by its side till told to get in: towels, sponges, old boots, gloves—all sorts of things he always brought together. His devotion to his master was most touching, and he seemed to understand if he was ill.

Another pet we had then, who was equally devoted to him alone, a green paroquet with a beautiful rosy ring round its neck. I saw a *Times* advertisement one day from an old soldier who had got a place as railway guard, and wanted to find a home for his parrot, not having time to look after it, so I answered it, and bought it. It called itself Jewy, and as it had always been in a regiment among men it hated the sight of a woman, and always flew at me, but attached itself at once to its master. It would fly to him whenever he came in, sit on his shoulder, kiss him, feed him, and say all it could say over and over again. If he went to sleep, it would walk gently over him, saying "Hush! What's the matter?" to any one who moved. We had a third pet and gem of all that year at Hastings—my sister's little girl Janet, the most fairy-like little creature, with grave thinking eyes which brightened up with extraordinary intelligence when one talked or sang to her. I can never forget the pretty pictures she made, poking her way through the long uncut grass stalks, great ox-eyed daisies, red sorrel and clover, and the odd little flower-arrangements she used to make, pulling off the heads and buds, with no stalks, mixing them up with feathers, pebbles, and shells. She wanted no expensive toys; the garden was her bazaar, and all nature her delight.[1]

After this came two more short journeys to the Italian Tyrol and to Mentone and the South of France. Then in November 1868 George Waldegrave's resignation brought on again the worries and work of a contested election. My father and Mr. Brassey came in with a large majority, but a petition to unseat them was at once lodged by the opposite party.

It was most galling to my poor old father, who had been all his life fighting against bribery. It is best to write no more on the subject, as it would only bore readers and injure my own temper. The very implication of corrupt practices broke his heart, though he knew his name was only dragged in from his connection with Mr. Brassey, who had the reputation of great wealth, but who always behaved most thoughtfully towards him in the whole matter.

It was a wretched year at Hastings, though all our kind old friends came down to cheer us with their company—the George Normans, Sabines, Benthams, etc. Then my father resumed his old work in Parliament, but his spirit was broken and his health declining. However, the suspense came to an end at last, and after five days' trial Mr. Justice Blackburn dismissed the petition with costs on the 17th of April 1869, our excellent old tenant at Rougham, Mr. Ringer, paying in his half-year's rent some months beforehand, as he said he "feared the Squire must have had some expense about the petition.

1869.—On the 4th of August we started for Gastein by way

[1] Janet H. Symonds died at Davos, April 1887.

of Frankfort. That journey is so full of painful remembrances that I shall make the note of it as short as possible. After a few days' rest at Salzburg we posted on to Gastein, and got our old rooms at the Hirsch. He grew so strong in a fortnight that he planned walking over the hills to Heiligenblut—eighteen hours! We went up an Alp 3000 feet above Gastein to try our powers. He came back so well that he went up another hill the next day, leaving me to rest at home, but it was too much; his old disease returned. We hastened down to Salzburg, where the doctor advised us to get home. At Munich he arrived in the greatest state of suffering. The people at the inn were kind, and persuaded me to go for Dr. Ranke. At last I got him safe home. He was so glad to be there, and to see Catherine and his friends again, that they would not believe how ill he was. Even his old friend and doctor, Mr. Ticehurst, did not discover it at first. After a last three days of exhaustion and sleep he ceased to live on the 29th of October. The last words in his mouth were, "Come and give me a kiss, Pop, I am only going to sleep." He never woke again, and left me indeed alone. I wished to be so; I could not bear to talk of him or of anything else, and resolved to keep out of the way of all friends and relations till I had schooled myself into that cheerfulness which makes life pleasant to those around us. I left the house at Hastings for ever, and my affairs in the hands of our kind old friend Mr. Hunt of Lewes.

*She took with her our old servant Elizabeth, and started for Mentone, then after a while by slow journeys along the Riviera to Sicily, where the whole of the following spring was spent and many sketches made. In the summer of 1870 she returned to take up life alone in the flat in Victoria Street, which was henceforth to be her home.*

## CHAPTER II

### CANADA AND UNITED STATES

### 1871

I HAD long had the dream of going to some tropical country to paint its peculiar vegetation on the spot in natural abundant luxuriance; so when my friend Mrs. S. asked me to come and spend the summer with her in the United States, I thought this might easily be made into a first step for carrying out my plan, as average people in England have but a very confused idea of the difference between North and South America.[28] I asked Charles Kingsley and others to give me letters to Brazil and the West Indies, his book *At Last* having added fuel to the burning of my rage for seeing the Tropics.[29]

1871.—On the 12th of July I joined my old friend Mrs. S. at Liverpool, the next day we packed ourselves into a comfortable cabin on board the Cunard steamer *Malta*, and moved away westward. It was rough, and a young French officer thought he was dying, sent for Mrs. S., and asked her to take his last will and testament to his betrothed at Boston. He wished her also to ask the captain to stop and let him out, and he would go on by the next ship. He also wanted her to make the steward bring him some pudding he called "by and by," which the latter was always promising and never brought. My friend promised to do all he wished, and that consoled him, and he didn't die.

The very sight of Mrs. S. did any one good; her head was covered with little curls of pure silver, her complexion was very fair, and she wore a purple knitted cobweb pinned on the back of her head, and diamond earrings. She was full of jokes and continual fits of laughter, her quaint American accent making her talk all the more amusing. She had a very pretty, but perfectly useless little French maid, and an enormous quantity of luggage, which was the plague of the little maid's life, for she could never find anything that was wanted, and used to wring her hands and exclaim "quel horreur!" at everything her mistress required. The people on board were a very uninteresting set of Yankees first class, and Irish second. The latter made a terrible noise all night, and the ship's officers did not keep any sort of order amongst them. The last night both ends of the ship were in such a state of uproariousness that we shut both our door and windows to keep out the row, and were nearly stifled in consequence.

After the usual scares of fog and icebergs we arrived safely in Boston harbour, and F. S. was soon on board, coming out with the pilot to meet us, and accompany us in through its many islands, crowded with all kinds of sailing and steaming vessels and many pleasure-yachts.[30] He employed his leisure moments in cramming me with stories about the inhabitants, aborigines, etc. etc., till we were sore with laughing. Yankee stories cannot be written; it is the dry peculiar way a clever American tells them that gives them their charm. The Custom-house kept us a weary while. The officers were all smoking and in no hurry, but they liked to see others work. As there were no porters, F. and I had to collect and carry all the boxes, which were distributed about a barn-like building in hopeless confusion, having to be hunted up out of its holes and corners. It was warm work, and I rather wished for a little less independence and more activity on the part of the gentlemen who were smoking and looking on at our exertions. There were no cabs, but we got into a great lumbering thing like a mourning-coach, and sent on the luggage by an "express," which means "come some time." Our horses, after shying right round at a railway train which whizzed past just across their heads, were driven on to a ferry-boat, which seemed to consist of two long square tunnels with a steam-engine in the middle, which with much puffing and groaning crossed the harbour, and landed us on the other side, together with a whole string of carriages, and we drove over the badly paved streets to the Somerset Club, from which F. brought out a most delicious bouquet of roses and jasmine and other sweet flowers for me, besides several old friends who came to welcome his mother home again. Then we called at her old house in Beacon Street, a sort of Park Lane looking over the public gardens. It was built, like all the other houses, of a beautiful dark red sandstone, with silver handles and knockers to the doors.

Then we drove out to Newton, about six miles into the country, where Mrs. F. and the baby welcomed us, and next day she drove me into the lanes, where I found many new plants; one of the Sweet Gale tribe, called Comptonia, was very common by the road-side, and had a delicious scent; they called it "Sweet Fern," and indeed its leaf had a brown furry back, and was much like our ceterach; its leaves are sometimes dried for smoking instead of tobacco. Large-leaved oaks, white pines, hemlock spruce and arbor vitæ hedges, wych-elms and maples, all showed one was not in England.

We finished our drive by a visit to "Jamaica plain" and its famous confectioner: tied up our pony to a post, then went inside and ate the largest "ice-creams" in a small and perfectly undecorated room. I was also introduced to cocoa-nut cakes and iced gingerbread, all first-rate, and then taken to see Mrs. S.'s garden, also a model dairy, with great lumps of ice slowly melting in it, and a regular stove and hot-water pipes for keeping it to the same temperature in winter, a new idea to English brains.

In the evening we dined with the parents of my hostess

at Brookline in the midst of another pretty garden: the whole piazza round the house was covered by one creeper—a fruitless vine, a dense mass of foliage. After dinner we strolled through the near garden and plantations to the top of a little hill to see a flaming sunset amid great thunder-clouds, and returned to the piazza for tea, when other neighbours joined us. The dinner-hour was generally four or five in these country houses, so that one could always have a stroll or a drive afterwards to enjoy the beautiful sunsets and evening coolness before tea-time and darkness came. I found very little dressing-up necessary, far less than in England.

After three days Mrs. S. and I settled ourselves in the house she had taken near West Manchester. It was built on the foundation of a fort or tower on the rocks, against which the sea washed on three sides at high water; the rocks were tinted with pink, red, brown, and gray, and above high-water mark with soft gray, green, and yellow lichens, wild grass, and scrub: there was no garden, and we wanted none. On the west side the sea ran up into a little sandy bay, the very ideal of a bathing-place; a few steps would take us down into it from the back door. On the east side were holes under the steep rocks, where we could find water at the very lowest tides, and these were almost as easily reached for bathing. So we bought some stuff to make bathing-dresses at "John Loring's, one price-dry-goods-warehouse" in Boston. Mrs. S. had chosen one of scarlet serge, I one of dark blue-gray, so that we looked much like two large lobsters in the water, one boiled, the other unboiled, but spectators were not common; we had three houses within sight, but none of them within half a mile. There were endless islands on the coast, and a lighthouse a mile off, which used to keep its bell perpetually tolling when there was any fog or mist; it was deliciously wild and quiet, with a beautiful mixture of rocks and green, and even a bit of marsh near, with tall bulrushes, reeds, and ferns, butterflies and wild flowers. Boats, with their clear reflections, were constantly passing over the bay, and the sea was an exquisite blue in the hot noondays, when I used to sit in the balcony or piazza which ran all round the outside of the house, as in most New England houses: on one side there was always shade and cool air, even on the hottest days. Our landlord's Newfoundland dog used to pay us many visits, and stand any length of time in the water watching me, in hopes of a stone being thrown in and giving him an excuse for a swim. The house had only just been built, and was furnished with clean new-polished fir-wood or basket-work, with Indian matting on the floor. Mrs. S. used to go up to her town house and bring down pretty things, brackets, and books, until she soon made it look most home-like. We had a fat cook, who had imported herself from Ireland twenty years before, but had not yet exchanged the brogue for the twang; a housemaid, who left her husband to come back to her old mistress; and Marguerite of France, who looked pretty and waited well, saying continually, "que je suis bête moi!" and "quelle horreur!" at everything new, especially grasshoppers and spiders, occasionally jumping on chairs to avoid them.

We used to go for a drive of an evening, and bring home great bunches of scarlet lobelia, which they called the "Cardinal Flower," white orchids, and grand ferns, smilax, sweet bay, sumach, and meadow flowers, to dress up our pretty rooms. The railway station was about four minutes' walk from our house—a shed with three chairs and a red flag, which we stuck up on the end of a bamboo placed there for that purpose if we wanted the train to stop. Newspapers and letters were thrown out by the guard as he passed; whoever happened to be going that way picked them up and distributed them, tossing ours in at any door or window that happened to be open: we had also a post-box, No. 115, at West Manchester, in which letters were sometimes found and brought over by friends.

Every day some fresh things were sent down to our little house; first a huge piano of Chickering's which filled two-thirds of the parlour, and, our captain said, "almost made the fish jump out of the water to hear it!" This captain was our only man, and a sort of king of the coast fishermen: he used to tie his boats to our rocks, and do "odd chores" for us, and I soon made fast friends with him. One day Mrs. S. arrived at the station with seventeen large parcels of pillows, books, pictures, etc. etc., and I asked the captain to come and help me to fetch them home: "Wall now, you hain't got a wheelbarrow, have you?" And I answered: "Wall now, if you had only mentioned it in time I'd have brought one over from England with me, but I didn't know you'd want it." On which the captain winked at me, and did without it.

We had a carriage with skeleton wheels, and used to drive along the shores to a real forest, which had never been anything else, though the trees were not very high. Kalmias and magnolias grew there, but the flowers were over. The cottages too were often 100 or 180 years old, and all were built of wood.

It was an idle enjoyable place, but the heat was too dry and glaring for much work. F. kept us supplied with American papers, coming backwards and forwards himself for a night at a time, boiling over with jokes and stories, and making us laugh till we cried. Sometimes I also used to take a hot day in Boston, sorting and packing with my friend; once I persuaded her to rest there quietly till morning, and started home myself, with a lot of her odds and ends, in the evening. I "got along" all right, merely getting into a wrong train, having to wait a while and change at Beverly, which was clear gain, as I could study the manners and customs of the natives. The guards took most kind care of me, and lifted my odd parcels in and out of the trains for me, including a huge basket of crockery, full of ornamental "vayser," a tin box of butter with ice gradually melting its way out, six parcels tied to two baskets of fruit, one of which came open on the platform and discharged peaches with marvellous rolling powers, one red bag, and a parcel of umbrellas, that was all! . . . There are no porters in America, but every one is courteous and helpful if you are civil too. When F. sent letters or papers to his mother by our landlord, who went to and fro to Boston every day, he always wrote outside, "By the extreme politeness of Mr. B.," and the old gentleman or his son turned out of their way to bring them up to our piazza with many bows, and we gave them many thanks and curtseys in return. Curtseys are still practised in New England, and one soon got into the habit of the thing too; Americans on first landing among us must be much struck by our want of manners.

The trains used to mark the hours at West Manchester, and on Sundays, when there were none, I never knew how time went, and wondered that we got our meals as usual. The food on those days was always extra good—huckleberry-puddings with cream were quite divine, and corn-cakes and chowder, a most glorious compound of codfish, soup, and crackers, not to be tasted off that coast from the St. Lawrence down to the Hudson. There were no "classes" in the train; but one night the ladies were invited to go into another car, as the workmen would be coming in and there would be plenty of smoking, "which ladies did not always like," the conductor explained to me. Tickets are taken and paid for in the cars. One day we went in them to Lynn, a town which is entirely inhabited by shoemakers; indeed all the country round is famous for that work, as the ground is so dry that agriculturists get but a poor living from it alone. Nearly every small farmer has a shoe-shop for spare hours and winter work. From Lynn we drove over a mile or two of sandy causeway, with sea on both sides, to the former Islands of Nohant, now a fashionable watering-place. Even there the houses were all detached, standing well apart, unpretentious wooden buildings with verandahs round them like ours. Longfellow was living in the house of his brother-in-law, Mr. A. The latter had invited us to dinner, and then gone out on a yachting excursion, which every one said "was just like him"; but the grand old poet with his daughters was expecting us under the piazza, and his kind sweet gentleness of manner and pleasant talk quite fascinated me. We spent some most delightful hours listening to him, then missed our train, and had to return in a slow luggage train. Another day we dined with some friends in the country and sat all the afternoon watching the little green humming-birds darting about the nasturtium flowers trained over the verandah. They used to build their nests in the apple-trees near, and come back year after year to the same trees, but my friends had never marked them particularly, and could not tell if they were the same birds.

On our way back that night we met a wooden house of three storeys being moved some hundred yards on rollers by means of a windlass. It entirely filled the road, and we had to drive over a field to get out of its way. The Bostonians have even moved large stone houses bodily in the same way for some small distance.

I paid a visit to Mr. and Mrs. Adams at Quincy, one of the oldest houses in America, full of curious family portraits and furniture, large low rooms, big open fireplaces, many windows, and old brocade hangings. Except for the outside being entirely of wood, it might have been taken for a two hundred years old manor house in England.

Mr. A. had lately added a stone fireproof library, detached, in the garden, to keep all his precious books and manuscripts in, for he had whole volumes of Washington's letters, and many besides written by all his greatest countrymen. His father and grandfather had both been Presidents; he might have been one himself, but he never would allow himself to be put forward as a political candidate at home, though he was long his country's representative in England. He was a remarkably quiet man, but his good wife made up for it, and her genial chatter used to make him sit and shake with laughter. It was a very pleasant family to be in, all the sons and his daughter had such a thorough respect for their parents, and when he did speak he was always worth listening to. The two elder sons had built houses within ten minutes' walk of their father's, and came in every evening to have a talk with him, going into Boston in the morning to their business and back by rail. Their children and wives were in and out all day long. From Colonel A.'s house there was a glorious view over sea and land, the former being about a mile off, and the whole coast broken up by estuaries, islands, and points connected by low isthmuses, so that one could never feel quite sure where the sea began and the land ended, being in that respect much like the island of Rügen.

The floors, staircases, and chimney-pieces were of different sorts of wood — black walnut, butternut, hickory, ash, and pine — beautifully put together, with very little ornament, sometimes a line or simple geometrical pattern cut and filled with blue or red, and the rich natural colour of the wood kept as a ground-work. Though the house was on the top of a hill there was abundant water everywhere. In the old house at Quincy was one room wainscoted with polished mahogany.

I heard an American story of Lord Grosvenor, who, when travelling, met a Yankee in the train, who asked him how he got his living and what trade he followed. He said, "he didn't do anything, and his father supported him." "What a dear old gentleman! How will you ever manage to live when he dies?" An American cannot understand that a son succeeds to his father's property; they are all expected to make their fortune for themselves, and it is considered almost a disgrace for a young man to have "nothing to do." I went one day with the S.s to see the free library in Boston, a splendid building. Any one who is introduced by a note from any householder in the city may not only read, but take home the books. I saw quite young girls and boys, as well as numbers of men and women, reading there. Another good institution in Boston was the ladies' room attached to the Somerset Club, at which I often dined or lunched with Mrs. S., F. or one of the other members passing us in. There were nice dressing-rooms, and a reading-room, as well as a refreshment-room expressly reserved for ladies belonging to the members.

The C. F.s were spending the summer near us at West Manchester, and were very good to me when Mrs. S. was away. He was partner of Ticknor, and editor of the *Atlantic Monthly*, and she a pretty poetess who went into floods of tears at the mere mention of Charles Dickens, whose name resembled that of his own "Mrs. 'Arris" in their mouths, and their room was hung all round with portraits of their hero.

[31] I enjoyed my expeditions with him and his wife. He invited me to meet Mrs. Agassiz at a picnic one day, and called for me in his pony carriage, picked her up at the railway station, and drove us to one of the many beautiful high headlands on the coast; then we walked over the cliffs to find a most curious old cedar-tree, perfectly shaved at the top like an umbrella pine by the sea winds, with its branches matted and twisted in the most fantastical way underneath, and clinging to the very edge of the precipice, its roots being tightly wedged into a crack without any apparent earth to nourish it. It was said to be of unknown antiquity, and there was no other specimen of such a cedar in the country; it looked to me like the common sort we call red cedar. We sat and talked a long while under its shade. Mrs. Agassiz and I agreed that the greatest pleasure we knew was to see new and wonderful countries, and the only rival to that pleasure was the one of staying quietly at home. Only ignorant fools think because one likes sugar one cannot like

salt; those people are only capable of one idea, and never try experiments.

Mrs. A. was a most agreeable handsome woman; she had begun life as a rich ball-going young lady, then, on her father losing his fortune, she had started a girls' school to support her family, and finally married the clever old Swiss professor, whose children were already settled in the world.[32] She made an excellent stepmother as well as travelling companion, putting his voyages and lectures together in such a manner that the Americans had a riddle, "Why were *Agassiz's Travels* like a mermaiden?" "Because you could not tell where the woman ended and the fish began!" The Professor was a great pet of the Americans, who were then just fitting up a new exploring ship for him to go on a ten months' voyage to Cape Horn and the Straits of Magellan to hunt for prehistoric fish in comfort. She told me much of the wonders and delights of her famous Amazon expedition, and promised me letters there if I went.[33] After a delightful morning we drove on to the woods behind Mr. F.'s house, and found luncheon spread for us, Mrs. T. and her sister, in white aprons and caps, acting servant-maids and waiting on us. Mr. F. let off a perfect cascade of anecdotes, and then I was taken into the house to do my part of the entertainment and sing for an hour, which I grudged much, as I preferred listening; but I suppose they liked it, as one of the ladies wept bitterly. After this we had tea on the piazza, and looked down on the great wild cliffs and deep blue sea a thousand feet below us.

Another day I went by street-car from Boston to Cambridge, and met two pretty girls, who spoke to me and told me they were the Miss Longfellows. When we got to the end of the journey, their father came and took me for a walk round the different Colleges, and home to have lunch with him in the house Washington used to live in. It was quite in what we English call the Queen Anne style, with plenty of fine trees round it, and large wainscoted rooms full of pictures and pretty things. The luncheon was worthy of a poet—nothing but cakes and fruit, and cold tea with lumps of ice in it; he was a model poet to listen to and look at, with his snow-white hair, eager eyes, and soft gentle manner and voice, full of pleasant unpractical talk, quite too good for everyday use. He showed me all his treasures, and asked me to come and stay with them if I returned to Boston, after which he showed me the way to Mrs. A.'s house. I found her and the Professor even more to my mind; he spoke funny broken English, and looked entirely content with himself and everybody else. They showed me photographs and told me of all the wonders of Brazil, and what I was to do there, then gave me a less poetical dinner. Then Mrs. Agassiz took me to the Museum and made Count Pourtalèz take us up to the attic to see the most perfect collection of palms in the world (all mummies), intensely interesting, as illustrating the world's history. Mrs. Agassiz showed me the great sheath of one of the flowers, which native mothers use as a cradle and also as a baby's bath, it being quite water-tight. The flowers of some of the palms were two to three yards long. She said, though she had wandered whole days in the forests, she had never seen a snake nor a savage beast. One day she heard a great crashing through the tangle and felt rather frightened, when a harmless milk-cow came out. After seeing the palms she caught a German professor and made him show us a most splendid collection of gorgeous butterflies: I never saw any so beautiful; they were all locked up in dark drawers, as the light faded them. Then came corals and madrepores.

I missed my train and had to wait at Boston for the last. I was rather astonished by the conductor putting his lantern up to my face and saying, "You are Miss North?" He only laughed when I said "Yes," and I found Mrs. S. considered me lost, and had been raising a hue and cry after me. Another day I went by invitation to see Miss Cushman, who was staying with friends near Beverly; she was very entertaining and kind, and sang me three songs in a most impassioned way.

Before we left the coast the sumach was turning geranium colour, and one little hill near looked as if it were burning with it. The red berberry bushes were also a beautiful deep tint. I found lots of creeping moss with corkscrew shoots above ground, and the root creeping underneath for twenty yards or more, sending up its pretty branches at every joint.

At last Mrs. S. made up her mind to the long-talked-of journey to Canada, and we started with an enormous quantity of luggage. We were five hours in the train, passing through a prettily wooded country dotted with bright autumn colours, and saw many varieties of people: as there are no classes in the long American carriages, all sorts are mixed up together. One girl had a large tom-cat on her knee, who did not like travelling, and panted with his tongue out like a dog all the way, every now and then giving most dismal mews.

Iced water in a tin kettle with mugs walked backwards and forwards through the cars; we bought a right to drink this with our tickets; apples, pears, popped corn and cakes, as well as Isabella grapes (with the strawberry flavour), were also brought for sale; I believe these grapes are the original wild vine of America. At Alton we left the cars, and a steamer brought us over the beautiful lake of Winnepiseogee, through its 365 islands, under the light of a great full moon. The lake was ten miles long, and the hills sloped gradually one over the other up to the white mountain-tops of 6000 feet. The views by daylight were very curious, owing to the gorgeous colouring of the maples and sycamores; nothing but our most brilliant geranium-beds could rival the dazzling variety of reds and crimsons, and the blue Michaelmas-daisies made tiny pyramids of colour in the foreground, the white ones looking like miniature fir-trees loaded with snow. The lake was particularly lovely on that gray rainy day, with all its countless islands in their gay autumn dress; its beautiful

Indian name, "The Smile of the Great Spirit," seemed to suit it well. Its hotel was my first experience of the regular American boarding-house; from six till nine there was an endless breakfast, twelve to three dinner, and five to eight supper, with an enormous list of dishes for each individual to choose from and order for himself—piles of hot cakes like pancakes, pumpkin-pies made with treacle and eaten with Cheshire cheese, a huge fish called holibat, and chowder and ice-cream. The season was nearly over, and the few remaining guests were much like old German boarders, and not interesting.

The Glen House had been shut up for ten days, and the landlord had to be much persuaded to take us in at all, and said we must be content to have such fare as he and his family lived on (he looked particularly sleek and fat), and he showed us into a dining-saloon intended for 400 or 500 people. He gradually thawed, and said we might stay the night if we liked and take the chance of the weather being fit for driving up Mount Washington the next day. There was a railway up on the other side, but it had ceased to run for the season. Meantime I went down among the river boulders and got an exquisite subject, with some orange and carmine maples bending over the water, and lemon-coloured feathery birch amongst the dark pines (spruce firs loaded with cones) and cypresses.

At one of the cascades we visited there was a table laid out with birch-bark baskets of popped corn, gingerbread nuts, and apples, and a slate on which was written, "Visitors are begged to help themselves and to leave five cents for each article taken" . . . which showed the neighbourhood to be honest. We passed one little settlement of about three cottages, a meeting-house, and an inn, which called itself Jackson's City; but with that exception, and one or two isolated farms, we saw nothing but forest all the way. We passed over several rivers on most crazy bridges, built for the purpose of being carried away in winter by the ice. The coachman said: "Yes, we has enough hunting in winter; there's big de-ars, and be-ars and foxes, and coons, and sometimes we traps them, and sometimes we puts the dogs arter them." Then we jolted on again to the railway and secured sleeping-shelves for the night. I had a good talk with the conductor, a coachman from Bridgewater, who had come out to make a fortune, had ridden races and driven a coach, and finally got a berth on the railway, and thought he should stick to it if it did not kill him; "but railway accidents was common in that country." After which agreeable information I slept well on my shelf, in spite of a most suffocating stove and no ventilation, and woke up in Canada the next morning near Quebec.

The city seemed to stand up on its hill like Corfu, and the river was almost hidden by the enormous quantity of floating timber or lumber. The last boat was going up the Saguenay in a day or two, but it was far too cold to enjoy such an expedition. Quebec seemed to me a mongrel place, with English-looking streets, and French quarters, Irish villages, and Indian settlements, and the climate was odd. I was freezing in a cloth jacket, and I found my beautiful friend Mrs. D. in a white muslin dress. She went round her garden in the same costume, and seemed to think the air quite genial. She showed me posts stuck up across the fields to mark the road by which they sleighed over the tops of the fences in winter when they were all deeply buried in snow. She was very charming, and though brought up so luxuriously had no maid, and attended most minutely to her household and children. Her house was quite covered with deep claret-coloured Virginian creeper, and a scarlet and crimson maple-tree shaded it on one side. I never saw anything more gorgeous than the colouring of her garden hanging over the high cliff above the St. Lawrence, four miles from Quebec, near the place where the English troops under Wolfe mounted and surprised the French in the war. I stayed two days with the D.s, and Mr. D. took me down before breakfast by a lovely path through the wood, from his garden to the cove below, to see his men at work on the timber, shaping and smoothing the ends in the neatest way with their axes. He had quite a small colony of his own down there. He took me also for a drive in a native carriage to the pretty little golden lake of Beaufort. The hills were quite dazzling with colour, dark tall pines standing up against the rounder foliage. There was a little inn and a ball-room in which they had picnics in summer and sledging parties in winter. It seemed difficult to say which season the Canadians enjoyed most, but I am sure the winter would be far too long for me. However, Canadian people seemed very happy, and we had a merry dinner-party of fourteen that night.

I resisted all invitations to stay, and went off to see the Falls of Montmorency. It was an afternoon of perpetual storms, then bright sunshine, and then storms again. The great fall looked particularly fine under those varying lights and shades. A river as big as the Avon falling sheer 250 feet, and yet every particle of water seemed to fall separately like grains of sand; it was very fascinating. I walked through the woods and across the fields to the natural steps where the same river tumbles in a narrow crack through layers of "stink-limestone," the guide-boy called it. It had the strongest sulphur smell when freshly broken, and the formation was very odd. I lingered long on the heights, seeing the storm-clouds gather over Quebec, and curious rainbows which formed and melted away again. It was an enchanting spot, and I sent the boy to fetch my things just for the pleasure of lingering there alone. Close by was a poor used-up dead horse, closely watched by his friend, a black dog. The poor thing would not leave it, but nestled close to it, licking it and whining, then came up to me with tears in its soft brown eyes, and its tail tucked tight between its legs, asking me to help its friend as plainly as a dog could ask anything, seeming to know how willingly I would have helped if I could. It made me cry too. At last the storm came; the good old landlady of the small inn came out to me with her umbrella in case I had none, and I went home and gossiped with her. She had three rooms to let, and artists often came to stay with her.

Parkman's book made me anxious to see the Indian village of Loretta, so I drove over to the chief's house first, who, though said to be of pure blood, looked more like a well-bred Frenchman. He sold me some moose-hair work. Then I went to see some of the less civilised and more interesting people, making friends with one young man with long lank hair and high cheek-bones. I got him to take me into the school. The children were a sight worth seeing; plenty of genuine Indian faces among them, mixed up with French. They had beautiful large black eyes, and they seemed very happy. They sang me several wild Indian hymns with soft-sounding words, in minor keys, with regular rhythm. The schoolmistress said they had never been written down. In winter there were as many as 300 Hurons in the village, but in summer they were spread about, some hunting, and some doing small pedlaring at watering-places in the way our gipsies do at home.

We started that same afternoon up the river to Montreal in a big top-heavy river steamer with grand saloons and comfortable sleeping-cabins opening off them. There was a supper-table so long no one could see the end of it. I sat next the captain, a modest practical little man, who went with me into the baggage-hole the next morning himself to dig out Mrs. S.'s tremendous boxes.

I found friends, the De L.s, who had once dined with us in England, and who were one of the oldest French families in Canada, still retaining all the characteristics of their nation. They were most hospitable. The next morning we went up by rail to La Chine, and then were carried over the rapids back again in a steamer. The white waves stretched across from shore to shore quite two miles, and we went over them, and glided past a huge bell-glass of water through which we saw the dark rock. The waves danced so fiercely against one another, it took my breath away for a moment or two, and then it was over. We soon came under the huge bridge over which our old friend Mr. Brassey spent so much time and thought. It is truly one of the wonders of the world, but very ugly. Then we went up the other great river to Ottawa, with the wood-ashes from the burning forests, a hundred or more miles off, falling on the steamer's deck and making us sneeze all the way.

Mr. and Mrs. D. and myself had set our hearts on seeing the Thousand Islands. The result was not interesting, but we did not repent waiting for the steam up the St. Lawrence, which was glorious, with its endless islands, some mere rocks, some miles in length, and all covered with trees in their gay autumn dresses, beautiful both in cloud and sunshine. For three days the air had been thick and hot with smoke; our eyes smarted, our lips cracked. People said it was from Chicago, but it was really from the forests, and had come even nearer, some twelve miles off from us; the sun only showed itself as a red ball of fire every now and then through the smoky atmosphere. We rushed through Kingston and Toronto, and arrived at ten o'clock at night to find comfortable rooms at the Clifton Station Hotel, kept by an old Swiss courier, "Rosli."

I was fairly tired out the next morning, but the quiet homely quarters suited me, and I determined to stay quiet at least a fortnight so as to enjoy and sketch Niagara at my leisure. It was so cold that I was glad of the two miles' walk back from the falls, after getting half-frozen over my sketching all day. The season was over, the big hotels nearly closed, and the wooden shops were being moved away bodily on rollers to other and warmer quarters for the winter. But the natives took the greatest interest in my work, and made several offers to buy it. A woman at a toll-gate, near which I had been sketching a marvellous group of coloured maples two mornings running, refused to let me pay toll when returning in a carriage, as she said I worked too hard for her to take anything from me.

The falls far outstretched my grandest ideas. They are enormous, the banks above and below wildly and richly wooded, with a great variety of fine trees, tangles of vine and Virginian creeper over them, dead stumps, skeleton trees, and worn rocks white with lichens; the whole setting is grand, and the bridges are so cobwebby that they seem by contrast to make the falls more massive. From my home I could walk along the edge of the cliff over the boiling green waters all the way to the falls, and if they had not been there at all I would willingly have stayed to paint the old trees and water alone. Mr. Rosli gave me wonderful accounts of the falls in winter, when great masses of ice came down from Lake Erie, got jammed between the rocks and banks, and gradually froze the water between them, then more ice slipped under and it was lifted up like a bridge; he said it was a most marvellous sight, and he had known carriages driven across on the ice under the bridge, but that did not often happen. It is much milder at Niagara than in Lower Canada, grapes and peaches ripen better; the old arbor-vitæ trees are splendid, as scraggy as any old silver firs, and the oak trees are drawn up into grand timber, the trunks rising without a branch for over fifty feet. It was difficult to choose out of so many subjects where to begin. The Horseshoe Fall tempted me much, standing close to its head, with the rapids like a sea behind, and the rainbow dipping into its deep emerald hollow; the tints were endless in their gradations, and delicious, but I got wet through in the mist.

Another tempting bit was below my home, looking down on the whirlpool, where the savage green boiling water seemed piled up in the centre like some glacier; there were foregrounds of great arbor-vitæ trees almost hanging in the air like orchids, with long twisted bare roots exposed against the edge of the cliff, from which all the earth had been washed. The rapids about Goat Island on the American side were also full of wonders. One day it blew such a gale that I had to sit down and hold on tightly to the bars of the bridge on returning; no carriages attempted it that day. There are thirty-five minutes difference in the time on the two sides of that bridge, and passengers are charged 40 cents for walking over.

I talked to a good many of the regular tourist Yankees; they were of a very different sort from my friends near Boston. (The Cataract House is a famous honeymoon resort.) "Now I guess you'll get a long price for that thing when it's done. What are you going to ask?" They seemed to have no idea of work being done except for dollars.

The Head Guide of the Falls, who came out from Scotland forty-seven years before, patronised me, and told me if I got chilled at any time just to go and ask his missus to give me a good cup of coffee, it 'ud do her heart good to make it for me. He showed me some lovely views at the bottom of a rickety old tower about seventy feet high, with a corkscrew staircase winding round one noble pole in the centre. The tower is fastened half-way down to the side of the cliff by an iron bar; it shakes and trembles with every step of persons going up or down. When I had settled to my work on the boulders below, between the two huge roaring falls, I began to think what would happen if it were to tumble down, and they were to forget my being there. But I had plenty of company passing and repassing after the first morning hours. Strange figures in suits of yellow oilskin came and looked at me at intervals. When I had got my sketch in, and myself sufficiently soppy, I went farther under the spray of the American fall and saw three quarters of a circle of rainbows on it, and watched the yellow oilskin people scrambling over the huge boulders in and out of the clouds of spray; they had left the paths and bridges, and were tempting death from the mere love of danger, but with that steady nerve and strength which showed them to be beef-fed islanders, and fit compatriots of Tyndall.

I was too excited to do more work on that day, so I took a carriage, drove over the bridge and up the rapids on the Canadian side, watching their lovely lines of dancing surf and their white horses. After a couple of miles we came to the Sulphur Springs; they are close to the edge of the mad waters, and a building is raised over them. I went into a dark room, and the guide set fire to a pipe at the end of a sort of wooden extinguisher over the spring; a flame of nearly two feet high blazed up, with a large space of blue vapour between the point of the flame and the mouth of the tube, so that one could put a piece of muslin over the opening of the pipe without burning it. The guide let down his torch into the actual spring which I saw bubbling up, and the flame ran about over it. The water did not taste very nasty.

While at Clifton I got a letter from two old Norfolk servants of my father, John and Betsy Loades, who had settled in Pontiac in Illinois. He was one of our Rougham boys, and eventually became our coachman, and could turn his hand to anything; he married our cook, and they had both helped me to nurse my father, and would have no other master when he died, but emigrated with their two girls to America. Betsy Loades now wrote, "that after knocking about at Chicago and other places, they had settled at Pontiac on the Vermilion River, and John had work for the winter on a new railway; they liked the place, and hoped to buy land there in time; that a man might rent a good farm (as soon as he was able to get a team and a few machines to work his land with), on which he could make a great deal of money in a few years to provide for old age."

I could not resist the temptation to go and see them and something of the prairie country besides, so I left my portmanteau with Mr. and Mrs. Rosli, and went off with my small hand-bag and sketch-book to Toronto, to refill my purse and see my cousin Dudley's friends, Judge G. and his family.

All the fashionable people at Toronto live out of sight of the lake, and its edge is taken up by warehouses and wharfs, yet its banks are very lovely and well wooded. All the trees, except the cedars and large-leaved oaks, were then bare of leaves; the oaks were a rich copper or purple brown, with now and then a sumach shining out like a carnation in front of all the neutral tints. The small curling waves on the shore remind one of a real sea, but Toronto is not the least attractive or picturesque.

The G.s put me into a Pullman car the next morning, and for 75 cents extra I was in solitary glory till 8 o'clock at night, with only the occasional society of three guards and the black man in charge, who now and then came up to say, "Wall, how are you? Quite comfor-table?" At Sarnia we had to cross a ferry to the other side of the St. Clair river, and then get into crowded cars the very reverse of the solitary luxury I had had all day. Such a rough lot,—they could not have been rougher,—but they seemed to know I was a lady, and gave me a seat to myself, and no annoyance.

We were turned out at Detroit, and the guard warned me to be quick, for the Great Western Express would barely stop, and indeed it was rolling on again before I was fairly inside the door. After that I fell into the hands of another black keeper of sleeping-cars, who persuaded me to take a whole compartment and pay an enormous sum; but I was tired of roughs, and enjoyed having room and peace, with two windows to look out at the burning forests on each side. It was a curious and fearful sight. Every mile or two we came to blazing roots and pillars of fire, often the tops alone blazing like giant torches, scattering sparks all around, which made passing dangerous. It looked a most ghastly sight all through that dark night. At last we came to the white sand-hills round the great Lake Michigan.

It was pouring with rain, and we went over endless bogs. Damp farmers and their families came in and out all the way. We reached Joliet just in time to see the train I wanted to go on by leaving the station and going slowly out of sight. There was no other passenger train till eight at night, so I decided to go by the freight train in two hours, and having had nothing but coffee and biscuits for twenty-four hours, I went to a baker's and had more coffee and bread for dinner. All those country towns seemed very poor, as if they were making a struggle to exist at all; with little houses dotted over a large space of ground, and wooden boards laid down between them to keep one from sinking in the soft mud. The nice baker's wife was rather doubtful about the freight train. If it was full they might be a rough lot, she said; women hardly ever went that way. However, I risked it, and was most comfortable, and hospitably treated in the one carriage, the guard's van, with three windows and three doors, a great stove in the middle, a divan all round, and an arm-chair and two stools, quite a cosy little room. There were three guards—one from some other line who was out for a holiday with his wife and sister, two very pretty women. Some played at cards, and the others looked on, and it was altogether a picturesque scene, but rather slow progress, as we had to stop to let everything else pass. The guards kept perpetually going in and out in most juggler-like ways through windows and doors, coming down feet first off the roof. Now and then strange men came in for a stage or two, all decent but very damp people. The elder guard told me he did not believe it was a bit better for a working man there than in England; for though they might get better wages, the living was so costly, and as soon as he had made enough he should go straight home to England.

At last we reached Pontiac about ten o'clock, and the guard

lighted me on to a pavement (boards), and told me to follow "that gentleman," he would show me the way to the hotel; so I tramped after him through the mud and rain, and he (a mere labourer) showed me the way most kindly, found the landlord out of a crowd of people in the shop below, and I soon had a capital supper and bed, which, though only a bag of straw, was a great luxury. The paper was all in tatters; there was no handle (but three bolts) on my door, but I felt quite safe, and slept soundly. The next morning the black cook brought me my breakfast at seven on a tray with nine saucers on it, containing one egg, one fish, one cutlet, one hot roll, one pat of butter, toast, cakes, biscuits, corn-cakes, a cup of coffee, a glass of milk, cup of sugar, and saltcellar. I never saw so much on one tray before, and felt equal to anything after such a sleep and such a meal, and started to find "Big John."

First I went to the post-office. The postmaster had never heard his name; he was a new-comer himself; I had better ask next door. Next door was a shoemaker and watchmaker combined, and he had his eye fixed in a magnifying glass over the anatomy of a damp clock. He was a thorough Englishman, and remembered both John and his watch, and described them, but could not say where he lived. Other gossips dropped in who also knew him, but not where he lived, and they advised my going to the new railway where he worked. So I tramped on again through the rain and the mud outside the town to the new station, and the stationmaster told me if I walked up the line I "should find him in a fur cap," which I did, and John straightway took off that fur cap and dashed it on the ground, and said, "Laws, if that beant Miss Maryhand!" Then went and told his "boss" he must have a holiday, and took me home to see Betsy. Poor fellow, he was the ghost of his old self. So thin from constant attacks of dumb-ague, but he said he meant and hoped to live it down, and thought he should get on in time. The ague only attacked new-comers.

He had a good boss, and got nearly two dollars a day. His wife made the best of everything in the same way. They took me for a walk through the fields; all the land was black but rich, and magnificent crops of corn and even grapes grew without manure. The soil had only to be turned over, without any harrowing, or cleaning, or manuring; the corn was thrown on it, and it yielded 80 to 100 bushels per acre. Grapes also yielded enormous crops the second year after planting. The ground was always moist, though they had had no rain for months, and no water fit to drink. Fencing was the great expense, as there was no wood. They had good coal quite near. Every "respectable" man carried a revolver in a pocket made on purpose, and John said he would not like the neighbours to know he had not a gun in the house. He was full of schemes for buying land and growing crops, as well as tobacco, but meant to try a year first if he could beat the ague, and if he had been a little more of a "scholard" he might easily have become a "boss" himself. The men always said he looked like one 'cos of his leather gaiters.

We saw grand fields of Indian corn-stalks on which the cattle feed in winter, and weeds as high as the corn—iron-weed and cockle-berries. The farming was most wasteful; one field was quite white with the shed beans left on the ground. The Reformatory School was a grand building with a model farm attached, kept up by Government; but the black, swampy, spongy ground and John's ague depressed me. Betsy baked capital new bread, and roasted a chicken for dinner, and while we were out the girls had swept up and made the house look smart. The chickens cost little to keep, as they feed on all the neighbours' ground; but the neighbours when they wanted one never hesitated to kill yours as well as their own, expecting you to do the same. They also had a habit of walking in when it suited them, and making themselves quite at home, in a manner that was not always convenient; but if you were ill they loaded you with kindness, so that you could not resent their cool ways. At the last place they had lived Betsy often heard a curious rattling going on when washing in the cellar, and one day saw a large rattlesnake come in and disappear in a hole in the floor. The neighbours advised her not to anger it, as where there was one there were sure to be more, but she left off washing in that cellar. One day she saw the cat dragging in a long snake to give to her kittens. She chased both beasts away, and supposed that the cat ate the snake, as she returned licking her lips and purring very happily. Betsy made her own soap and carpets.

At four o'clock the next morning I heard everybody moving, and saw a great light; it was a fire in the midst of the wooden houses, so I dressed and got myself ready to fly if necessary, but it was soon put out. Four houses were burned, and nobody doubted it was done on purpose. Many of the Chicago scamps were about; they had been shooting chloroform in at the keyholes, and then, when the victims were quieted, and the dangerous fumes dispersed, they got into the houses and robbed them. They were now on their trial, and the four houses burnt belonged to persons who had witnessed against them. Betsy said the burners would be hung if they were caught. She came to me at eight, and soon after the landlord's son brought his buggy to drive me to Chenoa, only a ten miles' drive, to catch the twelve o'clock train to Logansport.

Another long day was passed in skirting the southern shore of Lake Erie. I passed more pretty country, full of snake-fences and ague, and saw the big lake with waves like a real salt sea. I met many nigger "gentlemen" that day in the train, in full evening-dress coats, rings on their fingers, and gold chains, with their hair oiled and straightened as much as possible, and the full extent of possible dandyism. They were extraordinarily polite, lending their newspapers, and giving up their seats to any lady looking for one, afterwards sitting with their feet above their heads, and talking the grossest slang with some Irish roughs, or the news-boys. I took these black gentlemen for ex-Chicago swells, or billiard-markers.

Toledo seemed a busy manufacturing place, and Cleveland was even bigger. They are monstrous places, as black as Manchester; most of the people in the cars were quiet country people, but they had an anxious worn look one does not see in the same class at home.

At Buffalo I had to drive in an omnibus to another station, and then on again by rail. I was put out on the American side of the suspension bridge, and had to walk across in the dark starlight over the roaring river, while a train rolled over my head at the same time, shaking every iron bar. The Canadian toll-taker rubbed his eyes, and said, "I was wondering what had become of you!" and refused with indignation to look in my bag for tobacco. Mr. and Mrs. Rosli shook me by both hands, and sent me up buttered toast for tea; my little room looked quite like home again, and, but for the cold icicles hanging round the window, would have tempted me to stay on. I took a last stroll to say good-bye to all my pet views of the mighty waters.

I started by the night train so as to get to Albany by daylight and see the Hudson river afterwards.

The Hudson seemed to me like a very mild Rhine minus the castles. A clever talking woman travelled part of the way with me; she said she had been sent for her education to Leipzic among strange languages, dishes, and ways, had had no letters for the first month, and felt just like Columbus! She was very good-natured, and on our arrival at New York put me into a fly with my luggage. That quarter of an hour's drive cost me eight shillings! New York is not cheap. At the Hofmann House they gave me a very good room, looking on a deep well, with windows all round it, hot and cold water laid on, cupboards and all sorts of nice furniture, and five dollars printed on the door, a pound a day for room alone!

My food came in as I ordered it, from a restaurant, and was good and cheap.

I found a heap of letters to answer, and took a day's rest; then I went to call on Doctor Emily Blackwell, waiting in her back room till all the patients had had their turn, when she came out and we had a long talk.[34] She was a most jolly woman. She showed me her infirmary; about a dozen women and babies were all in one room, very clean and airy. She took out a day-old gingerbread baby to show me, such a funny little object with huge eyes, and the mother looked after it just as I have seen a cat do when its kittens were interfered with. Then she took me to see the female students being lectured to by a man; elaborate curls seemed the fashion among them, rather than the prevailing chignon.

I went home to luncheon and packed my bag, returned to the ferry, and by it to the railway. In another half-hour Mr. S. met me: he looked an ideal of benevolence and philanthropy, one of New York's most respected merchants; his carriage and ungroomed horses were shabby to the last degree; he drove himself, and I held the old patched reins while he did various errands in A., which was one of the oldest settlements in America, and quite an historical place; it was situated on an arm of the sea which looked like a lake, near which was the house of Eagleswood, a comfortable but not showy dwelling. That morning the old gentleman with his son and a fisherman had saved the lives of three boys from drowning, their boat having turned over in a squall.

Instead of going straight back to New York I got out at Newark, and went by horse-car to Orange, where I left my bag at the office, and walked off in search of Sydney Clack, the young gardener my father had had at Hastings, who also emigrated after his death. After a few false starts, I found my way through the woods to Mr. M.'s house, prettily placed on a long wooded hill with a view of New York and the Hudson river fifteen miles off. I found Sydney in his greenhouse, looking well and happy, with two or three men under him, and forty dollars a month, his board and lodging; he had several nice greenhouses, and beautiful flowers. He asked me if I would go in to see Mrs. M., who had told him to be sure to ask me to go in if I came. She was very hospitable, and pressed me to stay and come again. Then Sydney came back dressed like a young gentleman to see me safe home; he said he was very glad he came out to America, that any one with a distinctive profession and the will could do well there, and he asked me to write him a character saying what he could do. My agent at Hastings had written him one about being "honest and industrious," but that was just the thing they did not care for in the States. The particular line a person excels in was what they cared to know; idleness and dishonesty did not matter half so much. He thought John would have done more wisely to have taken a good coachman's place, instead of losing health in those horrid prairies.

When I got back to the Hofmann House, I found a kind invitation from Mr. Church (the first of living landscape painters) to come and see him and his wife at their cottage at Hudson.[35] They never got my answer, and I missed my train, and only reached Hudson in darkness and rain at half-past eight. I got into a fly and told the man to drive to Mr. Church's. 'All right,' said the man, and put two other persons in (a way they have in the States), and on we went. Presently he opened the door: 'Where did Mr. Church live?' How should I know? but the other passengers said six miles off; so I went to an inn for the night, and then started in a buggy, and met Mrs. C. in the road coming to hunt for me, and she took me home. She and her husband were quite ideal people, so handsome and noble in their ways and manners. They had four children. The eldest, Fred, had a supernaturally wise look, and told long stories to his brothers with the greatest gravity. Sometimes Mr. C. made him spin yarns in the same way to us, interrupting him with questions, and trying to put him out and make him contradict himself; but the boy always had a ready answer and a reason for everything. They were still living in their old cottage-farm; but the new house on the top of the little hill above them was already roofed in and approaching completion. Mr. C. had designed it himself after the pattern of a Damascus house, with a court in its middle paved with marble, having a splashing fountain in its centre. He had also had bricks and tiles made of different oriental patterns and ornamented the outside with them, but the floors were not yet laid down. The view from the arched entrance was fine, of the Catskill Mountains (then white with snow) and the winding River Hudson.

The studio was a detached building, with a picture in progress of Chimborazo, which seemed to me perfection in point of truth and workmanship. He showed me other tropical studies which made me more than ever anxious to go and see those countries.

In my own tiny bedroom were three pictures in oils—one of the Horse-Shoe Falls of Niagara, a study of magnolia flowers, and one of some tropical tree covered with parasites. They had imported two white asses from Damascus for Mrs. C. and the children to ride, and had also a gray South American donkey, quite curiosities in the United States, where the animal is almost unknown.

On my return I found a note from Mrs. M. (Sydney's mistress) asking me to come and stay with them, and as I liked to see as much as I could of life in America, I went back to Orange in the pouring rain, and Mr. M. met me at the station. We had a late dinner in the English style, with wine (which one does not often see on the table). Mr. M. told me a great deal of Chicago and the city of forty years before, its first rapid growth of poor houses on a most unattractive unhealthy spot, then how they raised first the roads, and then the houses (with a few exceptions) with screws to the level of the roadways, and finally built the great stone houses which replaced the others: how they bored a tunnel three miles under the lake, and then had a pipe put at the end so as to get pure drinking water—also of the wonderful way they drained the sluggish river.

At a quarter past seven we had breakfast, and the boys and their father went off for their day's work by rail, and after a gossip over the plants in the greenhouse with Sydney, Mrs. M. drove me through the park, not long before a natural forest, then sold in lots to rich merchants for building and farming. A speculator had made seven miles of winding road through it. The views were fine; there were wild rocks, glens, ferns, wild azaleas, and dogwood which in spring is covered with white flowers (like snow, they said). Some of the houses were pretty; some odd and unpractical ones had been designed by the original speculator when under the influence of "the spirits," who did not seem to excel in architecture, I thought.

On my way back I found the ferry crowded with smartly dressed people thronging to welcome the young Grand Duke Alexis of Russia, and I began to wonder how I should get on, when a lady told me to follow her, which I did, and she showed me the best omnibus. She was taking two pretty children to see the sight. Broadway was covered with flags and gay hangings.

Mrs. Botta was my best friend in New York, and soon made me leave the hotel and take up my quarters in her house. She was a most charming and cultivated person, had written one or two books on education, and brought up more than one set of orphan children just for love, having none of her own. She lived with her mother, was a lady of independent fortune, and had married an Italian professor.

She took me to many studios and exhibitions, where I saw some good paintings, but a great preponderance of French millinery amongst the favourite pictures. The prices of dresses in the shops were terrific, three hundred or four hundred dollars each, and gloves ten shillings a pair. No wonder the Americans come to do their shopping in Europe.

Mr. M. took me one morning over the "Equitable Life Insurance Company," a perfect palace of Pompeian frescoes, Italian marbles, and inlaid American woods. We went in the lift on to the roof, from whence is the best view of the city, river, islands, and sea. There was no smoke, but many white tassels of steam from the different steam-lifts and other small machines. I was struck by the absence of domes and ornamental towers in the huge place. The principal hall of the great building was like some Byzantine Church for gorgeousness, and to my thinking somewhat out of character with the clerks who were scribbling in it. We went to see the clerks' dining-room and kitchen, and then into the "Safe Deposit Vaults," a wonderful museum of locks and drawers, bolts and bars, hedged in with granite, iron, and telegraphic wires. If the door were moved the eighth of an inch it touched a telegraph and warned the police. It could only be opened by two of the head people; one could not do it alone. My head got quite in a maze over it all. Three men walk round and round all night between the inner and outer walls of this terrible treasure store, and look to the gratings. There seemed nothing combustible in the building. There was something horrible about the whole thing, with its defiance of evil-doers or accident. New York was altogether overwhelming in its constant movement, ugliness, and method; but still it had, like all big cities, much to attract and interest any one who could think.

The first night at Mrs. B.'s, after our supper-tea, we all walked out about nine o'clock to the horse-cars, and then on again and up a side street to Mr. R.'s, the literary critic of the *Tribune*. It was not a party, but just "a few friends," and most agreeable, for the friends were of the best sort: they wore only high dresses, and my old square-cut body felt almost indecent, the others were so decidedly "high." I was introduced to plenty of people whose names I did not catch, and had a long chat with M. du Chaillu, just home from Norway, where he had been as far north as he could get—an odd contrast to his last wanderings among the gorillas!

There were many artists there, including one who had painted the chief beauties of New York as the Nine Muses. I promised to go and see them. He had also, when in England, painted John Bright, Cobden, and Tom Potter. He was an Italian, and said Tom Potter was a good man, Bright was amusing, but Cobden he "loaved"! At eleven o'clock plates of cold salad, sandwiches, and oysters were handed round, with a napkin to put on the knees under them, with tiny glasses of wine, after that ice-creams and cakes, and then coffee and chocolate. The guests seemed to enjoy their picnic supper, after which we walked home again most of the way.

Mr. De Forest came one morning and took me to the Johnstone Gallery, a most exquisite collection of pictures. The great Niagara and a beautiful sunset scene in a swamp by Church were there. The latter is a wonderful picture. The four ages of life by Coles, and splendid Mullers and Cromes were there too. Every picture was a gem. We went one night to the opera to hear Nilsson as Marguerite. She was not appreciated, and the audience were most cold. All their applause was given to the French tenor Capoul, who was the fashion, and all the girls raved about him. The Grand Duke Alexis was there, a handsome well-grown middy. It was funny to see the audience rise gradually to the Russian hymn. We resisted as long as any one; but got up at last to do honour to the son of the Czar.

We were offered tickets to go in a steamer up the Hudson to Westpoint and back with H.I.H., but did not see the fun of being frozen on deck, or broiled in a cabin in high Russian society, so begged to be excused, and I went on to

Washington with a parcel of clam-sandwiches in my bag, which Mrs. B. made with her own dear hands, cutting a roll in half, buttering it, and putting the odd fishy things between; they come out of bivalve shells something like our scallops.

I did not stop at Philadelphia, but went straight to Mrs. Russell Gurney[1] at Washington at 1512 H. Street. The mode of distinguishing houses in America is certainly monotonous, but has the great merit of being easy to find, and the streets generally run at right angles to one another. Mr. G. had given up his comfort at home to come and try to settle the *Alabama* question, and was very weary of the task.[36] Month after month passed, and still nothing was settled. It was very cold, and I felt sorry for my friends there; they were most kind and hospitable to me. The first day I went to see Dr. and Mrs. Henry in their pretty museum building, built of pink stone with much-ornamented round archways, and creeping plants over it, and Miss H. showed me many interesting things. There was a large collection of birds' nests, and one trunk of a tree with holes made all over it by a Californian woodpecker in order to pick out its own pet grubs; then the chipmunk or squirrel puts the acorns in, which another bird steals again. We saw also the last of the auks, with its one odd egg; and a horrid little baby mummy which was tossed out of the middle of the earth by an earthquake in South America, and was supposed to be one of the very oldest of dead human beings.

We had a party at home of diplomatic people who discussed some of the new American ways. The young ladies have clubs among themselves, and give parties on alternate nights during the winter, every "Miss" bringing a gentleman. Mamma only has the privilege of giving the supper, appearing while it is being eaten, and retiring afterwards. Papa is allowed the privilege of paying for it, and does not appear at all. These girls go out to other people's houses under the escort of some young gentleman. Pas and mas have a dull

[1] Wife of the High Commissioner and Recorder of London.

life of it in U.S. society. When a man calls at a house he never asks for the mother, only for the girls, and the mother does not appear; if she did she would be snubbed, and made to know her place very quickly.

I had a card brought me the next morning, "the Secretary of State" and Mr. Fish followed it, to whom I had a letter of introduction. He was a great massive man, with a hard sensible head. He said he would call for me in the evening, and take me to the White House. So at eight o'clock in he came again after another big card, I being all ready for him in bonnet and shawl, and in no small trepidation at having to talk *tête-à-tête* with the Prime Minister in a small brougham. However, I found there was no need, as he did it all himself. We were shown in first to the awful crimson satin room which Mrs. G. had described to me, with a huge picture of the Grant family all standing side by side for their portraits. Then we were told to come upstairs, and passed from state-rooms to ordinary everyday life up a back staircase, which was the only means of reaching the upper storey allowed by the architect of seventy years ago. We were shown into a comfortable library and living-room, where a very old man sat reading the newspaper, Mrs. Grant's papa, who did not understand or hear any of the remarks Mr. Fish or I made to him. Then came Mrs. Grant, a motherly, kind body; then at last came the President, also a most homely kind of man.

We at first sat rather wide apart, and I had more of the talk to do than I enjoyed, and felt like a criminal being examined till Mrs. Grant hunted up a German book full of dried grasses to show me, and the poor withered sticks and straws brought dear Nature back again. I put on my spectacles and knelt down at Mrs. Grant's knee to look at them. They began to find out I was not a fine-lady worshipper of Worth, and we all got chatty and happy. Mrs. Grant confessed she had no idea "Governor Fish had brought me with him, or she would not have let me upstairs, but didn't mind now"; and she told me all about her children; and if I had stayed long enough would, I have no doubt, have confided to me her difficulties about servants also. The two big men talked softly in a corner as if I were not there, and I watched till Mr. Fish looked like going away, and then I rose. They were all so sorry I could not stay the winter there, and hoped I would come again, etc. etc., like ordinary mortals; and Mr. Fish showed me a water-colour drawing of the Grants' country house, took me into a blue satin room, which he said was very handsome, and conducted me home again.

I wondered if Gladstone or Dizzy would have taken as much trouble for the daughter of an American M.P. who brought a letter from the Secretary of an English embassy.

The next morning I found a big envelope with a huge G. on it, and a card inside from the President and Mrs. Grant asking me to dinner that night. The Gurneys had another, so we went in state and were shown into the blue satin oval room, well adapted for that sort of ceremony, and the aide-de-camp General Porter came and made himself most agreeable to us. Then came two Senators and the Secretary of Foreign Affairs, and then the President and his wife arm in arm, with Miss Nelly and a small brother, and grandpapa toddling in after. He had an armchair given to him, and General Grant told me he was so heavy that he had broken half the chairs in the house, and they were very careful about giving him extra strong ones now. After a terrible five minutes, dinner was announced, and to my horror the President offered me his arm and walked me in first (greatness thrust upon me). I looked penitently across at Mrs. Gurney, who looked highly amused at my confusion, and did not pity me in the least. I was relieved by finding the great man did not care to talk while he ate, and General Porter was easy to get on with on my other side. He seemed to know every place, inhabited and uninhabited, in America.

He gave me some curious accounts of the few remaining

Indians, some of whom are as near animals as mortals can be, too lazy to look for food till the strong pangs of hunger seize them, when they sit in a circle and beat down the grasshoppers with whips, gather them up and crush them in their hands, eating them just as they are, and then sleep again till the next fit of hunger seizes them. The President drank tea with his dinner, and had every dish handed to him first. He seemed an honest blunt soldier, with much talent for silence. His wife had a funny way, when shaking hands with people, of looking over their heads, and appearing to read off their names out loud from some invisible label there. I was taken out from dinner in the same distinguished manner, being made to stop in the red satin room and admire the family portraits and the youngest boy in a Grant tartan and kilt. I asked the President if he did not mean to go some day and hunt for his relations in Scotland, but he said he had quite lost all trace of them, four generations of his family having lived in America, and that he was "raised" in Ohio; and he sat down by me and was quite conversational. I told him about my visit to Pontiac. He said it was quite possible to live down ague, and that after seven or ten years of cultivation the prairies ceased to be unhealthy. How sad it is that the first brave men who make the country must be the victims to its climate.

General Cameron promised if I would come back in spring to take me to a place in Pennsylvania, only eight hours off, where they still talked pure Elizabethan English, and to another where they can talk nothing but Dutch, having kept themselves always apart from their neighbours. Miss Nelly got scolded for not playing the piano. She was kept very much at home, and not allowed to go with any of the fast girls of the day.

After that party we went to hear Santley sing. The Americans did not appreciate our great baritone any more than they did Nilsson, and I felt grieved that such a real artist should have thrown away his talent on such an audience.

The dear old Recorder slept, as only M.P.'s do, waking up after every piece to clap, and looking pleased too. The G.s were quite surprised (as I was) at the fuss the Grants had made about me, as they never gave dinners (they themselves had only dined there once before, when the High Commissioners first went over). I could not think what I had done to deserve all this; but after I left it came out. Mrs. Grant talked of me as the daughter of Lord North, the ex-Prime Minister of England. I always knew I was old, but was not prepared for that amount of antiquity.[37]

We drove out to Arlington, the late home of General Lee, a tasteless building of would-be classical style in a beautiful situation, with distant views of Washington; a one-armed ague-stricken soldier was its only inhabitant.

It blew a perfect gale of wind whilst we were in that dismal place, and we were glad to get out of it again. The large-leaved oaks were still holding on tightly to their brown-papery leaves, and kept up a continual crackle and rustle. In and about all the great towns of the States I saw little houses built for the accommodation of sparrows; the birds had been imported from England to get rid of a caterpillar which had been infesting the trees and eating up everything. The sparrows seemed to take kindly to their new homes and diet, but it was still a problem how they would endure the winter. The Potomac was frozen, and people were skating everywhere.

We went in the evening to a woman's meeting at which there were more men than women; all the men who would not go and hear men preach on Sundays seemed to make up for it there.

Miss H. brought me some beautiful dried specimens of the creeping fern with leaves like ivy which only grows at some place in Connecticut; it had been so much picked that a law was made and a heavy fine imposed on any one taking more of it. Writing of law reminds me that there was another State in America where divorces were so common that a lawyer would do the thing cheaply by the dozen, if you could take a sort of season-ticket for a set before you committed the folly of matrimony for the first time; one woman was pointed out to me who had been divorced eight times.

One morning we went to the opening of Congress; we drove to the Capitol after breakfast, a really handsome white marble palace with a large dome over its centre; then wandered on up and down, asking our way till we got to the gallery reserved for diplomats in the Lower House, and were told to take the front seats by Mr. P. the publisher to whom I had a letter, and who seemed to be a universal busybody and most important personage. The House looked twice as large as our House of Commons; all the names were read over to "the Bar of the House" (though there was no Bar). The oaths were decidedly calculated to keep truth-telling Southerners out, as they swore they had never counselled nor helped in any rebellion against the government of the U.S., etc. etc. There were two black M.P.'s particularly well dressed (not a general fault in the assembly), and there was a very ample supply of bald heads, as well as some preposterously young-looking men. There was a female reporter among the others in gold bracelets and a tremendous hat and feathers; the messengers were all boys, who dashed about continually amongst the members below, sitting between whiles on the steps of the Tribune. After a while a quorum of both Houses was declared, and a message sent to know if the President had anything to say to them. The House adjourned for half an hour, so we went out, and afterwards into the Upper House, where we stayed to hear the President's Message read, as it was done at the same time in both Houses. The Senate House looked dull after the other, and the Message was very long, it took nearly half an hour to read. The boy-messengers there were smaller than in the other House, some of them did not look more than eight years old; they sat on the steps of the Speaker's platform, and were very ornamental, reminding me of little boys in the foreground of old Italian paintings. We saw Sumner there, with a grand head; Butler, too, I saw in the other House. The lower one was filled with desks standing in pairs, and as they were distributed by lot, people who did not love one another must occasionally have been rather closer than they liked. There were only seventy altogether in the Senate, and each senator had his own desk, armchair, and spittoon. Both Houses seemed to have newspapers and periodicals on their desks, and could read through dull speeches openly, without having to creep up to dark corners of the gallery, as I have seen some highly respected M.P.'s in our own House (with Dickens's last number in their hands). Tobacco and cigars were selling in the lobbies. The central hall and passages were lofty, and full of fine marbles and frescoes.

On my return to New York, Mrs. Botta took me to some private theatricals in a friend's house. She had a real dramatic gift, and could make her audience either cry or laugh as she pleased. "She was a heaven-born genius," as a young German present called her. We had music, and I had to sing also; the amateur ladies had some very lovely voices among them, but chose Italian bravura songs far too difficult for them. It is odd how few people know the secret that a song cannot be too easy to sing well before an audience, and that the easier it is the more it generally pleases.

Roasted oysters are a great supper-dish at American evening parties, one oyster being as much as any one could eat at a meal. I think no food is better than those huge American shell-fish.

It was very cold before I left New York, and some snow had fallen, which made me very happy to go on board the Jamaica steamer on the 15th of December. I had a fearful cold in my head, and was nearly frozen to death, but we got warmer every day, and I always better as it grew warmer. We were soon amongst the mysterious festoons of floating gulf-weed. Even the sea-water was warm, and it looked such a solid black blue, and the weed as gold or amber on it, with the long streaks of floating white foam over all. A picturesque group of people were on deck when the warmth at last brought me up. A dark graceful half-caste woman of some sort, with her head on her still darker husband's shoulder, lay half asleep, while he was playing an accordion to a group of small children, all sitting in an admiring circle round them. A dear old American pair of people were going to spend the winter in Jamaica, and to return by the Isthmus of Panama and California in the spring to their home in Connecticut. They wanted me to go with them: he was eighty, she was seventy. The winter before they spent in Santa Cruz, but fancied seeing a fresh island that year. They had been a great deal on the West Coast of Africa, and I wondered what for? Had that mild old man ever bought and sold slaves? I looked at his feeble old face and began laughing, it seemed so impossible.

# CHAPTER III

## JAMAICA

## 1871-72

In the West Indies at last! Christmas Eve!

We passed Watling's Island and Rum Key, and after steaming through the crooked island passage we had a most exquisite sunset, the gold melting into pure blue so suddenly, and yet so softly, that one could hardly say where the beginning or ending of either colour was. What a contrast in one week! All the blankets were taken out of the cabin, and one sheet was almost unendurable, with both door and windows open. The next day we were within sight of Cuba, and the sunset had all the soft colours of a wood-pigeon's breast. I gave up the greater part of my dinner to enjoy it. The clouds closed in over it, till at last there was but one opening like a golden eye with red eyelashes, all the rest different shades of neutral tint, the land under it very green, while the sea looked like ink. The approach to Port Royal, with its long spit of sand and mangrove swamp, and then into the calm bay of Kingston beyond, was intensely exciting. Every tree was of a new form to me, the grand mountains rising gradually up to 7000 or 8000 feet beyond, all creased and crumpled with ins and outs, like brown paper which has been much used.

I landed entirely alone and friendless, but at once fell into kind helpful hands. A young Cuban engineer appeared from the moon or elsewhere, hunted up my luggage, paid my carriage and porters (for I had only American money), and saw me safe to the inn. The good brown landlady, having no other spare room, gave me up her own. It was not a quiet one, having the family on one side, the dining-room on the other, and only Venetian shutters between it and the traffic outside. Apparently all the dirty clothes of the establishment, as well as the stores, were kept in it, without much method as to their several arrangements. But the good woman meant to do her best for me, and she gave me my first mango to eat. Wasn't it good! I think no fruit is better, if it be really good of its kind. In Jamaica the best sort goes by the name of "Number 11,"[1] certain seeds having been brought over from India years ago with numbers attached and the names lost.

The next morning the landlady took me at daylight to see the opening of the new market. It was Christmas Day, and all the negresses went in their gayest ball-dresses; the transparent white muslin showing the black shoulders and arms most comically through. They were covered with pink, orange, and red satin bows, with artificial flowers, and feathers in their hair, a basket balanced above full of cakes or fruit. A band was playing, and all Kingston promenaded up and down.

On our return I found Dr. C., who insisted on carrying me off to stay.[38] One day Mrs. C. took me a drive up the Newcastle road; when it came to an end we walked on, and I saw a house half hidden amongst the glorious foliage of the long-deserted botanical gardens of the first settlers, and on inquiry found I could hire it entirely for four pounds a month. It had twenty rooms altogether, and offices behind, and had been a grand place in its day. So I did hire it, and also furniture for one bedroom. I put all but the bed and washstand in the long outside verandah, which occupied all the front of the upper floor, open to the lovely views (with occasional Venetian shutters), and pinned up my sketches on the opposite wall, keeping a little room at the end to sleep in, and another locked up for my storeroom. I gave eighteenpence

[1] Numbers in catalogue of a collection coming from East Indies to the Botanical Gardens, Martinique, when Rodney took the ship prisoner.—ED.

for a huge bunch of bananas, and hung it up instead of a chandelier from the roof of the verandah. The man who sold it to me could barely lift it; there were more than ninety bananas on it. They began ripening from the top downwards, and I ate my way steadily on, till one day the string gave way, and they came down with a crash and had to be given to the pigs.

Mrs. C. found me an old black woman, Betsy, to look after and "do" for me. She used to sit on the stairs or in the doorway and watch me, eating little odds and ends, and sleeping between whiles. She prided herself upon being "one of the Old Style Servants," which meant, I believe, old enough to have begun life as a slave; consequently she had a contempt for all newfangled notions about dress.[39] She wore *one* and a turban, and at night untwisted the latter article and put it on rather differently, that was her whole undressing. A second dress made her sole luggage. There was also a man attached to the house, old Stewart, a coal-black mortal with a gray head and tattered old soldier's coat, who put his hand up to his forehead with a military salute whenever I looked at him. I gave these old people six shillings a week to take care of me, and felt as safe there as I do at home, though there was not a white person living within a mile. I had a most delicious bath: a little house full of running water, coming up to my shoulders as I stood in it; it was the greatest of luxuries in that climate.

From my verandah or sitting-room I could see up and down the steep valley covered with trees and woods; higher up were meadows, and Newcastle 4000 feet above me, my own height being under a thousand above the sea. The richest foliage closed quite up to the little terrace on which the house stood; bananas, rose-apples[1] (with their white tassel flowers and pretty pink young shoots and leaves), the gigantic bread-

[1] *Eugenio jambos*, native of East Indies. Fruit the size of a hen's egg, rose-scented, with the flavour of an apricot.

fruit, trumpet-trees (with great white-lined leaves), star-apples (with brown and gold plush lining to their shiny leaves), the mahogany-trees (with their pretty terminal cones), mangoes, custard apples, and endless others, besides a few dates and cocoanuts. A tangle of all sorts of gay things underneath, golden-flowered allamandas, bignonias, and ipomœas over everything, heliotropes, lemon-verbenas, and geraniums from the long-neglected garden running wild like weeds: over all a giant cotton-tree quite 200 feet high was within sight, standing up like a ghost in its winter nakedness against the forest of evergreen trees, only coloured by the quantities of orchids, wild pines, and other parasites which had lodged themselves in its soft bark and branches. Little negro huts nestled among the "bush" everywhere, and zigzag paths led in all directions round the house. The mango-trees were just then covered with pink and yellow flowers, and the daturas, with their long white bells, bordered every stream. I was in a state of ecstasy, and hardly knew what to paint first. The black people too were very kind, and seemed in character with the scenery. They were always friendly, and ready for a chat with "missus." The population seemed enormous, though all scattered. There was a small valley at the back of the house which was a marvel of loveliness, bananas, daturas, and great *Caladium esculentum* bordering the stream, with the *Ipomœa bona nox*, passion-flower, and *Tacsonia Thunbergii* over all the trees, giant fern-fronds as high as myself, and quantities of smaller ferns with young pink and copper-coloured leaves, as well as the gold and silver varieties. I painted all day, going out at daylight and not returning until noon, after which I worked at flowers in the house, as we had heavy rain most afternoons at that season; before sunset it cleared again, and I used to walk up the hill and explore some new path, returning home in the dark.

I found no difficulty in walking, and could see the plants far better than when on a pony. I walked one Sunday down to the chapel two miles below in the valley: such a walk! The road in one place went through a gap in the cliffs just wide enough to let it and the rushing stream through together. Bridges crossed the stream more than once high above, with masses of the greenish bamboo feathering over it. There was no other white person in the church, and a black parson preached a good sermon, but not his own, and thereby showed sense.

People always ask how I fed there. I used to buy two pounds of beef from the soldiers' rations at the guardhouse a mile or two down the valley every Saturday. The meat was tough at first, but every day we stewed it up with fresh vegetables; then the black people brought me eggs and vegetables, and a woman went once a week into Kingston and brought me out any shopping I wanted. I was advised to buy some tins of turtle-soup, and was amused to find they were made at Glasgow. They are too indolent to make anything in Jamaica. The Seville oranges rotted on the ground, and sugar was growing close by, but they made no marmalade.

After about a month of perfect quiet and incessant painting at the garden-house, people began to find me out, and the K.s rode down and made me promise to come to their cottage for a night. Their home was a thousand feet higher than mine, with a most lovely view, and tufts of bamboo all round it, the first large specimens I ever saw; they made me feel in another world among their rattling, creaking, croaking, cork-drawing noises. Some of the canes must have been fifty feet high, thicker than my arm, and full of varied colour. There was a pretty garden, crammed with strange new plants. The cysak, which they told me was the sago palm, was very thriving. They get the sago from the roots of the young suckers; it has a number of scarlet nuts half hidden amongst its furry curly young leaves at the top, so wonderfully packed. I began a sketch of the bamboos the next morning, and then went on a mile along the ridge to stay with Captain and Mrs. H. and the old deaf General Commander-in-Chief, in a bare tumble-down old house, supported by two weird old cotton-trees and a sandbox-tree, built on the very edge of the precipitous wall of the valley.

I was taken to church on horseback the next morning, a lovely ride of half a mile to the most breezy spot on the south side of the island, on the very top of the hill. I knew every one in the church (with a white face), and the collection of "sorry nags" outside was very remarkable. We strolled up afterwards to Mrs. B.'s (the rector's wife) famous big tree, under which all the gossip and scandal of Jamaica were said to be manufactured every Sunday afternoon, enough to last through the week; but it was such a healthy spot, and she was such a jolly little woman, that I do not believe anything really spiteful proceeded from that locality. It was perched on a perfect pinnacle, and could be seen for scores of miles in every direction, out at sea as well as on land. The air was something worth living for; to breathe was a true pleasure. Captain Lanyon came up with the Governor's orders that I was not to go down the hill without coming to stay at Craigton; but I wanted more clothes and paints, so Captain H. promised me a horse at six the next morning to take me and bring me back; but when I got up I found the house like a tomb, not a creature stirring.

I got out of my window, only a yard above the ground, and went down to the stable: all asleep too, and the sun rising so gloriously! I could not waste time, so took my painting things and walked off to finish my sketch at the K.s. They sent me out some tea, and I afterwards walked on down the hill, among the ebony-trees and aloes, home. I passed one great mass of the granadilla passion-flower, with its lilac blossom and huge fruit, which is most delicious, and almost more than one person can eat at a time. Jamaica people scoop out all the seeds and juice, and stir it up in a large tumbler with ice and sugar, and nothing can be better for late breakfast, with the thermometer at 91°. The leaves are of a simple oval form. I found a Kingston doctor and his family had accepted my offer of rooms for a change, and had come up, furniture and all, for a week to a corner of my vast domain. So after a rummage and a bath I went up the hill again, and old Stewart carried my portmanteau on the top of his head as far as the little collection of cottages at the foot of the Craigton mound. Then he called out to ask "one of those ladies if they would carry this woman's trunk up the hill," and a lady did it, her woolly head being naturally padded for the purpose. Their heads are marvellously strong. When I first came there was a difficulty in getting the iron frame of my bed together. A carpenter was sent for. He first pushed at it and kicked it with his foot, then he thumped at it with his fist, and finally made a bull-like rush at it with the top of his head, and achieved it.

There was one of the great cotton-trees close to the path, and I went on zigzag, returning continually to the huge skeleton tree, and thought I should never get above it. The native cottages were generally hedged round with scarlet and double salmon-coloured hibiscus. The little children met one carrying flowers of it, and did not beg. All the people were sociable, with very gentle manners. I reached Craigton just after sunset; and the views over Kingston Harbour, and Port Royal stretching out into the sea beyond, were very fine. The house was a mere cottage, but so home-like in its lovely garden, blazing with red dracænas, *Bignonia venusta*, and poinsettias looking redder in the sunset rays, that I felt at home at once. The Governor, Sir John Peter Grant, was a great Scotchman, with a most genial simple manner, a hearty laugh, and enjoyment of a joke.[40] He was seldom seen till dinner-time, except sometimes when he came out for a game of croquet about five o'clock with any people who happened to collect themselves on the pretty green lawn which was always open to all the neighbours. Two sheep were kept tethered on it to nibble the grass and make it fine, and they had learned to stand on their hind-legs and beg for sugar at tea-time. After the anarchy succeeding the rebellion, Sir John was persuaded to leave Bengal and come to put things straight.[1] He worked enormously, pulling down old machinery and putting up new everywhere. He was never tired, and could work day and night without rest or exercise, trusting no one, and looking into the minutest details himself. His right-hand and secretary, Captain Lanyon, had no sinecure, and helped him gallantly, besides doing all the honours of the house; for the Governor hated "company," and never gave himself the least trouble to be civil to people unless he liked them. He told me to come and go just as it suited me, and to consider the house my home. He never took any more notice of me, and I did as I was told, and felt he had treated me in the way I liked best. He is always my ideal of a "Governor."

I begged to be let off formal breakfasts, went out after my cup of tea at sunrise as I did at home, and worked till noon. My first study was of a slender tree-fern with leaves like lace-work, rising out of a bank of creeping bracken which carpeted the ground and ran up all the banks and trees, with a marvellous apple-green hue. The native children used to take plunges into it as English children do with haycocks, and it was so elastic that it rose up after them as if nothing had happened. In the afternoon I could paint in the garden, and had the benefit of the tea and gossip which went on near me, sitting under a huge mango, the parson, his wife, and people coming up on business from the plains with three or four neighbours and idle officers from Newcastle. A brother of the Bishop of Oxford with his pretty daughter stayed a while also at Craigton. Orchids were tied to the trees, and all sorts of lovely bushes were on the terraces, the *Amherstia nobilis*, "Mahoe Yacca" tree, etc. etc., all wonders to be painted. In the evening Sir John always came into the drawing-room with the ladies (like all those who really do work in the tropics, he drank next to no wine). He used to curl himself up on the sofa amid a pile of books, kick off his shoes, and forget the existence of every one else, or he played a game of chess if he found a partner worth fighting. When he discovered I could sing he said he would have that other trunk up the hill, even if it took six men to carry it, so that they might find more songs to keep my voice going; and it was a comfort to think I could give him some pleasure in return for all his kindness.

The view from the dining-room was like an opal: the sea-line generally lost in a blue haze, the promontories of St. Augusta, and Port Royal with its long coral reef, stretching out into it all salmon-coloured, then the blue sea again, Kingston amid its gardens, and the great Vega all rich green, with one corner of purest emerald-green sugar-cane, the whole set between rich hillsides, with bananas and mangoes full of flowers, and the beautiful gold-brown star-apple[1] tree taking the place in the landscape which the copper-beech does in England. The mahoe is the hardest and blackest wood in the island, and its velvety leaves and trumpet-flowers of copper and brass tints made a fine study: all the flowers seemed so big. The poinsettias were often a foot across, one passion-flower covered two large trees, the dracænas were ten feet high, the gardenias loaded with sweet flowers. One day the captain started Agnes Wilberforce and myself on two horses with a groom for Newcastle, where he had arranged that Dr. S. should meet us and show us the famous Fern Walk. It was a glorious day. We rode up the steep hills straight into the clouds, and found rain in the great village of barracks, but we went on in spite of it. The scarlet geraniums and zinnias of former soldiers' gardens had seeded themselves all about, and above them we came to patches of wild alpinia, called by the English ginger and cardamom, with lovely waxy flowers smelling

[1] *Chrysophyllum Cainito.* Fruit the size of a large apple; the inside divided in two cells, each containing a black seed surrounded by gelatinous pulp.

like their names. Great branches of *Oncidium* orchids were pushing their way through the bushes, and creepers in abundance, huge white cherokee roses, and quantities of begonias.

At last we turned into the forest at the top of the hill, and rode through the Fern Walk; it almost took away my breath with its lovely fairy-like beauty; the very mist which always seemed to hang among the trees and plants there made it the more lovely and mysterious. There were quantities of tree-ferns, and every other sort of fern, all growing piled on one another; trees with branches and stems quite covered with them, and with wild bromeliads and orchids, many of the bromeliads with rosy centres and flowers coming out of them. A close waxy pink ivy was running up everything as well as the creeping fern, and many lycopodiums, mosses, and lichens. It was like a scene in a pantomime, too good to be real, the tree-fern fronds crossing and recrossing each other like network. One saw dozens at one view, their slender stems draped and hidden by other ferns and creeping things. There were tall trees above, which seemed to have long fern-like leaves also hanging from them, when really it was only a large creeping fern which had found its way over them up to the very tops. They were most delicious to look at, and, my horse thought, to eat also, for he risked my life on a narrow ledge by turning his head to crop the leaves from the bank, when his hind-legs slipped over the precipice. I said "Don't," and the Doctor and Agnes laughed, while the good horse picked his legs up again and went on munching in a more sensible position. We rode back by a lower fern walk, still lovelier because it was even damper.

Near Newcastle we found blackberries, furze, strawberries, and bracken on the drier hill-tops. Those were the only plants there the English soldier really loved, because they reminded him of home. We found it still raining there (we had been above it at the Fern Walk), so the doctor deposited us in two armchairs in his sitting-room, and went off to see a patient; we both fell asleep till he returned, and then went to see the view from Major W.'s arbour, and some Jamaica plants in his garden.

Two hours were enough of Newcastle talk. I rubbed my brains till they were sore to find recollections of Corfu, Quebec, or Gibraltar (which latter place I had only seen from the terrace at Ronda); but those places seemed the only ones which interested the "military." It was refreshing to get home again, and to hear the Governor's honest laugh when we told him how overworked and bored they all were on that hill above: he himself never knew what an idle hour meant.

One day the captain called me to the front door to see a black "lady," who had walked two or three days' journey over the hills with some appeal to the Governor. It was a funny scene: the petitioner talked of herself as "a lady who was used to the best society." (She had a long starched cotton dress trailing half a yard in the dust.) The captain suggested she should get some refreshment, which resulted in her being given a yam to gnaw sitting on the doortsep. I only left Craigton the day the Governor went down, and walked down at daylight in my usual way before any one was moving in the house but the old woman who brought my tea. I found old Betsy waiting for me alone, and was soon hard at my usual work again, painting the lovely *Alpinia nutans* or shell-flower, one of the most beautiful of tropical flowers.

Gertrude S., the Attorney-General's sister, soon rode down to see me; she lived only half a mile from Craigton, and was the person I liked best in Jamaica. As a young girl she had been taken out with her brothers and mother by a stepfather to Australia, where she had had no so-called "education," but had ridden wild horses and driven in the cattle with her brothers; had helped her mother to cook, wash, make the clothes, and salt down the meat; and till seventeen she could barely read; then her mother's health broke down, and she accompanied her back to England, nursed her through a long illness, and educated herself. Her eldest brother soon took great honours at the Bar, and was sent for by the Governor to help him in starting the new Constitution of Jamaica. I never knew a more charming brother and sister! so entirely happy together, and helpful to one another. Gertrude had taught herself German, French, and Italian in those few years, and still read much, though she did all the finer kinds of cooking with her own hands, and saw her horse and cow fed regularly. She rode like Di Vernon, and shocked the conventionalities of the country by taking no groom with her.[42] No one more thoroughly understood the management of a horse. She had a noble face and figure, with beautiful expressive dark eyes, and was a most perfect gentlewoman in spite of her rough training; another of the many examples I have known that a really distinguished woman needs no colleges or "higher education" lectures. Her brother was witty and bright, and when he went into the Governor's room at Craigton we were sure to hear the great laugh come rolling out over and over again. Three years later these dear people both died in the same hour, of yellow fever, and a letter to me was the last Gertrude wrote, telling how she and her brother had been nursing the master of the new college, who had come up for his Christmas holiday to them bringing yellow fever: he was better, and was on his way home, and I must come back with him and pay them a visit. The next day I read a telegram in the papers: "Attorney-General of Jamaica dead of yellow fever, sister dead also." It was too terrible! This is a long story, but I could not think of my friend without her curious history: her life was short, but I think a happy one, for she was always busy, and used to sing over her work, making all near her happy too. We took to one another at once with our whole hearts, and I well remember that afternoon when she rode down the hill to see me, and I walked nearly home with her afterwards, her horse following like a dog without any leading.

Some of the wild fruits were very good, though the English seldom eat them. The "soursop," or custard-apple,[1] was an especial favourite of mine; it was a green horny heart-shaped thing growing close to the stem of the tree, with a creamy pulp and black seeds, and an acid pineapple flavour. The avocado pear too was good as a salad; it looked like a pear, only sometimes it was purple as well as green, and had a large seed inside but the white part had the consistency of a very ripe pear without the slightest taste. I used to wander up the hill-paths behind the house in the evening and make friends with the logwood-tree, just then covered with yellow flowers: the anotha with pink or pearl-coloured buds and wonderfully packed crimson seeds in husks like sweet chestnuts wide open. One could hold these prickly shells upside down and shake them and the seeds never shook out, the prickles being curved over their surface, so that they were secured as with a network. I passed one evening through a cleft in the rocks so narrow I could touch them with either hand; they were covered with a scarlet lichen, pretty green and purple orchids growing among the moss. The allspice-trees were showing their white flower-buds, and the leaves were very sweet when crushed. I met a hideous old black woman, who told me she was Stewart's wife, a fact I knew before. I asked her if I could get a view of the sea higher up? "Oh dear no, no see sea, that very long way, very bad road." Five minutes more took me to the top, with a glorious view of the sea: why did she tell such lies? A nice lad up there gave me some peas to eat which he picked off a most unlikely-looking tree, and showed me the *Cassava mandioca* plant. Then I walked through a field of lovely waving green sugar-cane from which they make the coarsest sugar, nearly black, sending it to England to be refined and made into white. Nature has done everything, man nothing, for that beautiful island.

[1] *Anona muricata.*

There was a long ant-tunnel up and down my house and along the fence, but I found no outlet at either end: I broke it in several places and caused a great commotion among the ants; the next day it was mended. Near the house was an assembly I disliked much more than the ants, they called themselves revivalists, and used to howl and talk unknown tongues and foam at the mouth for hours together; sometimes it lasted all night, and Betsy and old Stewart used to go off together to see them, leaving me alone in my big house with the silvery banana leaves flapping against the shutters, the fireflies darting, and the glow-worms crawling all round, the crickets and frogs also having a revival and rivalling the bipeds in the noise they made, with probably more sense and meaning too in what they said to one another. I used to sit on the verandah writing, reading, and enjoying all these things, and never for an instant had the slightest fear; but I did not like the revival people any more than I do the so-called spiritualists of fashionable life, for they are both untrue and getting money on false pretences. Once I passed the camp as a man was preaching: he said, "A stranger is among us, if she will join us, we will bless her," but she didn't, and wondered if he cursed her? When I told the Governor about these things, he said he had no more right to prevent their amusing themselves in that way than he had to stop the white people from giving balls and keeping polkas and waltzes going till the small hours of the morning, preventing all near neighbours from sleeping; and that seemed just.

The principal palms on the hills were the cabbage, the young shoot of which is eaten boiled, for which the poor tree is killed; the "maccafoot" and the "groo-groo," whose great seeds take a high polish, and look like onyx stones in a bracelet: the mahogany-cones open in four leaves, and the seeds inside are packed like French bonbons in lace-paper. I was always finding fresh wonders. The sea-cucumber, a gourd which grew near the shore, had the most wonderful mat or skeleton sponge rolled up inside, which the natives used as a scrubbing-brush. The delicious star-apple got ripe, and was filled with blancmange flavoured with black currants.

My old American fellow-passengers came up one day, and I gave them a feast of fruits and made them very happy on my verandah. Then I went down to the plain half-way towards Kingston, to stay three days with the banker Mr. M. and his wife. Mrs. M. had a curious collection of odd pearls in an old patch-box. The pink pearl was only found in the great conch shell; only one in a shell, and none in most. It seemed a great sacrifice to break up such a noble shell for the chance of finding one little dot of a pearl. She had also a most beautiful gold-coloured pearl, and was in despair of ever matching it so as to make ear-rings. Formerly there were pearl fisheries near Port Royal, which was the original capital of the island but being built on a coral reef it was undermined by the sea, and one day it all tumbled in and was drowned; and now in very clear calm weather they say you can still see the city under the water.

One Sunday morning I walked up to Craigton, and on to Judge Ker's. I got up my 1800 feet before eight o'clock, and found his worship in an extra scarecrowish costume gardening. He was a very odd man, but was one of the people I liked, so original and honest, it was difficult to listen to his talk without laughing. His wife was a sister of the Poet Laureate, but could not live in Jamaica. He said that at last he had discovered what to do with cheese-parings: he threw them on the floor, and then the rats came in to eat them, the cat came in to eat the rats, and so there was no waste.

He lent me his good gray horse, and I rode up to the church, and asked Mr. B. to get me leave to go and stay at Clifton Lodge, which he did. The house belonged to a gentleman who had lost his wife there, and never cared to see it again; he did not let it, but *lent* it for a week at a time to different people, who wanted a dose of cool air, 5000 feet above the sea, beyond the lovely fern walk and in the midst of the finest and oldest coffee-plantations in Jamaica. It was a charming little well-furnished house, surrounded by a garden full of large white arums, geraniums, roses, fuchsia fulgens in great bunches, sweet violets, hibiscus, great pink and blue lilies, orange-flowers, sweet verbena, gardenias, heliotrope, and every sweet thing one could wish for. Opposite was the real Blue Mountain, with clouds rolling up across it as they do in Switzerland. There was a village just below, with a great coffee-growing establishment, and bushes of it for miles on the hillside in front—all pollards, about four feet high, full of flowers and different coloured berries. It seemed an ill-regulated shrub; its berries had not all the same idea about the time for becoming ripe, and the natives had to humour them and pick continually. It was a wonderful little house: I found plate, linen, knives, a clock (going), telescope, piano (best not to try it I thought), and a nice tidy woman and family in the yard to get all I wanted. I had brought old Betsy, and she did holiday and "lady out visiting," as all maids do when away from their usual homes. She said it was very cold, and shivered, but I did not find it so, though blankets and counterpanes on the beds looked as if it might be sometimes.

A great blue-bottle fly buzzed, and a bird whistled two notes, scientifically describable as the diminished seventh of the key of F, an E natural and B flat alternately, always the same and in perfect tune. A lovely little apple-green bird with a red spot on his breast also came into the garden, called the Jamaica Robin, which burrows a tunnel in the bank like a kingfisher; but after going in straight for eight inches it makes a sudden bend at an acute angle, and thus hides the actual nest from strangers outside. The Banana-bird, as yellow as the canary but bigger, and the Doctor humming-bird, with green breast and two long tail-feathers, used to dart about the garden in company with his wife, who was, like him, minus the tail, and the mocking-bird sang sweetly in the woods behind, having a vast variety of notes and trills. What nonsense people have written about the silence of the tropics; they only go out at noonday, when the birds have the sense to take their siestas. If they went out early, as I did, they would hear every sort of noise and sweet sound; then after sunset the crickets and frogs strike up, and a Babel of other strange talk begins.

I did one great study in the Fern Walk, sitting in my mackintosh cloak, and bringing it back soaking outside every day. Then one afternoon a dragoon arrived on horseback with a letter asking me for a week to Spanish Town Government House to meet the S.s only. The balls and heavy parties being over I could not resist, though sorry to leave the nice place I was in.

When I got home, I found no donkey had been sent for my luggage, and old Stewart had gone up the hills with the house-key in his pocket, so I got in at a window in a very bad humour, and then had to walk down a mile or more to tell Boltons the stableman to send up at once for the baggage and give me a carriage to Spanish Town. After which I crawled up the hill and in at the window again, and cooked some eggs a black neighbour gave me in a shallow pan without a cover, and made some tea, bathed, dressed, and packed before old Betsy and the things arrived, when I again started with my portmanteau on the head of penitent Stewart back to Boltons. I found all the Newcastle officers on their way down the valley to a dance on board one of the men-of-war; carriages were scarce, so I went round by Kingston, and shared one that far with them. We went at a great pace down the steep road and across the plain, with its tall candelabra-cactus hedges, varied by those of *Bromelia Pinguin*, a kind of bromeliad, with the centre leaves bright scarlet, from which lovely pink flowers wrapped in white kid bracts peeped out, but so hidden by the great rosettes of outer leaves that they are only visible when one is raised on a high horse or carriage above the hedge.

I reached Spanish Town in the dark, barely in time for dinner, and enjoyed all the more looking out at my window the next morning on the lovely convent-like garden below, full of the richest trees and plants. A tall spathodea-tree was just opposite, covered with enormous flower-heads, pyramids of brown leather buds piled up and encircled by a gorgeous crown of scarlet flowers edged with pure gold. They came out freshly every morning and fell off at night, making a dark crimson carpet round the tree. A great waxy portlandia was trained just underneath it, and cordias with heads as big as cauliflowers (Robinias, and *Petræa scandens* with wonderful masses of lilac-blue bracts). Larkspurs, with blue and white flowers and leaves like sandpaper, were in their fullest beauty. I never saw so many treasures in so small a space. Arches surrounded it leading to the different rooms of the ground-floor, where the Governor and his A.D.C. had their office open to the fresh air of the gardens on either side.[43] When the day's work was over, Gertrude and I used to go and sit there too: I still painting in a corner, she and her brother and friends swinging in hammocks and talking nonsense, till the great heat was over, and we could go for a ride or drive. The first morning she took me for a lovely walk down to the beautiful sandy river-bed, with the bamboos and big tree branches dipping into the waters, long-legged birds of the heron or stork kind walking in and out, fishing and pluming themselves with their long bills, and making their morning toilettes. There were many curious and grotesque old tree-trunks down there, with snake-like roots stretching over the ground, and arched buttresses, from which the floods had washed away the sand and earth at different times. Graceful little black children were running in and out of the water, bathing and splashing one another. Fish, too, were jumping out of the clear water, which ran rapidly over the golden sand. I went there very often afterwards to sketch, with the old bloodhound to take care of me; he used to gallop on in front till he found some solid yards of shade, where he would sit waiting till I came up, and then run on to another good halting-place.

I was told it was nonsense keeping the garden-house any longer, as I had so many other houses, so I resolved to go over and give it up.

I went by rail to Kingston over a rich plain of grass-land, dotted like an English park with magnificent trees, mostly of the flat-topped rosewood and allied species, passing also some giant cotton-trees, which adapt themselves to the flat land by growing in width rather than height: their buttresses were huge. Mrs. C. gave me breakfast, and arranged all my affairs for me. She was the universal referee for everybody, and quite untiring in her kindness. I got a carriage and drove up to the garden-house, paid off my two old retainers, and packed up my things. When the coachman refused to bring them in his carriage, I told him to go home alone and sent him off, cheered by all the villagers, who hated townspeople, and carried my things down to Boltons', who gave me horses which flew like the wind, and took me all the way back to Spanish Town.

One afternoon we saw the honey taken from some hives in the garden in a most primitive manner. Three blacks put nets over their heads and cigars in their mouths, sat on their heels and hammered at the two dial boxes which represented hives, till the bees all mounted into the upper box, leaving the honey in the lower one. The spoil was almost as good as English honey, but the bees were poor languid things, like all other imported creatures, and too spiritless even to sting.

Another day I mounted the Governor's famous old horse Blunderbuss, and rode out through pretty green lanes to a crack in the hills full of Broom Palm, growing like tree-ferns, with fan-shaped leaves on the top of a stem six or eight feet high: the plant delights in dry ungenial places. We turned in at a gate and climbed higher and higher through various fruit-trees, including the sapadilla or naseberry, whose fruit is about the size of an apple and tastes like a medlar.[44]

King's House was a most inconvenient building, internal comfort sacrificed to its classical outside, and to a huge ball-room which took up one wing of two storeys in height. The piano was there, and when the house was full we used to sit there as the coolest place. The Governor had a habit of waiting till the second bell rang, and then saying: "God bless my soul! I must go and dress." We used to get all sorts of strange and excellent dishes; everything seemed new. Fresh ginger-pudding, tomato toast, fried "ackee"[1] mango-stew, stewed guavas, cocoanut cream and puddings, and many other things not heard of in Europe, as well as roast turtle and other strange fish, the former rather unattractive food.

I had one delightful day in the Bog Walk with the M.s. There is a village half-way up the lovely valley nestling among large bread-fruit trees and cocoanuts, and huge calabash trees with fruit so large that nets were put over them to support them. I saw the great aristolochia trailing over the trees, as evil-smelling as its neighbour the portlandia was sweet. Tangles of ferns and orchids on every rock, with the clear river rushing among them, sometimes bounding between huge rocks, sometimes winding along serenely between great plumes of feathery bamboo. It was a glorious four miles of scenery, but too far from Spanish Town to work in comfortably. The trumpet-trees were lovely there, with their hollow trunks, branches at right angles, and great bunches of white-lined horse-chestnut leaves and pink shoots; it is one of the most remarkable of all tropical plants. On the road back we saw a splendid specimen of the "Scotchman hugging the Creole"—a fig-tree which begins as a parasite

[1] Probably the well-known Egyptian and Indian "okery."

and gradually envelops the original tree and strangles it to death. The air was sweet with the yellow-flowered logwood, another tree was full of clusters of bean-pods, with small red husks and black berries set in white, which they call "bread and cheese."

Two days afterwards I packed myself and my trunk into a two-seated box on wheels, with a flat waterproof top, and curtains tied up with bits of string. It had two horses attached to it; one pulled it and one ran beside the one who pulled. The former fell down and broke its knees at the first hill, which taught me a useful moral: never work too hard nor try to do more than your neighbours, or you may break your knees, which is unbecoming even in a horse! My driver was coal-black, and dressed much like an ordinary English scarecrow, but he was a good fellow. We passed over some pretty coast scenery, reminding me of parts of the Cornice; the sea quite as full of colour. Sometimes we were high above it, sometimes on the very sand at the edge, fording several rivers, one of which just made its way in over the floor of the carriage so as to cool my feet and damp my portmanteau-cover. The rivers are always bordered by palms, bread-fruit, and other fine trees. Cocoanuts are very difficult to draw, being so exceedingly high that unless one gets too far away to see the detail one cannot get both ends in; but they are the noblest of all, and after seeing them one fancies all other palms untidy and in want of combing.

We passed several sugar-plantations, with factories for crushing the canes, and groups of coolies about them: they were great contrasts to the negroes, being so graceful, frank, and intelligent-looking.

Morant Bay was too tempting to pass, especially as it had a good inn kept by Miss Burton, a large black lady with most amiable manners. The house was raised above the village; she gave me a nice corner room with a large tub in it—very acceptable after coming from the mosquitoes of Kingston—and

I began a sketch at once of a great cotton-tree half growing in the river, with the blue sea beyond, shaded by palms and bamboos.

After leaving the sea the atmosphere got more and more like a hot fern-house, till we reached Bath, where the inn was kept by a decent kind of white woman. It was really hot and without air; so I worked at home in slight clothing till four o'clock, and then walked up two miles of marvellous wood scenery to the baths, which were slightly sulphurous and very hot and delicious. Two large nutmegs, male and female, grew close to them, with the beautiful outer fruit just opening and showing the nut and the crimson network of mace round it. The flowers are like those of the arbutus. Lower down bamboos were growing in great magnificence, their great curves of cane arching overhead and interlacing like some wonderful Gothic crypt. Large marrow-fat palms were there too, with their whole trunks and heads covered with hanging ferns, and tangled up with creepers. The cabbage-palm was in abundance, with its leaves very much uncombed, and a yard or more of fleshy green shoots, the flowers and fruit under them, many of the former being then still folded tight in the green bract which sticks out at right angles from the stem: to cut open one of these palm flower-sheaths and shake out the contents like a tassel of the finest ivory-work was a great pleasure and never-ending wonder to me.

The town of Bath consists of one long street of detached houses, having an avenue down it of alternate cabbage-palms and Otaheite apples. The old botanical garden had long since been left to the care of nature; but to my mind no gardener could have treated it better, for everything grew as it liked, and the ugly formal paths were almost undiscoverable. The most gorgeous trees were tangled up with splendid climbing plants, all seeding and flowering luxuriantly; the yellow fruit of the gamboge strewed the ground under them, and the screw-pine rested on its stilted roots, over which hoya plants were twining, covered with their sweet star-flowers. I longed for some one to tell me the names of many other plants which I have since learned to know in their native lands; but it was delightful to have time to study them and not feel hurried.

I asked why I saw no snakes, and was told they had all gone up into the trees to drink out of the wild bromeliads! Those pretty parasites often held quite a pint of water in the cornucopias which form their centres, as I found to my cost one day when bending one down to look at its flower, and it emptied its contents up my sleeve. I drove into the more open country in the dusk, and saw a large acacia-leaved tree full of deep pink flowers and shaking leaves, and was told "Thorley's food for cattle" was made from it; the natives called it the guanga-tree. I saw the two marenga-trees, from the berries of which the oil of Ben used by watchmakers is pressed; they are both very sweet, especially the one with a lilac flower which they call Jamaica lilac. The chocolate plant is also much cultivated at Bath: it has large leaves which rustle like paper when touched; the younger ones are of all sorts of tender tints, from pink to yellow; its tiny flowers and huge pods hang directly from the trunk and branches under the leaves, and the pods are coloured, according to their degree of ripeness, from green to purple, red, or orange. The flowers, small bunches of gray stars about the size of a fourpenny piece, are scarcely visible close to the bark of the tree. The nuts are buried in a rather acid white pulp in rows inside the pod.

A mill had lately been established near Bath for the purpose of crushing the graceful bamboos and making a coarse kind of paper from them; this will soon rob the place of its principal beauty, but no one cares, as few strangers ever make their way to Bath.

The road eastward was very lovely, making short cuts from one beautiful bay to another, passing many little landlocked harbours of the very deepest blue, with cocoanuts fringing the very edge of the sea, and grotesque rocks hollowed out by the waves underneath, hung with leaves of maidenhair a foot long. We passed through rivers deep enough to oblige me to put my feet as well as my trunk on the seat (the floor of the buggy had holes drilled in it on purpose to let the water through). We rested during the mid-day heat at Manchineal, where I sat in the doorway to draw a palm, and the fattest hostess I ever saw sat beside me, cutting up guavas into a pot to make jelly of, while her little boy of four cracked Palma-Christi berries with his teeth preparatory to making castor-oil. The mother gave me some pretty shells as a keepsake, and white Frangipani flowers to smell (*Plumeria speciosa*). We passed long lines of its trees loaded with the sweet waxy flowers; they open before the leaves appear, and a caterpillar comes with them and eats them all off at once. We saw quantities of the creature gorging afterwards, about three inches long, black, with a red patch on his head.

We did not reach Port Antonio till after sunset, so many attractive things had tempted me to linger and sketch. The hotel was full, but they gave me a room out, in the house of a very beautiful brown lady. During the night the rats came and ate holes in my boots, which were very precious and not easily replaced, so I always put them on the top of the water-jug during the rest of my stay on the island. Port Antonio looked quite an important place from the hills above, where some friends of my hostess kindly allowed me to sit and paint it in its cocoanut setting. They gave us glasses of "matrimony," a delicious compound made of star-apple sugar and the juice of Seville oranges, like strawberry cream. It was very lovely, but airless; on the north side of Jamaica there were none of the refreshing sea breezes which made the Kingston side bearable.

My next resting-place was at Mr. E.'s, where I stayed a week. His home was perched on a rock like a fortress; one could see miles of cocoanuts on one side and of sugar on the other. The mountains came down close behind it, over which twenty miles of rough riding road could take one to Newcastle, and a very beautiful road it must have been. Thirty cocoanuts were used for drinking every day in that house. When one asked for water one heard a chopping, and a glass of cocoanut water was brought as a matter of course; but it used to hurt my economical feelings to see the way all the precious fibre and husk were wasted. Mr. E. had a large sugar farm, and enough to do—he was out from sunrise to sunset. At certain hours he waited below in his office to receive complaints, and made a rule he would not be bothered at home; so one evening when a lot of negroes came wanting to "peaky Massa," he told them to be off, and when they would not go he called the dogs (who were fast chained), and the way they all ran down the hill was an odd sight. Those bloodhounds have a hereditary dislike to a black, and though they no longer have slaves to hunt, are very useful as guardians to white houses, as no blacks will come near them if they know the dogs are loose.

The sugar-canes grew here magnificently, planted sufficiently wide apart to allow a plough to be worked between the rows. They threw up from fifty to eighty canes in one bunch, and were often fourteen feet high. Rats are their chief enemies, gnawing the cane near the ground so that it falls and dies. A penny was offered for every dead rat, and often 1000 were killed in one week. The governor had introduced the mongoose from India to eat the rats, but they preferred chickens, and rather liked sugar too, so were, on the whole (like most imported creatures), more harmful than beneficial. Cocoanuts sold for £3 : 10s. a thousand, a single tree often yielding one hundred in a year.

We stopped near the house of an old black man who hates blacks and will only speak to whites; he had a lovely garden full of rare flowers, and he came out and gave us a huge purple lily and a branch of double hibiscus. The dogs drove out a large land-crab from under my bed one night, walking sideways like an ordinary sea-crab: he was black, and as big as my hand, and I ate him afterwards for supper, all minced up in his own shell. Those creatures generally live on land, but at certain seasons go off to lay their eggs on the edge of the sea, and then nothing stops them; they go in a given straight line over everything that comes in the way. Two African niggers came one day to sell some "Obeah" or charm-sticks, and sat down in the verandah to finish and polish them, and Mr. E. made them talk.[45] They had been taken by the Spaniards for slaves, but were retaken by the English before they got to Cuba. They had worked hard, and now they had land of their own, ten miles off, on which they lived. They made twelve shillings a week each by carving those sticks for charms against the Obeah people, with "plenty snake and toadie on them." The men were very intelligent, and had the greatest contempt for "them Jamaica creole people. Dey work! Dem no 'ave tame teem in dem for work what we 'ave! Dem lazy brute nigger!" When asked if they were married, and why not, they said: "Me not marry dis lazy brute Jamaica creole girl, Governor send bring good nice African girl over me marry drekly." It was suggested he might find a yellow girl: "Yaller girl! me no marry yaller girl, she got all de brute lazy lying Jamaica nigger, and all de craft and dishonesty of Jamaica white too, she too cheap for me!"

One of the amusements at dinner was to play with a kind of small cockchafer, no bigger than an English house-fly, but so strong that he would carry a wine-glass on his back easily across the table. Two or three used to be set to run races in that way, and one of them once carried a small salt-cellar full of salt in the same way. The next house I stopped at was over another bay on a high hill-top, with most exquisite sea and land views over a park-like country, with groups of richest trees and palms; but they blew about too much to paint with comfort.

My host was one of the largest growers and makers of sugar, and managed seven other estates besides his own. His wife was a very nice woman. There were enormous parties of coolies working in the plain below, the women loaded with bangles and nose-rings, picturesque and apparently happy, but looking forward to their return to India just as our people thought of returning to England; no good people wanted to stay in poor Jamaica, even Mr. W. talked of ending his days on his estate in Cumberland (how cold he would be!).[46] One evening he took me for a ride with his two small children (also on big horses) up to the estates of Lord Howard de Walden, through the richest meadows dotted with large clumps of bamboos and a clear river winding through them. The mountains which surrounded these broad meadows were terraced naturally, and covered with guinea-grass, on which the horses principally feed; it had been imported from India with the Coolies and Mongeese: above the guinea-grass rose the virgin forest full of valuable timber trees. Once a huge bamboo cane cracked and fell across the road just before we reached it; it would have killed us probably if it had fallen a few minutes later: they constantly fall in this way in the sun, if left too long without cutting.

I went over the sugar manufactories and saw the great steam crusher at work, the green cane going in at one end and coming out a mere flat dry shell, which after being exposed for a day or two to the sun's rays becomes capital fuel for the engine: the great pans and gutters were all kept very clean. We were given some delicious sugar-candy to eat, and also tamarinds preserved in sugar, I saw the rum made and coloured with burnt sugar, and was told that its price depended more on the colour being good than on anything else.

In the afternoon we were climbing the 800 feet of

steep park-like road up to Shaw Park, the very gem of all Jamaica, where I was received with the heartiest of welcomes by Mrs. S. and her wild family. The house stood on a wide terrace of smooth green turf; wooded hills rose behind it; real forests of grand timber trees—teak, cedar, fiddle-wood, and astic, cocoanuts and cabbage-palms—came close to it; and on one side was a gully with masses of bananas and ferns, and a large fallen tree to act as a bridge over the stream, with a washerwoman in bare legs always ready to hand one across. That tree arched like the bridge on a willow-pattern plate, and always gave me a fit of nerves, though I bit my lip and said nothing. The stream was so deep below that when the wading woman held her hand above her head it merely reached my elbow, and I thought I should like to go back when half-way over, but had to go on, as there was no room to turn. The hills on the south, and the slopes up to the park from the sea, were covered with pimento, allspice, and orange-trees; the former, covered with white feathery flowers, scented the whole air; the latter had lately been stripped, and the fruit sent off to New Orleans by sea. But the glory of the view was to the north. From the edge of the small table of green a river tumbled past the house and over the hillside in endless cascades; one could watch it dancing down among extraordinary greenery to the sea. At one place, about one hundred yards below, a bath had been made some twenty or thirty yards in diameter, shaded all round with bananas, and a small dressing-shed under them; a perfectly ideal bath, which seemed too good for mortals.

The air was always fresh at Shaw Park, but there was little shade just round the house, as the trees had been cut away to make places for drying the spice—great floors of cement side by side covering as much space as a house. One tree yielded eleven shillings' worth of fruit in good seasons. The bay below was fringed with cocoanuts, and was very shallow: a long reef of coral separated its glassy water from the ruffled sea beyond; the white sand and corals coloured it with the purest tints—green, blue, and rosy lilac. The house was surrounded by cows, pigs, goats, turkeys, chickens, all feeding where and how they liked. Every one was welcome to come or go in their own fashion; neighbours continually dropped in from every direction, unsaddling their horses and turning them loose among the other odd animals to feed and roll. All called Mrs. S. "Mother," and seemed on the most intimate terms with the rest. If the bedrooms were all full, there were drawing-room sofas or hammocks in the verandah for them. Some arrived in the night, and did not come in until the next morning; but all seemed welcome in that primitive establishment. Few of these young men were burdened with much education.

Mr. S. lived on another property, and hardly ever came to see his family; but he and his wife wrote to one another every day. The cocoanuts near the house were sixty or seventy feet high, and had notches cut in them all the way up the trunk, into which the negroes put their great toes, and ran up like monkeys with the help of their hands. The butler was sent up one day; he flung down about three dozen nuts off one tree, and then came down, sat on the ground and chopped them open, and we all had a feast round them, he supplying us also with spoons, which he cut from the shell at the same time as fast as possible. The rest of the nuts were cut open and left for the fowls and pigs to finish without spoons. One tree will give one hundred and fifty in a year. Those fresh-cut spoons stained our dresses and hands; but there was a tree close to the house called the Blimbing, whose juice took out all stains; the fruit was about the size of a date, and hung close to the trunk or branches of the tree, with tiny bunches of red flowers. Many tropical fruits grow in that same way out of the bark of the tree, including the cocoa and jack fruit.

There were some curious lizards sunning themselves on

the walls, with green heads, reddish tails, and a sort of flapping lung-apparatus outside, lined with orange, like a leaf of Austrian briar. On our way down to St. Anne's we passed a dead cow, who had gone to drink in the heat of the day (the cow was a fool) and had dropped down dead, and the John-crows were eating it. Those birds were a great blessing, but most hateful to look at; they used to sit digesting in rows on the branches of a dead cotton-tree, with their wings spread out to dry in the sun like the eagles on German soldiers' helmets, and they looked most uncanny. The fireflies were a sight to see, particularly the large one with two green lanterns on its forehead and one red one on its tail; they were so bright that if half-a-dozen were put into a bottle one could read by the light they gave.

There was a small black imp called Ida—its mother had been with Mrs. S. as a child, so this was the pet of the house; though only half-witted, it ran in and out with as much freedom as the dogs, was supernaturally solemn, and when told to dance or laugh made hideous faces and antics, but showed no natural merriment of any sort. It liked riding on horseback, and was not averse to strong spirits, and some of the wild boys made it tipsy with rum. Poor solemn little atom, it was more hideous than ever then. Rum is the curse of the country, and in that house a large jug of it mixed with water was always on the side-table, being emptied and refilled all day, ruining the health of all those poor boys in the stifling climate.

One night while at dinner we heard a great screeching of hens and cocks; a black man was sent out to see if it was a snake, and soon returned breathless: "Him bery big yaller one, him wait for Massa Jim, come kill him." We all jumped up in a great commotion. Jim seized a great old sword from the wall, I headed the party, and we found that the niggers had driven the snake up into a tree after it had killed one chicken and nearly caught the old hen. And now the black people were dancing round and round the tree, and singing out: "heh! heh! him bery big," at the tops of their voices to keep him up there. Jim quietly pushed his way through the ring, climbed up the tree, and after a St. George and the dragon fight cut off the snake's head, the big beast hissing and spitting at him to the last. The butler brought the great body in wound round and round a branch six feet long, and as thick as my arm; they said his wife would be sure to come to look for him the next night.

One afternoon they took me for a wonderful gallop over some twelve miles of rough forest, meadow, and road, fording rushing rivers and limestone springs, a party of nine of the wildest young people on the face of the globe, trying to frighten me if they could; but they had mounted me on a strong sensible old hunter, and I just let her choose her way and have her head free, and enjoyed the scramble as much as they did. One place they took me to was a perfect fairy hall, with the clearest emerald-green water, trees with bunches of glossy leaves two feet long, and stalactites piled up fantastically against them; the leaves of these trees stung like nettles, and the stems grew straight out of the stalactites and water: all those springs come out close to the scar's edge, and are soon lost in it. The road along the coast to St. Anne's was shaded by bread-fruit and mammee-trees; the *Broughtonia sanguinea* orchid was hanging like a string of rubies from the rocks among the fresh green ferns. I was sorely tempted to take a small vacant house there called Eden Bower for £3 a month, with endless cocoanuts and grass for the horses, and enough allspice to pay my rent (fever also in plenty).

Prudence, however, drove me back to the civilised side of the island over the Monte Diabolo. Ascending by the veryferniest gully I ever saw, where the banana leaves were absolutely unbroken by any wind, we came to a kind of alpine scenery—a wide waving table-land of grass with trees dotted about it, oranges, allspice, and different timber trees hung with orchids, but not in flower. They were harvesting the oranges in one place in the usual way when the "Massa" or "Busha" is not by, that is, sitting in groups under the trees and eating them. Jim had spent the last week riding about seeing that that was not done, and had sent off 5000 during that week: only making ten shillings a thousand (and 1100 go to make a thousand), it seemed hardly worth the trouble. I stayed a night at Linstead, a pretty village at the head of the famous Bog Walk, and the next day drove all the way through Spanish Town, with its big deserted Queen's House, to Kingston, and climbed the hill to Bermuda Mount to stay with Gertrude S. and her brother the Attorney-General.

It was delightful to be with such people again—people who read and thought, and enjoyed a joke too, and were never idle. We were very happy together; though the summer heat prevented me from working out of doors, I always found abundance of flowers to paint in the cool verandah. In the evenings Gertrude and I took long rides or walks about those lovely hills; we often had orders to drive with the Governor when her brother could not come home, and used to walk home by moonlight accompanied by the great bloodhound, who divided his time between the two houses equally. On those nights I had a good sight of the beautiful night-flowering lily, with a pink edge, which was wide open; but as soon as the sun had thoroughly risen its head hung like a windless standard. The moonlight looks whiter on the smooth wet surface of the banana leaves; no native will go near them then for fear of "Duppies."

1872.—On the 24th of May Gertrude took me on board the *Cuban*—a roomy ship, with delightful deck cabins, and a jolly captain. We touched at three of the harbours of Hayti, with fine hills wooded to their very tops. At Port au Prince Mr. St. John, our chargé d'affaires, came and paid me a visit on board for a couple of hours, and told us much that was funny of the Black Republic, and of its army, which wore the lids of sardine-boxes for epaulets.

On the 16th of June we landed at Liverpool, and two days after I was at home.

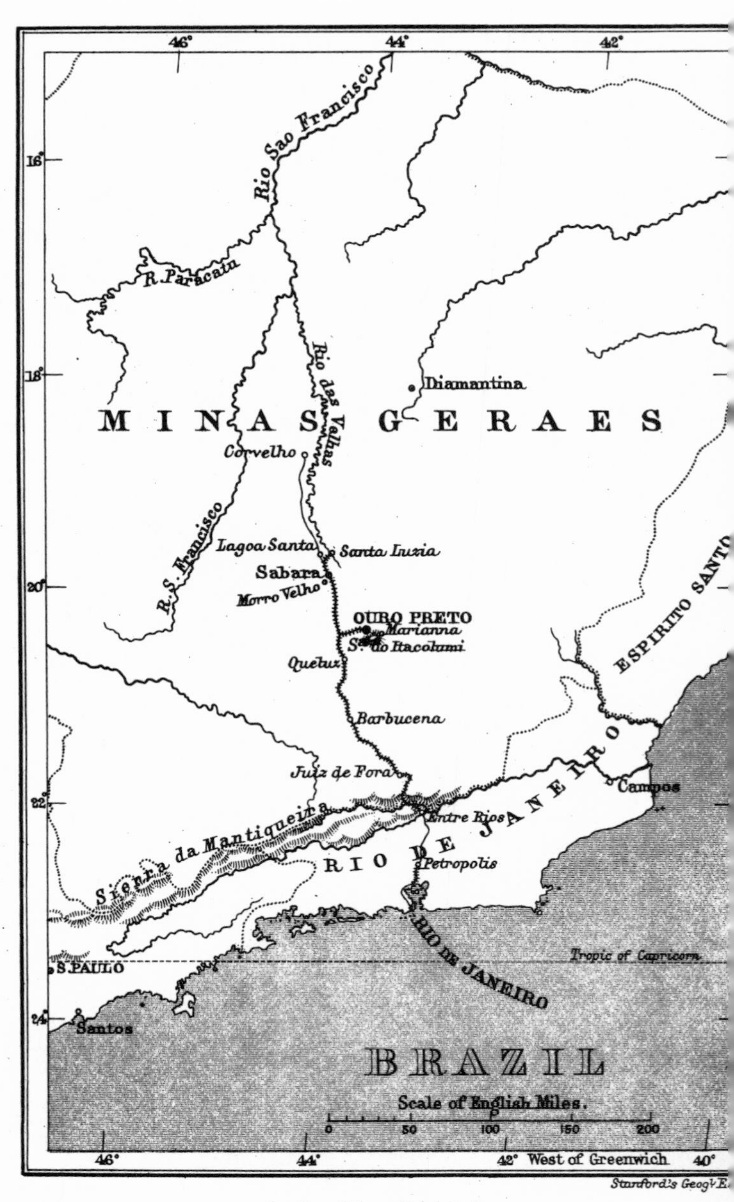


London: Macmillan & Co.

# CHAPTER IV

## BRAZIL

### 1872-73

FOR the next two months I enjoyed the society of my friends in London, and then began to think of carrying out my original plan of going to Brazil, to continue the collection of studies of tropical plants which I had begun in Jamaica.

1872.—I started in the *Neva* Royal Mail Ship on the 9th of August with a letter from Mr. R. G. to the captain. I had a most comfortable cabin, quite a little room, with a square window, and the voyage was most enjoyable. Lisbon was our first halt, which we reached on the 13th at sunset; the entrance to the harbour is striking, with the semi-Moorish tower and convent of Bela in the foreground; the domes and tall houses of the city gave me a much grander idea of the place than it deserved when investigated nearer: on the 19th we stopped to coal at St. Vincent. I did not land on that treeless island, which looked like a great cinder itself; but the boats which surrounded the ships were full of pretty things from Madeira, baskets and inlaid boxes, feather-flowers and fine cobwebby knitting, as well as monkeys and love-birds from the coast of Africa.

On the 28th of August 1872 we cast anchor at daylight off Pernambuco, and I saw the long reef with its lighthouse and guardian breakers stretching out between us and the land, and wondered how the crowd of ships with their tall masts ever got into the harbour. Seen through my glass, the buildings of the town looked much like those of any other town, but beyond were endless groves of cocoanut-trees, showing clearly in what part of the world we were. "Friend, a walk on shore will do thee good; my husband hath work to do there, and where he goeth I can go, and where I can go thee canst also," said a dear old Quakeress of New York to me; so I fetched my umbrella and prepared to follow the leader of our landing-party (a Belgian) down the ladder into the boat, but he went too fast and far, a wave went right over him, and we had to come up again while he changed all his clothes, for he was completely soaked. Our next start was more fortunate; we all watched till the boat was on the top of the swell and then dropped ourselves in cleverly one by one. It is often quite impossible to land at Pernambuco for many days together, and yet in this stormy sea, which is full of sharks, one sees the native fishermen floating about on the rudest kind of rafts, like hen-coops, with their legs in the water. The planks which form these rafts are so much more under than above the water that the men seem to sit on the actual waves as one sees them in the distance, and being black they fear no sharks. Our row over the surf was easy enough, though the white breakers on the coral reef looked angry on either side of us. Inside, the harbour is calm as a millpond, and we soon stood under the great umbrella-trees in the principal square.

It was Sunday, and the shops were shut with as much rigour as in Glasgow itself. I saw little to buy but parrots, oranges, and bananas; no ladies were about, they were all in church, and as my Quaker friend had told us nothing should induce him to take his hat off in those temples of idolatry, we did not attempt to enter the somewhat tawdry-looking buildings. But though the upper class of women was wanting, there were plenty of negresses in the streets, whose gay-coloured striped shawls hung over their heads and shoulders in the most picturesque folds; and in the suburb gardens we saw grand palms and other tropical plants new to me. The fan-palm of Madagascar was perhaps the most remarkable, with its long oar-like leaves and stalks wonderfully fitted together in the old Grecian plait, each stalk forming a perfect reservoir of pure water, easily tapped from the trunk; thirsty travellers had good reason for naming this palm or strelitzia their friend. The Frangipani-trees were also in great beauty, covered with yellow or salmon-tinted waxen bunches of sweet-scented flowers shaped like large azaleas, but as yet almost leafless. The flowers go on blooming for many weeks, then come the leaves, and with them a huge black and orange caterpillar with a red head, which eats them all up in a very short time; in spite of this the vitality of the tree is so great that it soon flowers again. The natives say that the moth lays its eggs in the very pith of the wood, and that if a bit is taken as a cutting to any part of the world and a young tree grown from it, the caterpillar will also grow, and appear in time to eat up its first attempts at leaves. Ants seem to abound about Pernambuco, and I noticed that all the rose-trees or other choice plants in the gardens had a circular trough of water round them, which I have little doubt is a protection till the clever little creatures learn to tunnel under them.

We drove out to the country by "the Bonds" or street railways which are now established in all the principal towns of Brazil, and are a great convenience and economy of time and money. These carriages are drawn by mules, and go at a great pace; the sides are open, and a substantial awning keeps the sun off the roof, so that one cannot well have cooler quarters at midday, obtaining at the same time a good sight of the country and its people.

At Bahia we also landed, and after mounting the steep zigzag to the top of the cliff, had another drive into the country, which is wild, hilly, and covered with rich forests. The market was most entertaining, and full of strange pictures. Huge negresses in low embroidered shirts (tumbling off), a gaudy skirt, and nothing else except a bright handkerchief or a few flowers on their heads, were selling screaming parrots, macaws, and marmosets, gorgeous little birds, monkeys and other strange animals, including a raccoon with a bushy tail, and a great green lizard as big as a cat, which they said was very good to eat.[47] I saw one girl quite covered with crawling and scratching marmosets; she never moved, but they did incessantly. One of the children on board bought a very tiny marmoset, so small that he hollowed out a cocoanut shell, put some cotton wool in, and used to keep his pet in it, having cut off the small end to let it in and out; its tail was eight inches long and very bushy. The oranges at Bahia are large and sweet, and they pack all their seeds into a kind of bag at one end, which renders them particularly easy to eat; the piles of this fruit, as well as of melons, tomatoes, egg-plants of different sorts, and pine-apples, make grand masses of rich colour, while bunches of sugar-cane, great whorls of bananas, and heaps of cocoanuts form a fine background. Lazy people were carried up the steep streets sitting on chairs in a kind of crazy palanquin, which was hung on a bent pole and carried on two men's shoulders; if the passenger were not a fidget he might arrive at the top of the hill uninjured. We did not try, but tired ourselves out in the usual British manner on foot, and were not sorry to get back to the *Neva* again. It took us in two more days safely into the beautiful Bay of Rio, which certainly is the most lovely sea-scape in the world: even Naples and Palermo must be content to hold a second place to it in point of natural beauty. I know nothing more trying to a shy person than landing for the first time among a strange people and language, I always dread it; so I asked the good Belgian merchant to help me, and he gave me into the care of one of his brothers, who not only landed me in his boat, but put me into a carriage which took me to the Hôtel des Étrangers at Botofogo, on the outside of the town.

I soon felt myself at home in Rio, and in a few days had a large airy room and dressing-room at the top of the hotel, with views from the windows which in every changing mood of the weather were a real pleasure to study; both the Sugar-loaf and Corcovado mountains and part of the bay also were within sight.

The house was wonderfully clean and comfortable, considering the people who kept it so; an American half-caste woman acted as chambermaid and did nothing slowly; a black man (a slave) did it quicker, and looked as if he enjoyed the work; he told me, "When you want nothing, call Auguste."[48]

The town of Rio has a great look of its relations in Spain or Sicily; the houses so full of colour, the balconies of such varied form, and the tiled roofs project in the same way, with highly ornamented and coloured waterspouts and terminals: the inhabitants have the same love of hanging out gaudy draperies and bright flowers from their windows and balconies, with the addition of parrots and monkeys screaming and scrambling after the passers-by, who are fortunately generally well out of reach. One day, however, I saw a tall slave-girl's tray of oranges robbed by a spider-monkey, as she walked underneath with a well-balanced pyramid of fruit on her head. The Brazilians are so fond of illuminations that there are permanent gas-pipes bent across their principal streets; these run perfectly straight from the sea to the hillside; the long vistas of flaming arches have a far finer effect than our isolated stars and cyphers. The shops in the streets seem very good, but the things are principally from Europe and exorbitantly dear. Brazil offers to a stranger few inducements for spending money, except its wonderful natural curiosities, its gorgeous birds and butterflies; "Even its bugs are gems," a Yankee friend remarked to me, and these latter are set in gold as ornaments with considerable taste and fineness of workmanship. To me the humming-birds were the great temptation. M. Bourget, one of Agassiz's late travelling companions, had a rare collection which he valued at 300 guineas, and I passed many happy mornings among his treasures hearing him talk of them and of their habits; but after the first few days I seldom went into the town.

The mule-cars passed the door of the hotel every ten minutes, and took me at six o'clock every day to the famous Botanical Gardens, about four miles off.[49] The whole road is lovely, skirting the edges of two bays, both like small lakes, to which one sees no outlet; the mountains around them are most strangely formed—on one side generally a sheer precipice, on the other covered with forests to the very top; and such forests! not the woolly-looking woods of Europe, but endless varieties of form and colour, from the white large-leaved trumpet-trees to the feathery palms, scarlet coral, and lilac quaresma-trees. Then the villa gardens along the roadside were full of rich flowers and fruits and noble trees; at one place a sort of marsh with masses of Indian bamboo gave the eyes a pleasant rest after the glaring gaudiness of the gardens. That drive was always charming and fresh to me, and I wished the mules had not been in such a hurry; but they were all splendid animals, and seemed to enjoy going at full gallop, after the first little scene of kicking and rearing which they considered the right thing at starting. They often went too fast, and would have arrived at the station before the appointed time if they had not been checked.

The gardens of Botofogo were a never-ending delight to me; and, as the good Austrian director allowed me to keep my easel and other things at his house, I felt quite at home there, and for some time worked every day and all day under its shady avenues, only returning at sunset to dine and rest, far too tired to pay evening visits, and thereby disgusted some of my kind friends. Of course my first work was to attempt to make a sketch of the great avenue of royal palms which has been so often described. It is half a mile long at least, and the trees are 100 feet high, though only thirty years old; they greatly resemble the cabbage-palm of the West Indies, though less graceful, having the same great green sheaths to their leaf-stalks, which peel off and drop with the leaves when ripe; about five fell in the year, and each left a distinct ring on the smooth trunk. The base of the trunk was much swollen out, and looked like a giant bulb. This huge avenue looked fine from wherever you saw it (and reminded me of the halls of Karnac).[50] There were grand specimens of other palms in the gardens: a whole row of the curious Screw-Pine, with its stilted roots and male and female trees; rows of camphor-trees, bamboos, the jack-fruit, with its monstrous pumpkin-like fruits hanging close to the rough trunks, and endless other interesting plants and trees. Beyond all rose the great blue hills. One could mount straight from the gardens to their woods and hollows, with running water everywhere. The garden seemed a favourite place for picnics, and tables and benches were set up under the wide-spreading bamboos and other trees. One day a most genial party settled near me, several of whom talked English; one of them brought me a saucer of delicious strawberries with sugar and champagne poured over them; he said they were not so good as those in England, but the best in Brazil; they were grown in his garden and picked by his children. The visitors were not all so well bred, and once my friend the director flourished his big stick and gave them his mind in strong German on the subject of standing between me and the tree I was drawing.

One day I was puzzled at hearing him continually calling "Pedro" in a coaxing tone of voice; at last up trotted a tapir, like a tall pig with a cover to its nose; he got something he liked out of the director's pocket, and a good scratch from the director's stick, and followed us as long as he dared. I found some difficulty about food for luncheon; if I put meat into a tin box it went bad, if I took it in paper the ants ate it up for me, even eggs they contrived to get into, and at last I came to the conclusion that oranges and bread were the best provisions to take. One day I asked the director if I could get a cup of coffee at the little inn near the gate. "Gott bewahr!" was his answer; he would not let his daughters even walk in the road alone among such people. Poor girls, they must have had a dull life of it; they were so thoroughly German and isolated, they had hardly ever been even into Rio. We had some pleasant scrambles together in the woods and up the hills; for they were nice simple girls, full of information about the plants and other natural curiosities of the neighbourhood. They collected marvellous caterpillars,—some hairy, some with the most delicate moss or feather-like horns on their heads and tails,—and fed them till they turned into the gorgeous butterflies or moths which abound in these gardens.

After a fortnight's daily work there the weather became cloudy, and I brought home flowers or fish to work at, my landlord kindly letting me go with him any morning I liked to the wonderful market, where the oddest fish were to be found, and where boat-loads of oranges were landed and sold all day long on the quay-side.

Almost all the menial work in Rio is done by slaves, either for their owners or for those their owners hire them out to serve; for though laws are passed for the future emancipation of these slaves, it will be a very gradual process, and full twenty years will elapse before it is entirely carried out. It would have been better perhaps if our former law-makers had not been in such a hurry, and so much led away by the absurd idea of "a man and a brother." I should like some of the good housewives at home who believe in this dogma to try the dear creatures as their only servants. One of my friends had been settled in Rio nine years with no maid-servant, only two black men (the lesser evil of the two), and some of her experiences were amusing. The blacks never kneel (except on the outside of illustrated tracts), and if they were told to scrub the floor they brought a pint pot full of water, which they poured over here and there, then put a bit of rag under their feet and pushed it about till the floor was dry again. If a black servant were spoken rudely to, or found fault with, he ran away back to the owner who let him out, and said he would not stay; his health would be ruined in that place, and his owner's property would be thereby injured in value. A good working man-slave could not be hired for less than £30 a year, though he might be fed and clothed (in slave fashion) for threepence a day: a girl for housework got £15 a year and two suits of clothes, besides sundry presents to herself to keep her in good-humour, and prevent her from running away to her real owners. It is a mistake to suppose that slaves are not well treated; everywhere I have seen them petted as we pet animals, and they usually went about grinning and singing.

The ladies in Brazil had the women well taught to embroider and make lace, doce, etc. etc.; they then sent them out to sell these things for them, which small trading was not looked upon as in the least *infra dig.*[51] The embroidery is some of it very fine, particularly the sort made by pulling out the threads of fine cambric, or even cotton stuff, and working different patterns on it; it takes much time, and is very expensive; this lace is coarse but effective.

At Rio I made my first acquaintance with a very common inhabitant of the tropics, a large caterpillar, who built himself first a sort of crinoline of sticks and then covered it with a thick web; this dwelling he carried about with him as a snail does his shell, spinning an outwork of web round a twig of his pet tree, by which his house hung, leaving him free to put out three joints of his head and neck, and to eat up all the leaves and flowers within his reach; when the branches were bare he spun a bit more web up to a higher twig, bit through the old one, jerked his whole establishment upstairs, and then commenced eating again. He had a kind of elastic portico to his house which closed over his head at the slightest noise, his house shutting up close to it like a telescope; and then when all was quiet again out came his head, down dropped the building, and the gourmand again set himself to the task of continual eating. He ate on for some months incessantly, using his claws to push and pull dainty bits down to him, and shifting his moorings in a most marvellous way. At last the sleep of the chrysalis overtook him, and he finally became a very dowdy moth. Some other caterpillars cover themselves in a much less artistic way with bits of their favourite leaves strung on a frame most clumsily, as a child strings paper to the tail of its kite. These creatures are very quick in their movements; I have often seen them cross the room and drag themselves up my dress and on to my knee in search of a bunch of rose branches I laid there to tempt them,—in a wonderfully short space of time.

The lady in whose garden I first found these caterpillars lived on the hill of Santa Theresa, and, instead of blinds, had her windows shaded with creeping-plants trained across and across them. Through the spaces left one could see the bay of Rio with its endless islands, strange Sugarloaf mountain, and many of the same odd form seeming to mimic it in the distance. The quivering haze and blueness of the whole scene was indescribably lovely, and the little terrace below was crowded with bright flowers. Daturas, bananas, cypress and palm-trees gave form to the foreground, whilst the orange *Bignonia venusta*, the blue petræa, bougainvillea, and rhynchospermum climbed over both trees and balustrades in great masses, the latter helping the gardenias, carnations, and jasmines to scent the air almost too deliciously. It was a small paradise, and though my friend grumbled at the nine long years of bad health and discomfort she had spent there, she will miss all this abundant beauty when she returns to foggy old England.

I spent some days in walking and sketching on the hills behind the city; its aqueduct road was a great help to this enjoyment, being cut through the real forest about a thousand feet above the town and sea. A diligence took one half-way up to it every morning; the road itself and the grand aqueduct by its side were made two hundred years ago by the Jesuits, and the forest trees near it have never been touched, in order to help the supply of water which is collected there in a great reservoir. In this neighbourhood I saw many curious sights. One day six monkeys with long tails and gray whiskers were chattering in one tree, and allowed me to come up close underneath and watch their games through my opera-glass; the branches they were on were quite as well worth studying as themselves, loaded as they were with creeping-plants and grown over with wild bromeliads, orchids, and ferns; these bromeliads had often the most gorgeous scarlet or crimson spikes of flowers. The cecropia or trumpet-tree was always the most conspicuous one in the forest, with its huge white-lined horse-chestnut-shaped leaves, young pink shoots, and hollow stems, in which a lazy kind of ant easily found a ready-made house of many storeys. The most awkward of all animals, the sloth, also spent his dull life on the branches, slowly eating up the young shoots and hugging them with his hooked feet, preferring to hang and sleep head downwards. Some of the acacia-trees grow in tufts on tall slender stems, and seem to mimic the tree-ferns with their long feathery fronds, whose stems were often twenty to thirty feet high. Mahogany, rosewood, and many less known timber-trees might be studied there; the knobby bombax, gray as the lovely butterfly which haunted them, were planted at the edge of the road in many places, and under them one got a really solid shade from the sun.

It was the favourite home of many gorgeous butterflies, and they came so fast and so cleverly that it was no easy task for a collecting maniac to make up his mind which to try to catch and which to leave; before the treasure was secured more came and tempted him to drop the half-caught beauties for other, perhaps rarer ones, which he would probably miss.

One happy mortal lived up in this neighbourhood and collected calmly, with his whole heart and time in the work, thereby gaining a good livelihood; he had drawers full of the different specimens, which were worth a journey to see: alas! when I went he had just sold the whole collection to the Imperial Princess, so I kept my money, as well as a most fascinating occupation for odd hours, which would have gone if I had, as I intended, done my collecting by deputy.[52] He lived on a lovely perch just under the Corcovado Crag, with a glorious view of the city and bay beneath, and a rare foreground of palms and cacti, one huge mamen tree in front of all, its thick umbrella of leaves supported by great pear-shaped fruit growing close to the stem. The common snail of Brazil introduced itself to me on that road; it was as large as a French roll, and its movements were very dignified. It had a considerable appetite for green leaves (as I afterwards found after keeping one as a pet in a foot-pan for a month), and its eggs were nearly as large as a pigeon's; the first I met was taking a walk on the old aqueduct amongst the begonia and fern-leaves, and moved on at least fifty yards whilst I made a two hours' sketch.

Of course (again), like all other visitors to Rio, I walked up to the top of the Corcovado and looked down on the clouds and peeps of blue sea and mountains seen occasionally through them, and on the splendid yellow and white amaryllis clinging to the inaccessible crannies of the rock; the whole way was a series of wonders and endless beauties.

On that expedition I met, for the first time, Mr. Gordon and his daughter, who asked me to come and see them in Minas Geraes, to which they were returning in about three weeks.[53] I liked their looks and manner of asking me, and it seemed a grand opportunity of seeing something of the country, so I said I would come for a fortnight, at which they laughed, and with reason, for I stayed eight months!

Meanwhile other kind friends asked me to pay them a visit

at a house they had taken for sea-bathing on the island of Pakita. I had already ordered my room at the hotel at Tignea, so I packed up my bag and carried it on board the little market-steamer, which took me in two hours across the bay to Pakita. The whole bay is sprinkled with islands and boulder-stones—some covered with woods and palms, some mere piles of dry boulders whose history is a sad puzzle to wise men. They are of hard granite, and some look as if they had dropped down violently from a high planet, and cracked in the process; others have orchids or aloes growing on their tops, and the tide level is marked by a small oyster not bad to eat.

Only two or three of these islands are inhabited, as there is no fresh water on them. But for this want Pakita, with its beautiful Indian name, would be the Island of Islands. It is so full of loveliness, indented as it is with many little creeks of silvery sand sprinkled with fine shells, the shores edged with drooping cocoanuts and other graceful trees and palms. There were about half a dozen villas on the island belonging to different merchants of Rio, and perhaps a hundred cottages, whose inhabitants make a poor living by fishing and burning shells for lime. The little house I stayed in was close to the edge of the sea, and any of the party who were inclined could run in and out of the water at any hour of the day, or stroll over the wooded heights or sands and enjoy the prettiest views possible with the least possible fatigue, for one could walk all round the island in less than an hour. After leaving this enchanted spot I went for a fortnight to Tignea, where the rains overtook me; but I had abundant work in the comfortable room which had been kept for me, painting different orchids and other flowers, with now and then a ramble in the hills and forests.

On the 25th of October I sent down my three portmanteaux in a return-cart drawn by eight oxen, and followed myself the next day, in pouring rain, to Rio. After some necessary shopping and other business, I crossed the bay and its lovely islands for Mawa, where a train was waiting to take us over the marsh to the foot of the Petropolis hills; in this same marsh were many fine plants, but the most conspicuous was the real Egyptian papyrus, growing with even greater vigour than it does at the source of the Cyane, near Syracuse. Tall white lilies and scarlet erythrinas also made me long to cry "stop" as we passed. At last we reached a more healthy-looking region, and stopped at Reiz da Serra, where I was put into a carriage with three Brazilians and conveyed up the ten miles of zigzag road, dragged by four mules, who kept up a continual trot, the rise of 3000 feet being well graduated. The mules were changed at a station half-way up, and the short stoppage gave one time to enjoy the magnificent view, the great mountains looking like ghosts through the mist and rain, the few giant trees which had escaped the cutting of the forest when the road was made, standing out all the grander for the background being veiled. As we rose higher the sun's last rays sent a red line through the openings in the clouds, and one or two of the highest points seemed on fire. From the top the view back towards Rio is perhaps as fine as anything I had yet seen, with the exception of its having no snow; the distant view of the city, with its two guardian masses of rocky mountains, as well as the bay full of islands, and the rolling middle distance shaded by floating clouds, was inexpressibly beautiful.

Two more miles at full gallop down hill took us to Petropolis, and I was soon in Mr. Miles's comfortable hotel, and again among friends, with whom I had a merry English dinner. Then came two days of rain and cold and loneliness, in which I worked and walked and soaked and froze, and came to the conclusion Petropolis was an odious place, a bad imitation of a second-class German watering-place, with its red roofs, little toy houses, and big palace in the midst, the river cut and straightened into a ditch, running down the middle of the principal street, with fanciful wooden bridges crossing it continually, and its banks planted with formal trees; though, when one came to think and thaw a bit, those very trees were in themselves a sight to see: umbrella-trees with their large heart-shaped leaves and pink fluffy flowers, and araucarias larger than any in England. My friend Mr. Hinchcliff had written me minute directions how to find one of his favourite walks, where he promised I should see ideal tropical tangles. I paddled through the mud and rain to find, alas! nothing but charcoal and ashes remained; some German women added insult to injury by informing me it was "verboten" to go further that way, so I returned to my packing in disgust. I was glad to see the Gordons arrive, and to hear them say they had taken their and my places in the coach for Juiz de Fora the next morning. Mrs. Miles took charge of my tin box and sketching umbrella, which, I may as well say here, is a perfectly useless article in the tropics; when the real unclouded sun is shining one requires a more solid shade than that of a gingham umbrella, and it is far too heavy to drag about in a hot climate, so I was glad to be quit of it.

It rained all night, and was still raining when we packed ourselves into the coach at six on the morning of the 28th of October, and four splendid mules, after their usual resistance, started suddenly at full gallop with the swinging, rattling old vehicle. A violent jerk brought us to the door of the other inn, and there our fourth place was filled up by a very important person in these pages, Antonio Marcus, commonly called the Baron of Morro la Gloria, who had been for forty years in the service of St. João del Rey Mining Company, to whose mines I was going. This old gentleman generally commanded "The Troop" which brought the gold up to Rio every two months at least; he was a great character, full of talk and pantomime, either grumbling or joking incessantly, or sometimes even doing both at once. Mary G. was his ideal of perfection, and understood how to stroke him the right way, so we had a merry journey through the most splendid scenery.

Such scenery! High trees draped with bougainvillea to the very tops, bushes of the same nearer the ground reminding one of the great rhododendrons in our own shrubberies in May at home, and of much the same colour, though occasionally paler and pinker. There were orange-flowered cassia-trees (whose leaves fold close together at night like the sensitive plant) and scarlet erythrinas looking like gems among the masses of rich green; exquisite peeps of the river, winding below its woody banks or rushing among great stones and rocks, came upon us, and were gone again with tantalising rapidity. My friends only laughed when I grumbled at the mules going so fast; now and then a peaked mountain-top pierced its way through the clouds for a moment and was lost again, then came a gray overhanging cliff sprinkled with bracket-like wild pines spiked with greenish flowers; the near banks were hidden by masses of large-leaved ferns and begonias and arums of many sorts, whose young fresh leaves and fronds were often tinted with crimson or copper-colour. The wild agaves too were very odd: having had their poor centre shoots twisted out, the sap accumulated in the hollow, and a wine or spirit was made from it; the wretched wounded things, sending up dwarfish flowers and prickly shoots from their other joints, formed a strange disagreeable-looking bush, several of which made a most efficient hedge. Under each of these flowers a bulb formed, which when ripe dropped and rooted itself, thus replacing the parent whose life ended at its birth. Another curious plant here abounded, the marica, like a lovely blue iris, which flowers and shoots from the ends of the leaves of the old plant, the leaf being often more than a yard in length, and weighed down to the ground by the bunch at its end. When the flower is over, a bulb forms under it which produces roots; eventually the connecting leaf rots off, so that a perfect circle of young plants succeeds round the original old one. When in flower the appearance was very peculiar; a perfect rosette of bent green leaves and a circle of delicate blue flowers outside them.

The grand coach road we went over had, of course, encouraged emigrants to settle near it; we passed miles of cultivated ground, and the long rows of tidily trimmed coffee and corn gave as much pleasure to my companions as the forest tangles gave to me. We stopped to dine at Entre Rios; here we came to the Don Pedro railway, and the real traffic of our road began.

There was no other way of reaching the rich province of Minas, or of obtaining its minerals, coffee, sugar, or cotton; so from this point we passed a continual stream of mules or waggons till we got to Juiz de Fora and its most comfortable hotel. The last part of the way was lighted by swarms of fireflies. We were two hours after our time, owing to the state of the roads and the overloaded coach; all the baggage was packed on the top in one high pyramid, and the outside passengers were clinging to every ledge, the whole machine swaying from side to side in the most frightful way. The Baron's head was continually out of the window, shouting directions to the driver and conductor, who of course knew him too well to take the slightest notice; they were both great characters in their way, two German brothers who had driven over that road ever since it was first made, nearly twenty years before.

Juiz de Fora is all one monument to the great and good man who founded it, Senhor Mariano Lages; even the excellent hotel was designed and built by him, and a college for agriculture, library, museum, his own pretty villa and gardens, and the grand road itself, were all made by him for the good of his country, as well as his own. He did not live long enough to see them prosper, but pined away after the death of his favourite daughter; and his college and other schemes will soon pine away too, for patriots are not common in his country. His garden was full of treasures, not only of plants, but of birds and animals; there was a fence of fifty yards at least, entirely hung with rare orchids tied together; every available tree-branch was also decorated in the same way, and many of them were covered when we were there with lovely blossoms of white, lilac, and yellow, mostly very sweet-scented. There was also a great variety of palms. I saw one huge candelabra cactus twenty feet in height, and the air was perfumed with orange and lemon blossoms. The village itself looked very comfortable, every cottage having its own luxuriant little garden and shady porch, under which the fair German women and children sat knitting with their hair plaited round their heads. Every one said the road to Minas was impassable from the late heavy rains. We heard of mules being smothered in the mud, a woman killed in it, etc.; but the more I heard the more I determined to see my friends safely through, if they were willing to be burdened with me; besides, people had said in Rio I should never really go, some had done their best to keep me from going, and one Scotchman had said I should "not find to paint any in Minas!"

The first loading of thirty-seven mules is not done in an hour; everything must be weighed and strengthened and hung with stout bands of cowhide, balanced well, or the mules will suffer. When once they are well loaded the things are numbered, and the operation on subsequent mornings becomes a much easier and quicker affair. All these arrangements were our Baron's glory; he had to think and be responsible for every little item, and made as much fuss as he possibly could, getting in and out of a score of terrible rages before midday. When the rain left off, his temper also cleared, and we finally started, forming a party which would not have shone in Hyde Park, but was admirably adapted for riding through Brazil in the wet season.

First went the loaded mules with their bare-legged black drivers, then the Baron in the shabbiest of straw hats, any quantity of worsted comforters, and brown coat and gaiters. Mr. G. on his noble gray mule, his daughter on her pretty little horse, and myself on Mueda, the steadiest and most calculating of mules. My dress was as good as any could be for such riding, namely, a short linsey petticoat and a long woollen waterproof cloak with sleeves. I had besides a light silk waterproof rolled up and hung on my pommel for extra wet hours, and my old black straw hat on my head. Behind us rode the two grooms—Roberto, the little bright-eyed mulatto boy, whose duty was always to look after Mary and myself, and Antonio, Mr. G.'s own particular attendant, in a gorgeous livery, glazed hat with a cockade on one side, top-boots, and a decidedly negro face. Alas! his magnificence soon disappeared; his coat was ere long splashed up to his shoulders, and, with his dear boots, had to be strapped and hung over his saddle, his trousers tucked up as high as they would go, and he was wading with the rest in front of us, feeling for holes in a sea of pea-soup, occasionally not only finding but falling into them, a wholesome warning to those behind. The road was one constant succession of holes and traps and pies of mud, often above the mules' knees, often worn by constant traffic into ridges like a ploughed field, through which the tired quadrupeds had to wade, or drag their feet from furrow to furrow of the sticky, soft, clogging mud. The only real danger was on the broken bridges, which are made of round logs or branches laid side by side, and liable to roll apart out of their places, leaving holes through which the mule's leg easily slips and breaks, or if the clever creature recovers it he may be thrown down and roll into the mud bath on either side. These "corduroy" bridges are constantly occurring, and when hidden up with mud are very dangerous traps indeed. Mueda was most careful, and seemed herself to know every inch of the road, and always to pick the safest places. When the difficulties began, my friends insisted on my taking the place of honour after our leader the Baron, whose track Mueda followed exactly (except when she had some good reason of her own for diverging); she seemed to put her feet into the identical places our leader's mule's feet had been in, and I believe the others almost always followed her example.

Every traveller we met delighted in magnifying the horrors they had passed, and said that as the rain had continued it was utterly impossible for us to go on; and one party which had started the day before were actually coming back in despair. Our progress through all this was slow; we were obliged to stop after only 3½ leagues of it, and put up for the night, while Mr. G. sent on a note to the chief engineer of the province to ask his help. An answer came the next morning, begging us not to start too early; he had set fifty men to work, and hoped to make the road passable by noon, which gave us time to enjoy and examine our present quarters. It was not a bad specimen of the ordinary roadside inn or rancio of the country—a small room with a table and two benches, and an earthenware water-jar with cups to dip into it, standing on a piece of wood which served for lid, the roof hidden by a mat of plaited palm leaves, and the floor made of clay taken from the walls of the great termite ants' nests and pounded down, a material which in its way is clean, though it does not look so. Besides this room, with its unglazed window and outer door, were two smaller rooms, also entered from the outside, and reached by stepping-stones set in mud; two beds were in each—mere wooden frames with a mat stretched over them, and a sack of well-shaken corn leaves, cotton sheets with embroidered or lace edges, and a gay painted cover. We took our own pillows and coloured blankets or rugs, for the nights are often cold. Near our inn was the shed, under which the men pile all the luggage and saddles cleverly and tidily, so as to make a substantial shelter from the wind; here they sit and sleep round a good fire, cooking, gossiping, and mending their clothes or harness, the animals tethered round them, feeding, or being groomed or shod, till it is time to turn them out to grass for the night.

Inside the house we fed right well, and as we had much the same fare everywhere more or less, I will here give our average rations. For dinner, soup, roast or boiled chicken and pork, rice prepared somewhat greasily, and Fejão, the staple food of the country—some English say "very stable, for it is only fit for horses," but I always liked it; it resembles the French haricot, only the bean is black instead of white; in Brazil it is always eaten with farinha sprinkled over it, a coarsely-ground flour of either Indian corn or mandioca. Then we had the country cheese, which was excellent, reminding me of the "fromage carré" of Normandy; this was always eaten with preserve of some sweet sort known by the general name of "doce," and followed by the best of coffee—the poorer the house the better the coffee. In the evening we had tea and biscuits, or bread and butter; but these biscuits, as well as wine and candles, we brought with us; and after tea a roast chicken was cut up, rubbed with farinha, and packed in a tin box for the next day's breakfast or luncheon, though we never started without a cup of hot coffee and a biscuit—a great security against the bad effects of a cold damp morning ride. The second morning of our journey it rained again, and we sat at the window watching the different passers-by, as they floundered about in the mud, with great interest, for our turn was coming next. There was a particularly bad place opposite our door; it probably had been particularly bad for years, and would be the same for years to come, it having apparently never come into the head of the landlord to mend it. Perhaps he thought it stopped people and brought custom to his house, as they were literally unable to pass his door. One by one we saw the poor mules go flop into the liquid mud-hole, have their loads transferred to men's heads, and themselves lifted out by tail and head, the lifters often replacing them in the hole during the process. We, however, all got safely over, and were soon met by "Beesmark himself," as our Baron called the great Prussian engineer, a large man with a magnificent white beard and tall horse, which I believe was once of the same pure colour. After many compliments and hearty greetings he took the lead, and we rode round the valley by the steep hill-sides, so as to avoid the muddy road and marsh, now powdered with lovely masses of the *Franciscea*, with its blue and white blooms. At last we were forced to descend again, and came to the worst place, from which the travellers had been turned back the day before. Here men were now at work throwing on turf and trying to make a causeway. The Graf and Mary passed over safely, then flop! in went a young engineer's mule in front of me, only his neck to be seen above the water, while his master tumbled cleverly on his feet beyond the danger, and every one shouted to me to stop, which Mueda had no objection to do. A big nigger was called up and ordered to carry me, and I submitted under protest. He had no sooner got the extra weight (no light one) on his back than he sank steadily in the spongy ground like a telescope, and would doubtless have disappeared entirely if I had not scrambled to my seat again on dear old Mueda, who stood steady as a rock, and seemed to grin to herself at the idea of any one but herself having the strength to carry me.

After we had done laughing at this scene I was allowed to walk over on my own feet from sod to sod, and Mr. G. followed my example. We afterwards rode on tolerably well till we got to the small town where we were to breakfast, the high-street of which was a torrent of mud. All the people had their heads and elbows out of the windows to see us pass; for many of them had not had a walk in the street for a month; they would only have tumbled into the pea-soup if they had attempted it. Our engineer and his party were lodging here, and after accompanying us a few leagues farther they turned back to give a few more despairing looks at the mud, and to tell the people nothing could be done till the wet season was over—a fact they already knew too well.

Our next night's quarters were worse than the first; for the landlord had not been out of his house for a month, and had not even a sack of corn for our poor tired beasts; but the night after that we passed in a fazenda or farmhouse, with a beautiful green grassy hill behind it, on which the animals did enjoy themselves, rolling over and over, cleaning their coats, and eating any quantity of delicious *capim* grass. This is almost as good as corn for them, growing in tufts like the tussock or guinea-grass of India, with a whitish downy leaf which is extremely sweet, and in the spring-time is covered with feathery lilac flowers, which give a glowing tint to all the hillsides. We also enjoyed ourselves, and ceased for the first time since we started to feel damp, as the dwelling-rooms were built on the second storey, the lower one being used as stables and servants' quarters. The family, too, were more civilised than any of the people we had been with before. The young daughter of the house delighted to hear about Rio fashions. She showed us all her finery, and her lace made by her own hands. Even the poor sick mother from her bed in the corner seemed to brighten up at having news of the outer world. She had a most conversational parrot on a perch. All the food he dropped was eagerly watched for and fought over by five cats and a dog. They had also the somewhat rare luxury of a dairy and herd of cows, brought up a great many calves, and made cheeses with the spare milk, pressing them with their hands in a primitive manner, with the help of a wooden ring and a board; butter they did not attempt.

Near here I first saw the araucaria-trees (*A. braziliensis*) in abundance; it is the most valuable timber of these parts, and goes by the name of "pine." The heart of it is very hard and coloured like mahogany; from this all sorts of fine carvings can be made; the outer wood is coloured like the common fir. This tree has three distinct ages and characters of form: in the first it looks a perfect cone; in the second a barrel with flat top, getting always flatter as the lower branches drop off, till in its last stage none but those turning up are left, and it looks at a distance like a stick with a saucer balanced on the top. During the first period the branches are more covered with green; but as it grows older only the ends are furnished with bunches of knife-like leaves, and the extremities alone are a bright fresh green, looking like stars in the distance among the bare branches and duller old leaves. Its large cone is wonderfully packed with great wedge-shaped nuts, which are very good to eat when roasted. These curious trees seldom grow lower than 3000 feet above the sea.

After crossing the grand pass of Mantiqueira we changed the general character of vegetation. I saw there masses of the creeping bamboo, so solid in its greenery that it might have been almost mowed with a scythe; also the Taquâra bamboo hanging in exquisite curves, with wheels of delicate green round its slender stems, reminding me of magnified mares' tails, and forming arches of 12 to 20 feet in span. I know nothing in nature more graceful than this plant. Over the stone fountain which marked the top of the pass was a palm-tree, three of whose branches were weighted at the end by the pendent nests of the oriole bird, at least a foot long, woven cleverly out of the fibre of the palm, and of the parasite commonly called "Old Man's Beard," which one sees hanging from the branches and waving in the wind, like masses of unravelled worsted from some old stocking. I have often taken hold of the end and pulled it out for yards; then, on letting it go, it returned again to its crinkly state. This fountain was a favourite halting-place in fine weather, and there could be few more inviting places for lingering in.

Every bit of the way was interesting and beautiful; I never found the dreary monotony Rio friends had talked about. Every now and then we came to bits of cultivation, green hills, and garden grounds. Once I saw a spider as big as a small sparrow with velvety paws; and everywhere were marvellous webs and nests. How could such a land be dull? Then we crossed high table-lands which seemed quite colonised by the "Jean de Barbe" bird; every tree was full of their nests—curious buildings of red clay as big as my head, divided into two apartments. The birds were flying about near their homes, and were of the same reddish colour as the nests they lived in. Roberto climbed a tree and tried to get me one of these nests, but broke it in the attempt; it looked like a half-baked and ill-formed earthenware pot. The ground of this same bleak region was dotted with the large wigwam-looking establishments of the termite ants, as big as sentry boxes, and with no visible entrances. The small creatures who make and inhabit them tunnel their way underground from openings at a considerable distance from the erections themselves, which are full of cells and passages made of a black sticky substance, much used by the natives as putty for stopping water-holes and fuel to heat their ovens; they also pound it down for the floors of their houses.

These highlands are frequented by a kind of small ostrich, about which many strange tales were told. I had often heard their call—a noise something between a quack and a bark. They are said to act in concert in many things, to form a large circle for the purpose of killing snakes, driving them nearer and nearer till they have them safely hedged in, when they seize them one by one by their necks and dash them against the stones till they are dead. They are said also to make sitting parties, building their nests close together in the ground near a stream or pool of water, and pulling up a circle of grass round their little settlements as a precaution against the inevitable fires; they then fill one nest with eggs laid promiscuously by all the party, then another nest is filled, and so on till all are full, the birds taking their turns at sitting as soon as each nest is ready. The reason ostriches manage their domestic affairs in this peculiar way is that they lay so many eggs that the first would be bad before the last was laid if they were to wait and sit on their own eggs only, like other more orthodox hens; so they become true Communists, devoting their energies to the general good of their kind. When the time comes for burning the grass (which is the Brazil substitute for manuring it), the cock-birds are said to walk repeatedly in and out of the water, shaking themselves over the nests and their surroundings, and repeat this operation till all danger is over.

After a long day's ride over these glaring plains, still sticky and slippery with mud, though the hot sun was shining on it, we were glad to find really comfortable night-quarters in the house of a gentleman who prides himself on producing the best cigarettes in Brazil. They are all rolled up with the greatest nicety in Indian-corn leaves, and tied together with coloured ribbons in pretty little bundles; the daughters of this house did them so neatly that report says they were forbidden to marry or to leave the work on any pretence whatever. We were received with a most hearty welcome, and lodged in their best rooms with every luxury—tubs of water, embroidered towels, and the best of coffee. Our dinner was also sumptuous, and here, for the first time, we persuaded the master of the house to sit and dine with us at the head of his own table, a post which was generally given to me as the greatest stranger. We had one dish for which the house is famous—a bowl of chicken-soup with a huge chicken boiled whole in the middle of it. There was a piano here, and we sang and played all we knew for the benefit of our entertainers, whose musical attainments were as yet very young indeed; but they formed a most enthusiastic audience, and the Baron declared, with tears in his eyes, he could not smoke while I sang. It affected him so much afterwards that he put the wrong end of his cigar into his mouth and burned it; no wonder he cried!

At this point in our journey Mr. G.'s carriage met us. Such a carriage! but if we had been ill I suppose we should have gladly submitted to its jolting; it was a sort of double sedan-chair, intended to contain two persons sitting opposite one another, and hung on two long bamboos, with a mule harnessed between them before and behind. Persons travelling in these littieras are very apt to be sea-sick from the swinging motion; but I am thankful to say none of us required to go through this ordeal, and the machine was sent on ahead with the baggage-mules. The sunshine continued as we rode on over the high country to Barbacena, the chief town of this district, beautifully situated on a hill about 4000 feet above the sea, with fine araucaria and other trees shading its garden slopes. Two tall churches made a finer show in the distance than they did near. The horrible paved road up to it was good neither for man nor beast, and reminded one of North Italy. These abominations seem a plague common to all Latin nations. We were entertained at the house of the agent of "The Company" most hospitably. I was shown a well-furnished and perfectly windowless room which I could have, if I liked to stay and paint flowers and scenery on my return. After breakfast we went to see the old chemist who was the naturalist of the neighbourhood. He had many valuable books, curiosities, and rare orchids, which he took the greatest delight in showing to us; but his chief pride was in one wretched little cherry-tree, which, after ten years of watching, had produced one miserable little brown cherry: he had brought the original stone from his dear native Belgium, and it reminded him of home.

The flowers on these high campos were lovely—campanulas of different tints, peas, mallows, ipomœas creeping flat on the ground, some with the most beautiful velvety stalks and leaves; many small tigridias, iris, and gladioli, besides all sorts of sweet herbs. There are many peculiar trees and scrubby bushes with brown or white linings to their leaves, and the stems powdered over with the same tints. I have never seen these elsewhere. When we descended into the greener hollows and crossed the swollen streams the vegetation became dense again, and wonderful in its richness. Gorgeous butterflies abounded, and seemed to be holding dancing parties on the gravelly water's edge. Birds, too, chirped and fluttered from branch to branch, canaries abounded, and small green parrots flew screaming across our path. Once I saw a great lizard nearly a yard long run along the road in front of us, with his tail held up in the air like a cat; he was very stupid about getting out of our way, and we had a good look at him. Gama was said to be the very worst house on the road, and it certainly was not what the Yankees call "handsome quarters." An idiot sat on the doorstep, pigs wallowed in the mud beyond; but the idiot was said "not to be often dangerous," and the pigs could not get past him into the house, so why should we mind either of them? Our next quarters made up for Gama; for they were in a friend's house, with a kind Brazilian lady and her children, who did all she could to show us hospitality, and came out the next morning before daylight, to give us our coffee, in her dressing-gown, with her long hair combed straight down her back; for we meant to make up for lost time now we neared the end of our journey. Mary had a threatening of diphtheria, and longed for home and her mother's care; so we toiled up and down the high ridge of Morro Preto, whose white sharp rocks stuck up like bleached bones, and whose cracks were filled with the brightest red, purple, or yellow earths. Occasionally there were fantastic earth-pyramids standing up, balancing balls of harder earth on the top, instead of stones, as in the Tyrol. At the top of this ridge I saw many strange plants for the first time, including the vellozia, a kind of tree-lily peculiar to these mountains. One of the varieties was called the Canella de Ema. It had a stem like an old twisted rope, out of which spring branches of the same, terminating in a bunch of sharp-pointed hard leaves like the yucca; out of these again come the most delicate, sweet-smelling blue-gray flowers with yellow centres, much resembling our common blue crocus in shape. There are many other vellozias, all having the same dagger-like leaves; some send up long stems with bunches of brown or green flowers.

It was most tantalising to pass all these wonders, but time was precious and my friend was suffering, and our next night behind a curtained alcove in an extremely draughty room after a good day's soaking did not improve her. The third morning found her voiceless, but she was determined to get home that night, though it was a full forty miles' ride; so on we came, and she bore it bravely. Suddenly a violent discharge of rockets in front warned Mr. G. he was coming among friends, and we stopped to breakfast at the house of a black man, whose late master had left him his freedom as well as house and property. There were many bits of curious old carved furniture here, as well as fine silver-work in the little chapel, and our host treated us as if he loved us (for a con-si-der-a-tion). Over the wall round his house were masses of bright scarlet-flowering euphorbia, from the juice of which the Indians poison their arrows, and of which the Jews say the Crown of Thorns was made. The journey was a weary one; for we were all anxious about her who was generally the life of our party, and when we reached the bridge over the deep river-bed where we were to change mules I thought she would have been suffocated. Soon, however, the hill of Morro Velho came in sight, and, though still far off, her spirits rose and her troubles grew less in proportion as the distance shortened. Every house we passed sent off rockets, and one enthusiastic man pointed his gun straight at Mr. G., and kept firing at him as long as he was in sight, fortunately only with powder. A fearful storm came on, and our waterproofs were of real use, and brought us in a comparatively dry state to the house of a very remarkable old lady, Dona Florisabella of Santa Rita, who hugged us all round in the heartiest way, and then led us up by a rough ladder to a set of handsome rooms, which had been frescoed in a most gaudy and reckless manner with every bright tint of the rainbow.

The open verandah attracted me at once. From it there was an exquisite view of the Rio das Velhas, winding through its wide green valley, surrounded by hills wooded to two-thirds of their height, and a noble ceiba or silk-cotton tree standing sentinel by the house, which I afterwards saw covered with the most lovely pink hibiscus-like flowers—a perfect mass of colour, looking in the distance like a large old cabbage-rose against the green hills. Across the river I now saw the pretty church and village almost hidden in groves of bananas and palm-trees. Above were the peaks of Morro la Gloria, the property of our old leader, and from which we gave him his title. From this view politeness required me to turn at last to our hostess and her abundant conversation. She was of good family, and had seen better days; her children were dispersed in the world, and had left her to make what she could of a small property. She had spirit enough to work that or anything else, and her power of talk and pantomime beat even her rival the Baron's. She wore a once-handsome silk dress, and a gaudy silk handkerchief bound over her head so as to hide every trace of hair; but, in spite of the disfiguring costume, showed remains of great beauty. Soon Mary rushed out to meet her brother, the clever young engineer. She found her voice at the same moment; and we all sat down to a grand dinner, excepting our hostess, who stood and helped us all, and woe betide any one who refused to eat or drink what she offered them. After she had filled all our plates she seized the drumstick of a chicken in one hand and a bit of bread in the other and took alternate bites at them, after which she washed her hands at a side-table, and began carving again, drinking all our healths separately, and making speeches to each as she did it. One of her dishes had a duck in it sitting upright as if it were swimming, with a lime in its mouth. Her "doce" were excellent, particularly a kind of sweet pudding made with a great deal of cheese in it.

It was no easy task to get away from this hospitable lady,

but at last we started, and about a mile farther crossed the great bridge over the river, and were on "The Company's" property.[54] About twenty of its officers were waiting to receive us, all mounted on mules, and there was a general hand-shaking, most of the party being English. The Baron was low-spirited, for he was no longer our leader, and his work was over. Mr. G. and I led the way and jogged over the muddy road up hill and down to the village of Congonhas, when the rockets and firing and hand-clasping began in good earnest, amid torrents of rain. The mules became quite unmanageable, either from the noise or from the nearness of their well-loved stables, and we all took to galloping violently up and down the steep paved streets, which were now torrents of liquid mud—such a clattering, splashing, umbrella-grinding procession! Mr. G.'s mule objected to a rocket-stick on his nose, and kicked his rider's hat off, after which the Baron galloped on ahead to stop the fireworks if possible; he looked very picturesquely wild, with his red-lined poncho flying out on the wind like the wings of a blue and scarlet macaw. At last we were stopped by the band awaiting us, and had to tramp solemnly behind it into the grounds of the Casa Grande—a mass of close-packed dripping umbrellas and damp bodies; and before I knew where I was I found myself dismounted, and hugged and welcomed by one of the best and kindest women I ever met in all the wide world, and called "dearie" in a sweet Scotch voice; no wonder Mary longed to be at home! And I felt that I was right and the Rio people wrong about coming to Morro Velho, and the only drawbacks to the journey left were blistered lips and slightly browned hands.

The Casa Grande of Morro Velho was indeed a rare home for an artist to settle in, and I soon fell into a regular and very pleasant routine of life. I had the cheeriest and most airy of little rooms next my friends, with a large window opening on to the light verandah, in which people were continually coming and going and lingering to gossip. Beyond that was the garden, full of sweetest flowers; a large *Magnolia grandiflora* tree loaded with blossoms within smelling distance; around it masses of roses, carnations, gardenias (never out of flower), bauhinias of every tint (the delight of humming-birds and butterflies), heliotropes grown into standard trees, and covered with sweet bloom, besides great bushes of poinsettia with scarlet stars a foot across; beyond these were bananas, palms, and other trees, and the wooded hillsides and peeps of the old works and stream in the valley below.

Mrs. G. was constantly passing my window, looking after her lazy blacks, who sat down to rest as soon as her eye was off them. She had also many pets besides her flowers to attend to. There were two macaws or araras (as they call themselves), one blue with yellow breast, the other red and green. The latter was called "the Mayor," and was very tame; he was much attached to Pedro, the oldest slave of the establishment, and would allow him to do anything he liked with him. He was also very fond of one of the cats, and the two strange friends used to huddle close together in the sun for hours, scratching each other's heads. Occasionally there was a row when one or other put too much zeal into his work, but after a rather noisy argument in the macawese and cat languages they again became friends. Then there were three green parrots, with blue foreheads and yellow waistcoats, and pink patches on their wings, who were extremely talkative, and sang and danced in the negro style; nobody near passed these birds without a talk. Numberless pigeons and doves, and a peacock which mounted a certain tree at six o'clock regularly every evening, announcing his arrival at his perch by a shrill scream; so that if the cook were asked why the dinner was not ready, she would very likely say, "It was not time, the peacock had not screamed yet." Two pretty gazelles lived below the garden in the poultry-yard, and cats and dogs abounded; but my chief friend was an old smooth-skinned dog called Lopez. He was large, and doubtless belonged to that famous breed called mongrels, but was full of intelligence, and used to sleep under my window and accompany me in all my walks.

We had delightful rambles together, and always found new wonders on every expedition. Just below the flower-garden was a perfect temple of bananas, roofed with their spreading cool green leaves, which formed an exquisite picture. Sometimes a ray of sunlight would slant in through some chink, and illuminate one of the red-purple banana flowers hanging down from its slender stem, making it look like an enchanted lamp of red flame. Masses of the large wild white ginger flowers were on the bank beyond this temple, and scented the whole air. This was a grand playground for the Hector and Morpho butterflies; here, too, I used to watch the humming-birds hovering over and under the flowers, darting from bush to bush without the slightest method—unlike their rivals, the bees, who exhaust the honey from one entire plant before they go to another. Farther down the steep path were masses of sensitive plants covering the bank with the brightest of green velvet and delicate lilac buttons. I never could resist passing the handle of my net over this, when instantly the whole bank became of a dull, dead, earthy tint, and only the dry twigs and stalks of the plants were visible, with their shrinking branchlets starting from them at most acute angles. Below this there were two or three old gray trees, on whose trunks or roots I never failed to find some new wonders of cocoons or larvæ, or odd spider's web, green, gold, or silver, as they glittered in the bright morning sun, often spangled with diamond dew. Lower still were the clear stream and rickety little bridge from which I used to watch the humming-birds and other small creatures bathing, pluming themselves afterwards on the leaves and stalks of the wild ginger or castor-oil plants. These latter grow to a great height in this country, and make fine foregrounds, with their large cut leaves and purple or green heads of flowers and berries.

The curious grass which bears the gray berries called "Job's tears" was also a handsome plant, and abounded here. Beyond the bridge was the kitchen-garden, in which several superannuated black people did as little as possible from sunrise to sunset, at certain very frequent intervals leaving off to light a fire and cook for themselves various sorts of savoury decoctions in an iron pot, taking as long as possible to eat it, after which exertion of course they required rest. In spite of this not very energetic mode of cultivation, the place was crowded with vegetables, including cauliflowers, peas, beans, turnips, carrots, asparagus, parsnips, potatoes, as well as sweet potatoes, mandioca, "quianga" (the okery of the United States —a kind of hibiscus with an eatable pod), many kinds of cucumber and pumpkin, and a large bush of real Congo tea. Lettuces and parsley were the most difficult things to raise; the ants had such a taste for them that the only chance was to grow them in boxes isolated over a tub of water.

Beyond this garden a slight scramble took me on to the path beside one of the aqueducts which brought water to the gold-mines and works; along that I could walk for miles, winding through the valleys, crossing and recrossing them over crazy wooden planks, startling enough at first, but which one soon learned to think nothing of. At first, too, my head was full of stories of poisonous snakes, and the dread of stepping upon them, particularly the rattlesnake or cascabella which was common there, but I soon forgot such creatures existed; and during the eight months I was in Minas I never met a live one, though I frequently saw them dead, and even heard occasionally the rattle near me. Lopez generally cleared the way for me, and gave all enemies notice to quit. One day he came back with his tail between his legs, giving me notice, in plain dog-language, to be careful, while he insisted on keeping between me and a kind of big slow-worm they call here a "two-headed snake," which is quite harmless, and was on this occasion dead. Another day he returned in the same way, and showed me two guinea-fowls who were pulling a snake to pieces between them. He had quite an idea snakes were to be avoided. There is one whose habit is to live in branches of the trees and drop down on the passers-by, which is not pleasant to think about. One day a black washerwoman was returning up my favourite path with her basket of clothes on her head, when this happened, and she did not hesitate to drop the freshly-washed load in the mud, snake and all, and to run for dear life, as did the frightened reptile also, after his wriggling fashion. However, deaths from snake-bites are very uncommon. Rewards are given for all dead rattlesnakes, for whose bite there are no cures.

It was an odd sensation living in an English colony which possessed slaves; but this company existed before the slave-laws, and was with some others made exceptional.[55] As far as I could see, the people looked quite as contented as the free negroes did in Jamaica, and, thanks to the new Brazilian regulations, they have the happiness of being allowed to buy themselves at a fixed price, if they can save sufficient money. The girl who brought me my coffee in the morning had bought two-thirds of herself from her own father, of whom she was hired by Mrs. G., as he was said to be such a brute that it was a charity to keep her out of his hands. After five o'clock her time was her own, and she did embroidery and other work to sell, so as to complete her emancipation fund. She and another girl used to squat on the verandah close to my window working, generally with their feet poked through the balustrades. Mary told them one day if they did so I should paint them, toes and all, which would be a disgrace; after which there was always a great scuffle to tuck away their feet whenever I looked that way. Every other Sunday there was a revista or review of the blacks in front of the house, and they were all dressed in a kind of uniform; the women in red petticoats and white dresses, with red stripes for good conduct, bright orange and red turbans and blue striped shawls, which they arranged over their heads in fine folds; the men had red caps, blue jackets, and white trousers, with medals pinned on for good conduct, and a general grin passed over the faces as Mr. G. and his officers passed by. I could not see much discontent or sadness in these poor slaves, and do not believe them capable of ambition or of much thought for the morrow. If they have abundant food, gay clothing, and little work, they are very tolerably happy: seven years of good conduct at Morro Velho gave freedom, which they had just sense enough to think a desirable thing to have.

Coppers were given at all these revistas for good conduct, Mr. G. thinking rewards went further than punishments in the management of these people; but the quantity of spirit to be sold to each black per day was limited, and these coppers were only taken at the Company's shop, so that until freedom was obtained, it was not easy for the negro to indulge his dear vice of drunkenness. All babies born were free, the consequence of which was that the mothers took no more care of them, as they said they were now worth nothing! In the "good old days," when black babies were saleable articles, the masters used to have them properly cared for; and the mothers didn't see why they should be bothered with them now. At Morro Velho every man has his garden—such gardens! With running water passing above them so that they could irrigate to any extent, and full of the richest fruit and vegetables. The leaves of their *Caladium esculentum* were often nearly two feet long. At noonday the beautiful banana-leaves lose all their fresh shining greenness, and shut themselves up tight like sheets of folded letter-paper, so as to keep their moisture in, and appear mere knife-like edges to the sun's scorching rays; as it sinks lower they again spread out ready to collect the evening dews. The blacks devote all these garden treasures to their pigs, which they fatten up till they are worthy of Smithfield, with almost invisible necks, little snouts, and short legs.

They were full of superstitions. Old Pedro, the macaw's friend, put bread and meat every other Friday night in the room his old master the Padre died in; and as it was never found in the morning, he declared the old gentleman himself came and ate it. But Pedro was the pet of the house, and had his own way in this as in everything. He would have been free years ago but for his infatuation at times for strong drink. He had now taken the pledge, and the only thing he could not resist stealing was tea, which always had to be carefully locked up. There were two fortress-looking piles of building on the hills opposite my window, where these poor creatures lived and were shut in every night.

On the other side of the valley stood the Cornish village—such a contrast! All its pretty cottages standing in their own well-fenced gardens, with pure water running through each, roses and other familiar English flowers, and fair English children with clean faces playing at hop-scotch and other British games in the road. All the head-work in the mines was done by those strong countrymen, who were well paid, and could soon put by considerable savings if they had the sense to content themselves with the food of the country. The beef was abundant and cheap, so was Fejão farinha, coffee, and sugar. Most vegetables and fruits were grown without difficulty; but the miners, who could not do without beer, champagne, horses, and other extravagant luxuries, of course soon found their pockets empty, and got into debt besides.

In a weedy garden near was a humming-bird's nest, hanging to a single leaf of bamboo by a rope of twisted spider's web three or four inches long, swinging with every breath of air. These wee birds generally build in this way about Morro Velho, often hanging their nests over the running water, on the ends of fern-fronds or on the blades of grass, where the eggs are safe from the attacks of snakes or lizards; but they always choose places with some protection above from rain or sun. They sit twice in the year, never laying more than two eggs, which are always white, and about the size of small Scotch sugar-plums. Some of these humming-birds were quite sociable. One pair had come regularly twice a year to one of the outhouses of the Casa Grande for many years, apparently using and repairing the same nest, which hung by a tiny rope from the matted ceiling.

Everybody "collected" for me, and the results were sometimes rather startling. One day a hideous black woman, without any previous announcement, poked her Turk's-head broom at me through the open window, grunting something unintelligible; and behold, a large specimen of the odd insect called "the praying mantis," clasping its hands devoutly on the Turk's-head. But collecting is one thing and preserving is another, and the difficulties of the latter process are great in so hot and damp a climate. Many of my specimens were eaten even when hung up by a string, apparently out of reach of all creeping things. Some of my collections were not pleasant to handle. One day one of the boys brought me a large black beetle in his cap which he said would bite. "Oh no, it never bites," said Mrs. G., scolding him for the very idea; then she screamed and dropped it. I got a bottle of restil to put it in, but screamed and dropped bottle and all as it hugged my fingers with its sharp-hooked feet, and Eugenio had the last word, after corking it up securely in the spirit: "He knew it *did* bite!" Another unpleasant creature was a stinging caterpillar, whose hairs were as dangerous as a scorpion's tail; a rub against them might cause the hand or arm to inflame so that amputation was necessary to save life. These dreadful things were common enough at certain seasons. I kept and fed one for a long while, hoping to see the kind of moth it turned into; but the blacks hated it to that extent that they pretended it crawled away when I was out one day—a difficult thing to accomplish with a glass shade over it! The varieties of spider were endless, and their works worthy of the old Egyptians. One huge colony formed a web from the roof of the house to the flagstaff opposite, dragging one of its sustaining ropes into an acute angle. We broke down the web, and released the rope to its old straight line. In less than a week it was again pulled towards the roof, forming a tight bridge for the enemy to cross and recross. This spider's body was no bigger than an ordinary green pea. Some of the webs were so thick and strong that they gave my face quite a cutting sensation as I rode through them.

About January the heat became more oppressive—86° was the average, though it was often 91° in the shade—but the nights were always cool enough for sleep at Morro Velho, which is about 3000 feet above the sea, and I was never uncomfortably hot. The Gordons, however, who had lived sixteen years in the climate, longed for a change; so they determined to go to pay a long-promised visit to Mr. R. at Cata Branca, taking a young Scotch lady who had been spending Christmas with them, and myself.

It was a beautiful day's ride of about 26 miles, the road winding for the greater part of the way along the high banks overlooking the Rio das Velhas, which eventually runs into the Rio San Francesco, and enters the sea above Bahia. The river we followed was about as broad as the Tweed at Dryburgh, running through wooded hollows. A good road was in course of making along this valley, on which some hundreds of the Company's blacks were working, who greeted us with hearty cheers. In the fresh clearings I saw many new and gorgeous flowers, as well as some old friends, including the graceful amaranth plant of North Italy, with which the wine of Padua and Verona is coloured.[1] How did it get to the two places so far apart? I longed more and more for some intelligent botanical companion to answer my many questions.

We rested a while at a collection of huts that have been put up for the work-people near some fine falls of the river, and the Head Man there told us of one curious fish he had caught,

[1] *Phytolacca decandra.*—ED.

which seemed to have a sort of inner mouth, which it sent out like a net to catch small fish or flies with. He showed us a rough drawing he had made, and was very positive about the story, which is not more difficult to believe than many other well-proved wonders of nature. After leaving this settlement, we mounted up bare hillsides another 1000 feet, and came to the green plateau of Cata Branca, with its groups of iron-rocks, piled most fantastically like obelisks or Druid stones standing on end, dry and hard, and so full of metal that the compass does not know where to point. Amid these rocks grow the rarest plants—orchids, vellozias, gum-trees, gesnerias, and many others as yet perhaps unnamed. One of these bore a delicate bloom,—*Macrosiphonia longiflora* (No. 67 in my Catalogue at Kew),—like a giant white primrose of rice-paper with a throat three inches long; it was mounted on a slender stalk, and had leaves of white plush like our mullein, and a most delicious scent of cloves. Another was a gorgeous orange thistle with velvety purple leaves. I was getting wild with my longing to dismount and examine these, when we met our kind host Mr. R. coming out to meet us, and in another half-hour we were in his pretty cottage, where he had been living for the last two years watching a dying mine, in almost perfect solitude, expecting to be released any moment. The once-famous mine of Cata Branca had long been filled with stones. All around were the ruins of fine houses which had helped to ruin so many people, and the small cottage we were in was the only habitable place on the hill, with the exception of a negro hut or two, and must have been a dreary position for so sociable a character as our host.

The summer of St. Veronica was endless that year, and we had the most glorious weather. The air was much fresher on the height and did us all good. Every day's ramble showed me fresh wonders. There was a deep lake near the house, said (of course) to be unfathomable; it was surrounded by thick tangled woods and haunted by gay butterflies. In it Ounces were said to drink morning and evening. I never saw them, but they had lately carried off two of our host's small flock of sheep, and I saw some skins of these small tigers, which were richly marked and coloured. One morning we spent on the actual peak, which rises a perfect obelisk of rock 5000 feet above the sea. Some of the more adventurous of our party mounted to the very top.

The earth had entirely disappeared from the crannies, leaving the huge ironstones loosely piled on one another. In spite of the want of soil, these rocks were loaded with clinging plants, bulbs, orchids, and wild pines. Tillandsias and Bilbergias of many sorts crowded round them; the latter were very curious—great green or lilac cornucopias with feathery spikes and many-coloured flowers, or beautiful frosted bunches of curling leaves, from the centre of which fell a graceful rose-coloured spray of flowers. There were also many euphorbias and velvet-leaved gesnerias, trailing fuchsias, ipomœas, and begonias with wonderful roots like strings of beads. It was impossible to carry away half I longed for, even if possible to climb over such rocks with two loaded hands. One day I rode with Mary and the Baron to visit a dairy-farm some miles off, where we sat and gossiped, ate toasted cheese, and drank enough strong coffee to poison any well-regulated English constitution; but our life was far too healthy to be hurt by such little luxuries. Another day we rode down to visit some people on the plain below. All these expeditions showed fresh beauties of nature and miseries of humanity. At last Mr. G. came to fetch us, and on the Sunday before we left he read the Service, three Cornish miners coming up from below to assist at it. Their captain afterwards made a speech to say "what a pleasure and a privilege it was, etc.," on which Mr. G. said if they would only come up, he would read it every Sunday in the same way. "Oh no, it warn't that, it war them four ladies all stannin' of a row; it war so long sin' he had seen four English ladies all at once, it war!" The poor old man almost cried over the extraordinary event. He went down to his expiring mine again, and we rode home, leaving our kind host to utter loneliness.

Soon after that a tragedy occurred which filled all our thoughts for some time afterwards. On my first arrival at Morro Velho I had found a visitor staying there—an old friend of the family, a singular old Scotchman with long flowing white beard and mane, a poet whose brain was full of his country's legends and genealogies, who had remarked to me the first night when I said I was half Scotch, "Why, woman, d'ye no ken ye're a Johnstone?" This Mr. B. was the superintendent of another mine about twelve leagues off. His niece had now returned with us from Cata Branca, and we were all to go home with her shortly to pay her uncle a visit; when one day news came he was dangerously ill, and before she and the doctor had gone far on the road, they met another messenger who said he was dead. She sent him on to beg Mr. G. to come and help her; before any of them reached home he was not only dead but buried, and there were strong reasons for supposing him poisoned. In Brazil it is almost hopeless to think of getting justice in such a case, though there was little doubt who did it; valuables of different sorts were found in the suspected hut, the owner of which had lately been making and selling "charms"—the national name for poison. As British Consul, Mr. G. could help the poor niece better than any one else could through her troubles; he also determined to have the old man's body taken out of the unconsecrated hole into which it had been hastily thrust, and to have it brought over and buried near his friends, with all proper forms and ceremonies, as it would not do to let the natives think Englishmen might be treated so. There was a great gathering of English from all quarters to the funeral, with black clothes and much solemnity; for the strange old man had made himself liked in the country. The cemetery at Morro Velho was a lovely spot, on the top of one of the small wooded hills which jut out into the main valley; and the last sunset rays reddened the wooded tops around as the old man was at last let down into his quiet resting-place. Kind hands dropped in pure white alpinia blossoms and roses over him, while his friend read the old service of our Church.

There was something very touching about that quiet service and its old-fashioned hymns and psalms in that far land, with the great mass of bougainvillea seen through the open door, and the sweet hoya trained over it. In that same bougainvillea bush there was an exquisite little nest made of the finest possible twigs and straws, so very fragile in its open-work that one wondered the small gray bird and her eggs did not fall through it as she sat there. She was perfectly tame, and merely fluttered away to an upper branch when we went to look at her treasures, coming down again directly we left the spot. Six sheep, some guinea-fowl, and the great peacock, were always in waiting at the church door for chance handfuls of crumbs or corn; and when the service was going on, a black man had to walk backwards and forwards to prevent them from coming in too.

# CHAPTER V

## HIGHLANDS OF BRAZIL

1873

About the end of March we all started up the hills with bag and baggage, crossing over a shoulder of the Coral mountains, and on to Sabara, the chief town of the district, where we took coffee at the house of "a most respectable brown woman," who hugged all my friends most warmly. Mrs. G. told me she was much to be pitied, having lost her only son, to whom she was devoted. "Was she a widow?"—"Oh dear no; she had never heard of her being married, but as Brazilians go she was a most respectable person!" Opposite her house we saw a room full of remarkably clean little black boys, all dressed alike, and an ill-looking, dandified man in charge of them. He was a slave-dealer, buying up well-grown boys over twelve years of age for the Rio market. The law now forbids the sale of younger children, and every year a year will be added; so that children of eleven are safe for life where they are, and all the next generation will be free. These boys looked very happy, and as if they enjoyed the process of being fatted up.[56]

Our road followed the banks of the river for some way; sometimes along the low banks amongst reeds and bushes of *Franciscea*; often over the higher sierras, among strange scraggy trees, which were covered with more flowers than leaves. On one especially the white lily-like flowers were very fascinating. The ipomœas and bignonias were in great variety. One large lilac ipomœa grew in massive bunches on the tops of the trees;

and a smaller white one had fifty or sixty buds and flowers on every spray, making the trees look as if they had just been covered with snow. The round-backed Piedade mountain got nearer and nearer, and we put up for the night at Caite, a village at its foot.

It was a perfect fairyland. The great blue and opal Morpho butterflies came flopping their wide wings down the narrow lanes close over our heads, moving slowly and with a kind of see-saw motion, so as to let the light catch their glorious metallic colours, entirely perplexing any holder of nets. Gorgeous flowers grew close, but just out of reach, and every now and then I caught sight of some tiny nest, hanging inside a sheltering and prickly screen of brambles. All these wonders seeming to taunt us mortals for trespassing on fairies' grounds, and to tell us they were unapproachable. At last we left the forest, and the real climb began amidst rocks grown over with everlasting peas, large, filmy, and blue haresfoot ferns, orchids, and on the top grand bushes of a large pleroma with lilac flowers and red buds like the gum-cistus, and beds of the wild strawberry, which some Italian monk had introduced years ago. Two old ladies, "Beate," lived alone in the old convent, which was still in good repair.

The Baron took charge of the two girls and myself over the hills; and at the edge of the Rossa Grande property its superintendent met us, showed us his trim little mine and big wheels, and gave us luncheon, then took us up the hill to admire the view, and accompanied us through two leagues of real virgin forest, the finest I had yet seen, to the old Casa Grande of Gongo—a huge half-ruined house which had originally belonged to some noble family.

The great gold-mine here had at one time yielded more than £100,000 a year. In that day there were a thousand miners there, and twenty servants in the great house alone. The superintendent used to drive a carriage with two horses over the tangled and stony path by which we had just come;

now, one old black man and his family alone inhabited the place to keep the keys (which didn't lock) and hold up authority. Gaunt ruins of the different houses stood around; but though their roofs were whole and unbroken glass in their windows, they were scarcely accessible or even visible, from the thick growth of tangled trees and greenery which had wreathed itself around and over them. All was thick mat and forest, except on one side where the grassy hills rose, affording abundant food for the flocks and herds which supplied the mines on the other side of the mountains. To this old deserted place Mr. D. had sent furniture, food, and slaves, and persuaded his mining captain to let his pretty little wife go and keep house for us, taking Baby Johnnie to amuse her and us too. She soon made us at home.

After tea we played a game of whist in the ghost-like old hall with its heavy wainscoted cupboards, and great gilt hooks from which the mirrors and chandeliers had formerly hung, and on which a late superintendent had once committed suicide. When the present one would have ridden back through the forest we begged him to stay and keep the Baron company and the ghosts out, and wishing the two good-night we began our retreat towards our own part of the house; but when we came to the grand staircase, behold! a gambat was coming down it very quietly. Now a gambat is not a fascinating quadruped. He only sees in the dark, and his wife carries her young in her pocket like a kangaroo. He is like a tiny bear with most human-looking hands, and a long prehensile tail so enormously strong that when once he has twisted it round some firm anchorage it would resist the pull of a strong man, and hold on though bleeding and torn. He has also the power of emitting a horrible smell like the skunk, thereby driving away his enemies. Once I remember Lopez was himself so objectionable after killing one of these creatures that he had to be locked up for a day or two; he was, unfortunately, not with us now, and we all cried out for help. The poor little beast, looking extremely puzzled at seeing his usually quiet premises invaded by strange creatures, with strange lights in their hands, was too brave to turn, and, I am sorry to say, was killed ruthlessly by our two knights, who had rushed to our assistance.

The next morning our Baron, who had begun life as a blacksmith, went round and mended the locks of all the doors we were likely to use, and our party dispersed, leaving me to enjoy a fortnight of perfect quiet in the great empty house and rich forest scenery, with Mrs. S. and her baby boy to keep me company : to her it was an agreeable change, to me the thing of all others I had longed for. I used to start every morning on my mule, with Roberto on another, for some choice spot in the forest, where I gave him my butterfly-net (which he soon learned to use very deftly and judiciously), while I sat down and worked with my brush for some hours, first in one spot and then in another, returning in time for a good wash before dinner. Washing and dressing were very necessary, as the abundant vegetation here was covered with Garapatas, the most intolerable of insect-plagues, and at this season in their infantine and most venomous stage. One blade of grass might shake a whole nest on to the passing victim, no bigger than a pinch of snuff, and easily shaken off then; but if left, the hundreds of tiny grains would diverge in every direction till they found places they fancied screwing their proboscis into, when they would suck and suck till they became as big as peas, and dropped off from over-repletion. Of course none but idiots would allow them to do this; but the very first attempts of these torturing atoms poisoned one's blood and irritated it for weeks after. When the insect grew older and bigger it was less objectionable, as it then could be easily seen and removed before it did any injury; it attacked one then singly, not in armies. But even this plague was worth bearing for the sake of the many wonders and enjoyments of the life I was leading in that quiet forest-nook.

I used generally to roam out before breakfast for an hour or two, when the ground was soaked with heavy dew, and the butterflies were still asleep beneath the sheltering leaves. The birds got up earlier, and the Alma de Gato used to follow me from bush to bush, apparently desirous of knowing what I was after, and as curious about my affairs as I was about his. He was a large brown bird like a cuckoo, with white tips to his long tail, and was said to see better by night than by day, when he becomes stupidly tame and sociable, and might even be caught with the hand. One morning I stopped to look at a black mass on the top of a stalk of brush-grass, and was very near touching it, when I discovered it to be a swarm of black wasps. When I moved a little way off I found through my glass they were all in motion and most busy. When I returned again close they became again immovable, like a bit of black coal, and I tried this several times with always the same effect; but foolishly wishing to prove they really were wasps, got my finger well stung. This little insect drama was in itself worth some little discomfort to see. The brush-grass on which these wasps had settled was itself curious, each flower forming a perfect brush—a bunch of them made the broom of everyday use in the country; scrubbing-brushes were generally formed out of half the outer shell of a cocoanut.

One had always been told that flowers were rare in this forest scenery, but I found a great many, and some of them most contradictory ones. There was a coarse marigold-looking bloom with the sweetest scent of vanilla, and a large purple-bell bignonia creeper with the strongest smell of garlic. A lovely velvet-leaved ipomœa with large white blossom and dark eye, and a perfectly exquisite rose-coloured bignonia bush were very common. Large-leaved dracænas were also in flower, mingled with feathery fern-trees. There were banks of solid greenery formed by creeping bamboos as smooth as if they had been shaved, with thunbergias and convolvulus and abutilon spangling them with colour. Over all the grand wreaths of Taquârá bamboo, and festoons of lianes, with orchids and bromeliads, lichens and lycopodiums on every branch.

I had one grand scramble in a neighbouring forest with Mr. W., and brought home a great treasure—a black frog. His face and all the underparts, including the palms of his hands and feet were flesh-colour; he had black horns over his garnet-coloured eyes, which he seemed to prick up like a dog when excited, and which gave much intelligence to his countenance. I kept this pet for three months, and then trusted him to a friend to take to the Zoological Gardens in London; but alas! he died after three days of sea-air. If he had been corked up with some moss in an air-tight bottle he would probably have lived. In the same woods we found several specimens of the exquisite little butterfly, the *Zenonia Batesii*, which appeared to come out twice a year here. The large semi-transparent green Dido was also abundant, but very shy and clever at eluding my net. A messenger at last recalled us to Morro Velho. Visitors had arrived, and Mrs. Gordon wanted us to help in entertaining them; so we obeyed at once, stopping by the way to breakfast with the Baron's family, to his great delight.

Ouro Prêto, the capital of the province, is full of convents, and one of them I was told had been built with the washings of the negroes' heads, after their day's work in the mines was done, their woolly heads being first sprinkled with gold-dust, and then sent to be ducked in the church fonts—an original way of paying tithes! All along the roads are old diggings—some deep, some shallow, but all deserted. This honeycombed valley reminded me somewhat of Ipsica in Sicily. After rounding the shoulder of the mountain, we came to the village of Passagio, also long deserted, and to the Casa Grande. Such a pretty house, and such pretty English children ran out to meet us—no children in the world are so pretty as English. The country about the Passagio Mine was very picturesque, in a narrow and deep valley worn by a noisy rushing stream bordered by rich green tangles. Great fern-fronds of the gold and silver varieties were often a yard in length among the rocks above.

We rode one afternoon to Marianna, where the Bishop lived, surrounded by numerous convents. The hills round it were bare; the place looked gloomy; of the Bishop himself but one story was told—of his saint-like simplicity of life. One of his admiring devotees sent him once four embroidered shirts made by her own hands. A beggar soon afterwards paid him a visit, to whom he gave two of them, as he said no man wanted more than two shirts—one at the wash and one on. He was also charitable towards heretics, and said he really believed Mrs. G. would go to heaven in spite of her faith. In general, Brazilians are not so kind, and would say of a Protestant friend, "Yes, she has good manners; what a pity it is she must burn in hell!"

It was dark when we turned back from Marianna and rode to the country-house of a rich man of good family, a Commendador or Knight of the Empire, who was giving a party that evening. His handsome wife wore a cotton dress striped with gay colours, and, wonderful to say, appeared without the usual handkerchief on her head, to bind up and hide all her hair; instead of this hideous head-dress she had a bunch of China roses stuck behind her ear coquettishly. We were first shown the state-apartments, in which the family never lived, with ounce and other skins laid about on the floors, and nick-nacks on the tables, some old portraits of Portuguese ancestors on the walls; a tawdry little chapel, and pretty garden; then we returned to the real living-room—a sort of back-kitchen full of litter and black children, with its door opening to the dirty stable-yard; pigs and cocks and hens strutting around the doorstep. By this entrance we and the other guests had arrived, and here we dismounted from our horses and mules. Gradually a curious collection of people assembled—Herr W.

and his wife (a perfect ideal of Voss's "gute verständige Hausfrau"), a clever lawyer, the local M.P., a perfectly round Canonico with red embroidery on the back of his gown and a redder face, an old Cornish mining captain with a grievance, and several Brazilians, who were soon seated round a large tea-table in the room I have described—half pantry, half passage. The lady of the house insisted on my taking the head of the table as the greatest stranger, while she waited on her guests, afterwards taking her own tea in a cupboard, from which Mr. M. dragged her, much against her will. The tea itself was peculiar—a mixture, I believe, of the native tea and strong green tea. I had two cups to make sure if I liked it, and was as much puzzled at the end as I was at the beginning. After this meal the men went into another room and played at loo, by which they often lose much money, while we women "did company" in foreign languages, German being a real rest after the difficult and unpronounceable Portuguese; till Mr. M. thought he had gambled enough for all purposes of politeness, and we rode home through the thick darkness, under the clouded sky, following the steps of Mrs. M.'s old white horse, who knew every step of the way. What a noise the crickets and frogs and other small creatures make at night! they are as bewildering to one's ears as the buzzing of a London "at home," and are one of the things which strike a stranger most in tropical lands.

One day we went up the big mountain whose shape is so unlike any other, sprinkled with rocks as big as houses, the two top ones, from which it takes its name, being seen from enormous distances: none but cats or Tyndall would think of climbing them, but we enjoyed our ramble at their base. The beautiful scarlet sophronitis orchids quite coloured every rock and tree-stem, shining out gloriously among their green leaves and the gray lichens round. It was difficult to make up one's mind to cease picking them, the plant came off so easily with such great satisfactory slabs of roots, and we knew how they would be valued in England if we could only get them there. There were two varieties; one, "coccinea," rather deeper-coloured and smaller than the grandiflora, which was about the size of the English peacock butterfly. A few hundred yards beneath the top there was fine pasture-ground, varied by groves of spreading trees. One wonders the rich people of Ouro Prêto do not build villas up in this lovely spot to pass their summers in; perhaps they fear the large boa constrictors which people say haunt these big rocks, occasionally attacking the cattle.

From Itacolumi there is a grand view of the fine massive mountain of Caraca, amid whose peaks the great Jesuit College lies hidden, and the other side of which we had seen from Cocaes. On the lower slopes we passed through pretty woods in which I saw heads of a white-flowered begonia with large velvety leaves poking itself through thick bushes with much pertinacity, its stalks often six feet high at least. The light-blue plumbago also grew in the same way. I had an example at Passagio of how quickly the leaf-cutting ants can work. A citron-tree in the afternoon was perfectly green; the next morning nothing but bare stalks remained, and a long stream of apparently walking bits of leaves was still moving off in a long straight line towards the nest. They are said to attack all lemon, citron, and orange-trees, but never to touch the lime, whose essential oil is too strong for their taste, and which has armed itself with thorns. Nothing seemed to turn these little creatures from their path; even pouring boiling water on them does not have any effect till many are killed, when they make up their minds to go another way. Often that desperate remedy had to be applied, or the whole house might have been invaded; but it always went to my heart to see the murder of so many intellectual beings, for ants are not as other insects.

At last Mr. R. came, as he had promised, to ride home with me; and the M.s accompanied us back to Ouro Prêto, where my friend Dona Maria had a pretty gift for me—a large spray of maiden-hair fern with a tiny little humming-bird's nest hanging from its end by a rope of twisted spider's web, lined with the finest silk-cotton down. The enthusiasm I showed over it diverted the family, and still more when I said I would not part with it for a hundred pounds; they thought me indeed mad.

Our last stop on this journey was at the house of a friend—a big farmhouse where crowds of little black children were playing round the mules in the yard, or packed away in a building on the other side of it, together with the pigs and the poultry, so that the living-house was clear of them. The garden, too, was a miracle of neatness for Brazil. Strawberries grew in it and Brazilian raspberries (very unlike ours). We sat outside for some time star-gazing, and talking of home. The famous Southern Cross is as unlike a cross as four stars can well be, but nevertheless is a pretty constellation, its two very bright pointers moving with it as our Bear moves round the Pole Star; and when we started in the dark the next morning they were above, instead of below the Cross, as they were when we had last looked at them. The southern end of the Cross touches one of the curious Black Clouds of Magellan which veils part of the Milky Way, and makes as odd a collection of tropical curiosities as any of the nearer wonders, either vegetable or animal. As we had more than forty miles to ride, Mr. R. mounted me on his own spare mule—a noble beast who kept close to his friend, and needed neither whip nor spur. My mule was abandoned to Roberto, who wore a big spur, though he had neither shoes nor stockings, the strings being held between the great toe and the next in the usual negro style. Poor Brissole did not gain by the exchange as much as I did over the long dusty roads.

A fortnight's work at home was very pleasant, with kind friends, occasional visitors coming to relieve the monotony of a family party. Amongst others, a gentleman came, a real genius, who combined the accomplishments of making false teeth and tuning pianos; he also excelled as a dog-doctor and maker of guitars. Once we made a picnic to the top of the Coral mountain, 2000 feet above us.

At last Mr. Gordon said he would start on the 21st of May for his long-talked-of holiday-journey to the Caves of Corvelho, and arranged to take his daughter and myself and an English gentleman. We were to have started at daylight, but many things came to detain our leader, so that we really did not get off till the full heat of midday. There were three baggage mules, six spare mules, five men, and Lopez. A. G. rode over the shoulder of the Coral mountain with us, and then divided his spurs between his sister and myself before returning—a fact the mules soon felt established, and they required no more use of them, but jogged on merrily at their natural ambling pace. It is quite a mistake to make mules walk; they do not understand it, and dawdle mournfully. Up and down the steep roads we met a dozen or more ox-teams, dragging great trees by means of two huge wheels made of solid segments of trees bound with iron. There were often twenty or more poor beasts in each team, struggling and groaning, with an army of slaves poking pronged staves into them, and howling at them. It was a most painful sight to see. They often come a hundred miles with their unmanageable loads over these roughest of roads.

We passed the night at a lonely farmhouse—the roughest of quarters, where we were glad to camp round a wood fire on the mud floor, sitting on our wraps. However, we were supplied with an excellent supper, sent down at midnight on the heads of three black giants, by friends whose hospitable house we had missed in the growing darkness. We reached their Fazenda the next morning. They were educated Brazilians of the upper class. We breakfasted there, then rode on through a wooded country, till at sunset we came in sight of the Lago Santo, a shallow sheet of water getting gradually filled up, and I could hardly see the likeness our old Swiss friend found between it and the Lake of Geneva. It was certainly long since he had seen the latter, he said.

An air of indescribable dulness seemed to hang over it and the poor straggling village on its banks, and we wondered more and more what the charm was which had kept the famous Danish naturalist Dr. Lund here for more than forty years. He was now nearly eighty years old, and had made several collections of natural curiosities and plants, which had gone to Copenhagen. He had corresponded with many of the scientific men of Europe, but always lived entirely alone, and since the death of his secretary he seldom had an educated man to speak to. Once the Danish Government sent a man-of-war to Rio to fetch the doctor home, and he rode as far as Morro Velho, then lost courage, and returned to his dear lake. In former days he used to pass the heat of the day in a room he had built and fitted up as a laboratory over his boat-house on the lake; but now his habits were those of an invalid, and he seldom went outside his garden, and never left his room till after midday, when he liked to sit in his arbour and talk, which he did well in many languages. His English was astonishing, considering he had only learned it from books. He had a good library, and several times in the course of conversation he hobbled off into the house to seek some book, and to show us the authority for what he was saying. He seemed full of information on general subjects and on what was going on in Europe, as he read many foreign journals. His garden was full of rare plants and curiosities, collected and planted by himself. The trunk of one large date-palm was covered with a mass of lilac Lælia flowers, and a beautiful night-blooming cactus hung in great festoons from another tree, or climbed against a wall like a giant centipede, throwing out its feet or roots on each side to cling on by—it seemed to change its whole character by force of circumstances. I had made a painting during the morning of a rare blue pontederia which the doctor had persuaded with considerable difficulty to grow on his lake, and he was much delighted with it, and declared I was "one very great wonder" to have done it. The *Philodendron Lundii* was another of his most curious plants, a sort of great cut-leaved and climbing tree-arum, whose leaves are almost as good as a sundial, showing by their temperature the time of day.

Thirty years before our visit Dr. Lund had discovered the stalactite caves of Corvelho, and, as we were now on our way to them, was much interested in giving us directions how to find parts of them unknown to any but himself. Few persons had been far in since he first found them. He told us of the large apes, lizards, snakes, and other antediluvian beasts whose bones he had found there, as well as those of men with retreating foreheads, whose teeth showed they lived on unground corn and nuts.

It was late in the afternoon when we said good-bye and rode off, our saddle-bags well filled with cakes and oranges, a parting gift from our old friend. In a few hours we reached the Fazenda of Commendador O., a huge building like a fortress, with its enclosure of slave-houses, and its one closed gate, at which we knocked loudly with a stone, our admission being heralded by the barking of some score of mongrel dogs. Their owners matched them well, and a most miserable collection of beings received us. "The ladies" (about a dozen) conducted Mary and myself into the state bedroom, and then brought in chairs and sat down to have a comfortable stare at us, their slaves standing behind them and staring likewise. After some attempts at conversation, we sat down and waited patiently for them to go. After a time they departed, to our great relief, but took turns to look through the keyhole and the chinks in the window shutters, till we stopped them up with pocket-handkerchiefs.

The men seemed to spend both night and day lounging in the verandah. Whenever our shutters were opened they came and stared in too at us without the least ceremony. We had a ewer and a basin of solid silver, but the candles at dinner were stuck into empty wine-bottles. I never saw any family pretending to gentle blood so dirty; the Commendador himself was absolutely unwashed and unshaved. Fortunately those people do not consider it polite to sit down and eat with their guests, but feed afterwards at the other end of the table.

We did not leave this charming family till the sun was high in the heavens. Poor old Lopez was as glad as we were when at last we started; for he did not take to the four-footed company more than we did to ours. He kept close to us all the time, and gave a series of barks and jumps as he heard the great gate swing behind us.

A couple of hours through the woods brought us to the house of a friend, the Baron R. de V., a great ally of Mr. Gordon's, where we were received with the most boisterous geniality. His house looked thoroughly practical and thriving; his wife and daughters were lady-like and neat, but very quiet. They gave us coffee and cakes, ale and fruit, on gorgeous silver trays said to be worth £150 each, in a long room with windows nearly all round it, shaded by banana-trees with their fresh green leaves, and black grapes hanging from the vines trained round the frames. Portraits of the old Portuguese royal family were on the walls, chairs were arranged round the room, and arm-chairs surrounded a hearthrug at the end, to which we were conducted as the place of honour on entering. There was a round table in the middle of the room, on which was a massive silver-branched candlestick, with smaller candelabra round it.

About a league from this were some pretty stalactite caves, which our host took great delight in showing us; for he had once hidden in them for three months in time of political troubles.[57] One of them was 1500 feet long. There were caves beyond that, but lower, and too damp to enter. The thing which struck me most in these caves was the exquisitely finished margin left by the different water-levels, like the rims of marble fountains. The cliffs outside were much like those of Saxon Switzerland in their singular forms. Some of them formed fine old gateways, making, with the different creeping-plants which hung over them, rich pictures of form and colour. Indigo was cultivated among the sugar plantations, and there I saw for the first time quantities of large loose hanging nests made of twigs and sticks. Some of these were two feet long at least, and were the work of a comparatively small bird, in order to guard its young from squirrels and snakes. The Baron was a famous hunter, and his table was well furnished with excellent fish, paca, partridges, quail, etc.—everything in his house was superexcellent and abundant. The next morning he himself called everybody before light, and gave us a grand hot breakfast at six. We started with a large party, including the great man, his bailiff, and pet pointer, who, he said, was entirely English (though it had forgotten its mother tongue).

We had nine long leagues to ride, and passed only three houses all day, stopping at one to change mules and drink coffee. The mistress of the house gave me three pink eggs, and one large ostrich egg, which she thought she had improved by dyeing it a bright blue. Eggs are not the easiest things to carry on a jogging mule when hung on the same pommel with a sketch-book and a waterproof cloak, and I was glad to get them safely to our night's quarters. Cedros, where we were to stop for two nights, was a collection of houses built round a small cotton-mill by two brothers who had been possessed by the rare wish to do something for themselves, though they belonged to a rich family, and their old father divided his income equally between all his sons every year, giving each sufficient to do nothing genteelly on; but these two eccentric youths thought they would like to make more, and started this cotton factory, getting the best machinery they could, and importing an English mechanic with his American wife to teach the natives to work it.

Bernardo took us out of our road to show me the most magnificent specimens of Buriti palms, with great fan-like leaves, and noble bunches of red fruit full five feet long, which Bates describes as "quilted cannon balls," from the embossed markings on them; they make a kind of butter from the fruit, and hammocks and ropes from the fibre. These noble trees were standing in a kind of marsh amidst long reeds and stagnant water. Oxen were feeding near, and they told there the old story of the great boas killing and swallowing these beasts whole, leaving the horns sticking out of their mouths till they rot off. We slept that night at the house of Bernardo's father, a toothless old gentleman of eighty; his old wife looked very happy with all her big sons round her, and her home was a good specimen of its kind. Her verandah was edged with one long trough of growing plants, fringed with carnations hanging down outside a yard or two in depth—thick masses of foliage dotted with bright flowers.

Beyond this garden of sweet flowers we had a good view over the usual large enclosure of the Fazenda, at the other side of which was the sugar-making machinery. Part of the court was covered with freshly picked Indian-corn heads, which the slaves were shelling by the light of several bonfires, these being fed with the husks and dry remains of the sugar-canes. One man was leading a kind of monotonous chant, which the rest followed in a series of howls (not the kind of negro melodies we hear in London streets). After dinner we followed the old lady into a room, apparently the laundry, which seemed also to be used as a nursery for numberless black babies. I asked how many there were, and was told, "Oh, she had never counted them, there were always more born every day! Always the same bother of finding names for them." They are said to be fully black from their birth, and do not darken with age and light, like some other varieties of blacks. There were several looms in the passage, in which the women work beautiful counterpanes of a brown-coloured cotton, ornamenting them at the same time with different fanciful patterns in bright-coloured wools. Near this place is a very curious old altar, probably Indian. It is made of a single block of ironstone, shaped and grooved rudely, and hollowed into a sort of font. It stands on another block of stone, and when tapped with any hard substance produces a ringing metallic sound. It is placed quite alone at the top of a hill, and no legend is known about it. The Brazilians have little curiosity about things in their country, though these people had plenty about the reason of our wanting to see the caves of Corvelho.

It was a good ten miles' ride to the caves, and we stopped on our way to explore some smaller ones, the entrances of which were beautifully draped with cacti and other parasites and creepers, from under which flew a troop of beautiful white owls. The cave was quite dry, but bore the marks of a considerable body of water having flowed through it at times. The stalactites were very perfect. Outside was a lake, also quite dry, which refills every year at a certain season, like that of Zirknitz in Carinthia. A league beyond this was the great cave we had come so far to see, whose entrance was reached by a steep climb of a hundred yards from the stream below. It too was quite dry, but the steps or terraces which marked the different water levels were edged and banked up as regularly as the fountains of old Rome, and the grand hall at the entrance was like a bit of fairyland. Great masses of stalactite stood up from the ground, or hung from the roof, tinted with delicate blues and greens and creamy whites. Within the cave our scanty supply of light did little towards showing the endless halls and passages, and as Mr. Gordon was bent on making a minute measurement of them all, the progress was slow. The width of one of the great elliptical roofs was over fifty French metres, and it had no pillars or supports, only the elegant pendent stalactites hanging in groups. Of course there were plenty of mosques, fonts, pulpits, curtains, and Milan-cathedral roofs turned topsy turvy and every other way, and every variety of cinnamon, creamy, and white alabaster; but the only water we found was scarcely deep enough to wash a baby in. It would have made a lovely font for any cathedral, with its hanging canopy of pure white lace-work over it. We found abundant footmarks of ounces, gambats, and pacas, but only one place where bones were buried in the stalactite, and they were too much broken to be worth moving. Four long days we passed in this wonderful cave, our illuminations improving each day. We had bull's-eye lanterns, torches of *Canella di ema* dipped in tar, half orange-skins full of oil to place in certain niches and mark the road. Then we had ladders made of bamboo-canes bound together with lianes or stalks of climbing plants; but in spite of all this I felt we were groping in the dark, and the air was hard to breathe and very hot so far in the earth.

Our rides home were by the light of a very young moon and the fireflies. We tried to keep the cloud of white dust kicked up by the mule in front at a sufficient distance to guide us, and yet not get into our eyes. It was tedious work, and at the end of our journey we found little rest. So after the second day underground we determined to try another Fazenda. Mounting the wooded heights above the cave, then over open downs dotted with silk-cotton and other trees, one caught fine distant views of the Diamantina mountains, dark purple in the sunset, which was a rare gold and vermilion tint that night. The show was scarcely over when we reached "Once." The name sounded ominous, but the people were kind, and though somewhat astonished at such an invasion, received us hospitably as we rode into their enclosure.

A most jolly fat lady came out, lightly attired in the usual embroidered chemise and a red petticoat. She took Mary and myself at once to see our "rooms"—a large barn with many tiles wanting in its roof, and well ventilated walls, half filled with looms. There was a low wall in one corner to keep back the rice and corn on the floor. This had been swept back to make room for two little beds. Another part of the barn had been partitioned off by a higher wall, behind which the two gentlemen were to sleep. These were our quarters, though some obstinate old hens and pigs wanted to persuade us that they had engaged them beforehand; but Lopez soon settled that matter. Our men camped in a shed close by, and the farmer and his jolly wife lived in a small house on the other side of the yard, to which we went for dinner, returning to "Mr. G.'s room" (the tidiest part of the barn) for a game of whist and coffee, putting a tray on a stool for a table. Mr. G. and his daughter sat on one side of the rickety beds, and Mr. B. and myself on boxes, with one candle stuck in a bottle for our light. Our host and hostess and all their family witnessed the novel entertainment at a distance, seated on the edges of the different beds, for there were several in the room. It was very cold, and we put on all the wraps we possessed, including our own particular blankets, which we carried with us. We must have formed rather a strange collection of human oddities, and no doubt the natives thought we wore the national costumes of our country. At last they were tired out and retired, while we kept on at our game till nine o'clock, when we all drank hot sugar-and-water to keep out the cold (having finished our last drop of anything stronger), and dispersed to our several corners of the barn. Lopez, as usual, kept close to us, but was not content till he had driven all the poor chickens from their usual roostings on the looms or beams of the roof. He would also not allow a single cat or rat in the same room with him, and made sundry rushes at fancied intrusions during the night. How cold it was! How the wind whistled through the holes in the wall close to us! Mary said she should die if she stayed a second night, but she did not; for we had another long day in the cave and another night in the barn, and then rode over the windy sierras back to Cedros, where we were again loaded with Mrs. N.'s abundant talk and kindness.

Our host rode on with us and lost his way before we had gone three miles, though we were bound for the principal town of that district! It was very cold when we reached the top of the sierras, and the cold seemed all the stranger that we were passing through miles of burning grass. Setting it on fire at certain seasons is the only form of manuring it ever gets; after that it springs up with new life. The fire also kills many of the snakes, and insects who would devour it and the other crops. At night the country was quite illuminated by the burning hills. The name of Sette Sugons describes the place—a low village near a marsh and pools of water, which suggest a longing for quinine, but it was said not to be unhealthy. We stopped at a really decent little inn kept by an old black woman named Donna Anna. Our room was full of sacks of grain and bales of groceries; but the beds were covered with gorgeous quilts, the linen dazzlingly white, edged with fine lace and knotted fringe, also made by hand. We had even a looking-glass, and a basin and jug; but these luxuries had to go the round of the guests from room to room, including a strolling photographer, whose chemicals occasionally sent the water in rather black. His price was twenty millen reis a head (£2). I believe a good living might be made by any one who could take portraits (however indifferently) in these far-off countries. Many of the country people here who appear so poor are really rich, and only want the vanities or luxuries of life to tempt the money from their pockets.

Donna Anna was a famous cook, and did nearly all herself, her "helps" sitting down and grinning by her. The whole evening we vainly endeavoured to keep the doors shut and the draughts out; as fast as one was shut some grinning black head was poked in at the other. The free blacks were especially curious about us, and in honour of our arrival they put any amount of grease on their hair, stretching and straightening it with weights at the end till they fancied it looked "like any other gentleman's." Another cold day's ride brought us to a large farmhouse belonging to a very remarkable family, who would have made their fortunes at fairs. The farmer himself was perfectly round, with a bullet head and face, all over which grew hair and beard apparently cut with his wife's bluntest pair of scissors as close as he could with his left hand. His fat wife had a thick black beard and moustache (uncut), her grandmother the same in gray. The children were all perfectly round, like their fascinating parents, but as yet beardless. We had a grand but greasy dinner; the table quite groaned beneath the quantity of heavy dishes on it. A small pig cooked whole, and considerably over the usual size for making such a barbarous exhibition of itself, was among the dainties. After dinner we sat in a large unfurnished saloon and did "company." It was no easy task to keep up a conversation even for my friend, with her perfect knowledge of the language; for these people were absolutely without ideas, except the usual desire to know what everything cost.

It was a relief the next morning to hear the tremendous voice of our friend the Baron R. V. before daylight, shouting to the gentlemen in the adjoining room. He seemed to bring a more genial world nearer to us. He had ridden over in a wonderful peaked woollen hood to make sure we did not pass his house without going in. Now his house was not in our road at all; but it was impossible to refuse our friend's positive determination to take us there, and we resigned ourselves to his will. After having my face scrubbed by the old grandmother's gray beard, my mule took to fidgeting, and I escaped more adieux and went on ahead. The old lady said she should always have "Sandades" of me, and I am sure I shall never forget her, though the beautiful Portuguese word is untranslatable. It was very refreshing to sit a while with the quiet Baronessa and her silver coffee-tray and Minton cups, all so bright and clean; but after a rest and a chat we went on to Dr. Lund's again, and were obliged to stop another night at the grocer's, as the old gentleman never came out of his shell before midday. After a while we rode on a little farther to the house of our old Swiss friend, which he had turned outside in for us. He would let no one wait on us but himself, had cut down twenty dwarf palm-trees to make one dish of cabbage for us, and hung up various branches and bright flowers about the verandah—a decoration no native would think of.

We had but a short journey on to the "City" of Santa Lucia, a most picturesquely situated village on the top of a hill, looking over a long stretch of the winding Rio das Velhas, which again reminded me of the Tweed, and except for a few palm-trees looked not a bit more tropical; while the churches, with their metal pepper-pot towers, and the tiled roofs of the one-storeyed houses, suggested Hungary. A few houses stood up out of all proportion to the others. To the largest of these we now rode, and found a friend there who was staying on a visit to the poor old Baronessa of S. L., who was then a terrible sufferer, paralysed even to her tongue. After having been a generous and sociable woman, she was now neglected and left to the entire care of slaves, while her relations were fighting over her property. She had built hospitals and schools and churches, and had brought up numerous nephews and nieces, not one of whom seemed to take any charge of her comfort now. The lady we found there was, like ourselves, a passing guest. One nephew she had sent to Rio a short while before, telling him he might draw money if he wanted it from her credit; he stayed three months, drew £8000, and had nothing to show for it. Somehow we made out that the poor old lady liked music, so we dragged her chair into one of the bedrooms where there was an antique piano, and gave her as much as she liked of it. She gave us each at parting a piece of wide thread lace made by her slaves. The chests in which this lace was kept were covered with the skins of big snakes of the country; they looked like scale armour, and were said to be as strong as any leather.

From Santa Lucia our way was hot and dusty as we crept round the shoulder of the Piedade mountain, and came at last to a ridge from whence we looked down on the pretty town of Sabara, descending by a road so steep that walking was almost a necessity, after which we were glad to escape from the glare into the shelter of Donna Anna's roof, where we lingered till the cool evening and rode home by moonlight; what luxury "home" was after such a three weeks of wandering!

On the 2d of July I saw the last of dear old Morro Velho, and accompanied Mr. and Mrs. G. back to Rossa Grande, and on to Cocaes. The forest of Gongo had lost much of its beauty during this cold dry season; more trees had lost their leaves than I expected in a tropical country, and flowers were quite rare. I was rather glad of this, as it made me regret less that I was leaving so lovely a country, and I took away the hope of seeing my kind friends again in England; but in spite of this it was hard to say good-bye to dear Mrs. Gordon at Cocaes. Mr. G. was to go down with me to Caraca, so down we went to a bare hill country, and leaving the village of St. John to our right, soon came to the bridge and ravine of Caite. It was a fête day, and everybody was on the road dressed in their best.

The college of Caraca reminded me somewhat of the Great St. Bernard, minus the snow. We found the Superior Padre Julio expecting us, and after dismounting in the court he walked down to a lower building and introduced me to a stout old lady with a black silk handkerchief tied over her head, whom I afterwards found to be the Chief of the washerwomen; he left me to her care, taking his other guests to be entertained in the convent itself. The smallest little room I ever saw had been prepared, but after seeing me the old lady seemed to think it would be too tight a fit, and she moved the bed into her own comfortable apartment, where I spent a pleasant evening with a heap of valuable botanical books sent down from the library by the good Padre, who also took care to feed me well. He was a most fascinating character, full of general information and knowledge of the world; moreover, a thorough gentleman. The next morning he was down before seven o'clock looking at my drawings, and giving me the names of many of the strange plants I had been hunting for so long. He took me to see the library and garden, and told me I was the first woman who had entered there for seventeen years. There are about two hundred and fifty students in this college, and nine padres, besides my friend; but what a difference there was in those men! I was told there was one other who would have interested me, as he was a naturalist, but he was away. Padre Julio told me he wanted to start a museum and classes for natural history, but the Brazilians did not see the good of it, and did not care to inquire into such things. This same absent priest was a good carpenter, and they showed me a beautifully finished flageolet he had made out of the heart of the Araucaria pine, using up his old spurs for the silver parts. The boys at this college paid only £30 a year, and were taught French, English, and Latin, as well as mathematics and Portuguese.

The neighbourhood abounded in rare orchids and other plants, but the rain never ceased to pour, and at this time of year it generally did pour on these mountains; so there was little use in staying, and I resisted all the kind wishes of the Superior that I should stop on at the washerwoman's, and said good-bye to him and to Mr. Gordon, who had loaded me with such continual kindness and hospitality for the last eight months. He now returned to Cocaes, while I rode after the Baron in the opposite direction.

We crossed the high boggy watershed, every pool and river being bordered with a curious dwarf bamboo peculiar to this mountain, more like young cypresses than canes; and the rocks were everywhere covered with rare orchids. The descent was over the roughest of tracks, and we had to walk for quite two hours, and quickly too, for the darkness was creeping on.

The Baron grumbled incessantly at those who had kept him so late. The wild bromeliads were glorious; I saw acres of one I had given ten shillings for not many years ago at Henderson's —a nidularia with deep carmine nest and turquoise flowers in the centre. After much sliding, tumbling, and slipping, we arrived at Senor Antonio de Sonlea's, and were received most kindly by him and his young wife. Before eight the next morning we were riding over the smaller spurs and still under the wet clouds of Caraca, now and then getting a good shower-bath from some overhanging curl of bamboo or green tangle as we passed. Everything was dripping with moisture; how lovely those wet mornings were! And the huge spiders' webs all strung with crystal beads, so strong that they seemed to cut one's face riding through them.

About eleven we entered the principal iron basin of Minas, San Antonio de Pereira, where all the roads and rivers were black and all the rocks red. After winding for some time through this singular valley, we mounted high over the next ridge, and presently looked down on Santa Anna, Marianna, and finally on Ouro Prêto; Itacolumi being covered with cloud like Caraca. Instead of going into the capital, we turned aside through its eastern suburb and crossed the river to the new road. The town is wonderfully picturesque, though most inconveniently built for its inhabitants, stretching over the steep spurs of real mountains. At a distance, the roofs of the houses looked like a succession of steps. I longed to sketch, but did not like to stop the Baron, as we had had a hard day's work for the mules; so we went on farther and put up at a lonely rancha by a bridge, and secured the whole three rooms in the house to ourselves. All the sheds outside were filled with merchandise and people on their way from Diamantina to Rio. Now that the Cape diamonds are so much easier to obtain, the famous old mines of Brazil have been nearly abandoned; labour is so much dearer there that it hardly pays to work them, though the stones are of better quality.

More troops of mules and men from this old Diamantina arrived that same night, and the Baron was most mysterious about knowing them and not letting them know him, etc. etc. He insisted on starting before daylight to get ahead of them; so we jogged on through cloud and cold into sun and dust, and over dazzling many-coloured sandstone roads for five hours, only stopping once at a lonely hut to change mules and eat our bread and cheese, then on again by a road which was in many places "carriageable," but then minus the carriages. We passed instead endless processions of loaded mules, generally led by the "Madrina"—a horse with a peal of bells, feathers, and a curious little dressed-up doll in orthodox crinoline, doing duty for the Madonna, suspended over his headpiece.

We arrived at our night's quarters soon after midday, and got the rooms the Baron wanted, in a most comfortable house kept by people who had seen better days, and had done their best to educate their children. Three of the girls took me a stroll through the farm, and showed me many curious little insect-nests of different sorts, for in these countries the insects are even more curious in their home-architecture than the birds; one little cocoon I saw here seemed of the finest frosted silver. The small town of Che Luz is famous for its guitars, and I saw the work of making them going on in several of the houses. Ribera I shall never forget; it is notoriously the worst quarters on the road. I had a very tolerable mud-floored room to myself, and a quantity of pigeons pattering over the mat (which did duty for a ceiling over my head), cooing to one another, and kicking down dust and fleas, which last were also taking visibly all sorts of calisthenic exercises on the floor. From the window I had an uninterrupted view of the farmyard or general slop-pan of ages; but when I had rushed from all this up the hill to the wild country, I thought even Ribera was worth a journey to see. Birds like canaries were twittering in the bushes; armies of ants were carrying on their mysterious occupations; their great nests lined the road with such regularity one almost fancied they had been part of the design of the engineer who made it. The air was as delicious as air could be, and as I returned from my walk I almost envied the man who owned Ribera! *He* was a Justice of the Peace, and my beautiful room was his "study" on ordinary occasions. Besides its two beds, it had a table with legs of such very odd shapes and lengths that three of them required stones of different sizes to keep them standing at all; there was nearly a foot of difference between the longest and shortest leg; on the table stood his worship's law-dictionary, and inkstand full of black dry porridge. He and the rest of his variously coloured family lounged about and stared; and I was again assured, as on coming up, that the lunatic who sat on the doorstep was *generally* harmless: on the whole, one would avoid this place if there were any other within ten miles to stop at. I suppose the Baron was glad to leave it as early as he could, for we were in our saddles before daylight. It was cold, but very beautiful, seeing the full moon gradually fade into the more gorgeous tints of dawn, the hill-tops dipped in pure gold, the nearer and lower ones deep purple. My poor old leader was quite ill, and when we got to the half-way house he said he could go no farther, but changed his mind after breakfast.

I walked into the back verandah to get into the sun, for I was half numbed with cold, and found a huge copper pan (the usual bath of the country) placed on a great bonfire in the yard, full of stewing meat, stirred round and round by two men with long poles; on the kitchen-fire a huge caldron of potatoes was boiling, so we had little chance of starving. The fat landlady took a dish in one hand and a long kind of toasting fork in the other, with which she fished out dainty bits for us from the two steaming messes. We left again by moonlight, but this time had no sun to cheer us, and though I put on a pair of gloves for the first time for a year I could scarcely feel my fingers for the cold, and yet it was not freezing.

We stopped at another solitary house for breakfast, and I watched the woman making biscuits of mandioca flour and white of egg, each one separately rolled out like a ring on a large pleroma leaf, which was then put in the baking-tin (they generally used the banana leaves; these would have gained a prize among foliage plants at an English horticultural show). At Popoyas I declined the state apartments upstairs, where I should have had to sit bolt upright in a chair and smile blandly all the afternoon in return for the hospitable stares of the family, so I pleaded fatigue and stayed in the ordinary travellers' room. The next day we passed again over the fine sierra of Mantiqueira; the holes where mules were drowning in the wet season were now full of dust but *unmended,* the bridges were even more dangerous than they were then, and the half-eaten carcase of a mule surrounded by a tribe of Burinboos was tainting the air for miles.

My last night on this journey was an unquiet one, in another solitary house near the new railway works. It was Sunday, and half-drunken navvies came and thumped at the door all night. My room opened on the verandah and got its share of thumps too, but I knew if the Baron or Roberto wanted anything they would begin "O Dona Pop!" and not hearing that, I hugged the cold blankets and kept still till called as usual at four, for I knew there was a wooden bar across the door which would resist any quantity of thumping. But the mules had got into sweet pasture and would not be found, and the thick cloud made it no easy task to hunt for them. Four hours it took, when the poor men came in soaked and shivering, and the Baron stormed and grumbled: "There had been such a row he had not slept a wink; it was too cold even to take off his boots, and the coffee was burned," etc.; so he grumbled himself into high good-humour long before we entered the trim German suburb of Juiz de Fora. The next morning, after squeezing the good old Baron's hand for the last time with real regret, I packed myself into the crowded coach and was whirled away towards Rio.

The distant Organ Mountains peeped at us over the ends of the green valleys, and I again thought nothing in the world could be lovelier than that marvellous road; and then what a welcome the kind M.s gave me, and what a cosy little room in their house at Petropolis! It was rather pleasant too to see my old box again and its contents. Of what priceless value those shoes and stockings and paints seemed to me! And how I longed for them! I had intended starting for Para in a week, but was persuaded to give it up, as the yellow fever was still lingering all along the coast; and I had a longing first for rest in my pleasant, comfortable quarters, and then still more for a sight of home, friends, and books again.

Meanwhile I made two visits to Rio, the chief object of which was to see the Emperor, to whom I had a letter from my father's old friend Sir Edward Sabine.[58] The Emperor is a man who would be worth some trouble to know, even if he were the poorest of private gentlemen; he is eminently a gentleman, and full of information and general knowledge on all subjects. He lives more the life of a student than that to which ordinary princes condemn themselves. He gives no public entertainment, but on certain days he and the Empress will receive the poorest of their subjects who like to take their complaints to them.[59] He kindly gave me a special appointment in the morning, and spent more than an hour examining my paintings and talking them over, telling me the names and qualities of different plants which I did not know myself. He then took the whole mass (no small weight) in his arms, and carried them in to show the Empress, telling me to follow. She was also very kind, with a sweet, gentle manner, and both had learned since their journey to Europe (of which they never tired of talking) to shake hands in the English manner. They had both prematurely white hair, brought on by the trouble of losing their daughter and the miserable war in Paraguay.[60] On my second visit to the palace the Emperor was good enough to show me his museum, in which there is a magnificent collection of minerals. He took especial delight in showing me the specimens of coal from the province of Rio Grande do Sul, which promise to be a source of great riches to the country if his schemes of facilitating the transportation can be carried out. At present, though the coal itself is close to the surface of the ground, there are so many transhipments necessary in bringing it to Rio that it is cheaper to bring it from England or the States. I have not the slightest knowledge of mineralogy, but I blacked the ends of my fingers with a wise air, and agreed heartily with the Emperor's opinion, that if the precious stuff could be brought into consumption cheaply, it would be of more use to Brazil than all the diamonds of Diamantina. Then he showed me many of the most precious books in his library, some views of the San Francisco river, etc.

The palace is not in a good situation; but the Emperor passes a great part of the year at Petropolis, around which there are endless beauties. One spot there especially attracted me, where an old companion of Humboldt's had settled himself in an unpretending cottage.[61] He had planted all sorts of rare plants and palms around it, and the real virgin forest sloped down to it at the back, while a glorious view of blue mountains was seen from the front windows, with some few great forest giants left as foreground, their branches loaded with parasites and festooned with creeping plants. This little house was the highest inhabited house of the neighbourhood, the path up to it sufficiently steep to keep off ordinary morning visitors, though I am told it is a favourite walk of the Emperor's, who found the old German naturalist a pleasanter companion than many in the world below. When I was there this old man was dying, and his pretty place would soon be a ruin. Already his treasures of moths, books, birds, and butterflies were half destroyed by mould and devouring ants;

even the bridge which crossed the cascade and the path up to his house were falling away. I never felt anything more sad.

Petropolis seemed full of idle people and gossip, and it was thought rather shocking and dangerous for me to wander over the hills alone; wild stories were told of runaway slaves, etc. I felt out of place there, and got more and more home-sick, but determined to have at least a glimpse of the Organ Mountains before I went. I was told the way was most difficult, and even dangerous; neither mules nor guide could be got. Still I persevered, and finally heard of a mason at Petropolis who knew the way and would like a change of air and a holiday, but he could only spare four days. Mr. M. kindly lent him a famous old mule, and sent it on the day before to San Antonio, where I was also to find a horse; and, in spite of the persistent rain at Petropolis, I and the mason started by the Juiz de Fora coach at six o'clock, and were set down about sixteen miles on the road at a venda near a bridge, where we saddled our steeds and mounted, my small bag and paint-box being fastened to the crupper of my guide's mule. My horse was of the Rosinante order, very bony and old, with two great gaping wounds on her shoulders caused by the bites of vampire bats, into which the flies walked in the most distressing manner. After winding along two or three valleys, we began to mount in good earnest. The only danger on our path was from the hanging wreaths of bamboo, and the acacia called "cat's paw," which had been long untrimmed, and might easily do serious damage to the faces of unwary travellers. My guide used his long knife, and I met with no accident, and soon reached the top of the pass, having left all rain and humidity at Petropolis. It was a curious view, and well worth some trouble to see; but the "difficulties and dangers" we in vain searched for.

We arrived at Theresopolis by two o'clock, went on for another two leagues, and put up at a quaint and lonely house on the sierra. The boulders there had fallen all round it; they propped it up, and seemed to rest on its roof, and the stables were built under one huge hanging boulder. Great trees and all sorts of rich vegetation grew over and round these big blocks of granite. Beyond all were the most splendid distant views of Rio Bay and its mountains, and over our heads strange obelisks of granite. It was a spot for an artist to spend a life in.

Did I not paint?—and wander and wonder at everything? Every rock bore a botanical collection fit to furnish any hot-house in England. Then there was a real Italian vine pergola leading down through the banana trees to the spring, with picturesque figures continually fetching water from it, and troops of mules, goats, cows, and sheep always moving about; for the grass had failed in most parts of the mountains this year, but was unusually abundant here. I found it hard to leave the next day, and lingered over my work till nearly noon, when a gentleman came down the hill leading his horse, and spoke to me about the view I was taking, then went on and spoke to my guide, arranging with him that as the inn of the place where we were to stop the night was bad, he should take me to his house, writing at the same time a few lines to his wife, to take with us and explain who we were. Who were we? And who was he? We were both ignorant on these subjects, but accepted his kind offer of hospitality in the frank spirit with which it was given, and which one only meets in remote places far from the cautious rules of civilisation, which believes every one till properly introduced to be a rogue. We descended the glorious road to Barrera (another spot for an artist to settle in), rested a few hours of the extreme heat of the day, and I worked at the view from the shady verandah. A mad river made its noisy way through great purple and gray boulders of granite from the strange group of mountains beyond, which here seemed to open themselves out like the walls of an amphitheatre, the sharp points piercing the clouds which formed its roof, and the whole in a state of quivering blue heat most difficult to represent on paper, as the intense glare of the almost perpendicular sun's rays puzzled one. Was it all shade? or all light? Flies and tiny wasps with a taste for chemistry were anxious to ascertain what my colours were made of, and carried various fancy tints into my wet sky, producing effects that were startling but not artistic. The air was heavy, and there was every appearance of a coming storm, but none came.

At last we reached the sea, stopping every now and then to chat at different roadside cottages, where my guide bought different refreshments for himself, as he was always hungry when travelling, he said. Sometimes he bought a paper of boiled prawns very large and pink, then oranges or sweet lemons, or a beautiful sort of cornucopia of dazzling white made of the thinnest paste of mandioca flour rolled out and baked; it was a fit food for gods.

The sky was still red when we reached the little town we were to stop at, and inquired for the address our friend had given us. His young wife would not let us in till she had held a long conversation with us from an upper window, which ended in a good deal of laughing on both sides, she thinking she could talk English, and I Portuguese, and each of us thinking the other talked her own native tongue. But when I was at last admitted she was most kind, and gave me her best apartment—a cupboard inside her sitting-room. She walked up and down the room combing her beautiful "back hair," a mode of entertaining her guest which was certainly original. After that she and I passed an hour or two gossiping at the window, she constantly talking to friends in the street below, who, like her, had lately arranged their heads most becomingly, and stuck natural flowers behind their ears. She also sent out and bought some cups of freshly pressed sugar and rice made hot, which a man was crying in the street; it was very good, and about the consistency of barley water. After a time her husband came home, and she left me to give him his supper, after which he came also to gossip at the window. I found he was the chief of the police of that province—an educated man of good family. He was extremely curious to know why I was travelling alone, and painting. Did the Government pay my expenses? I certainly could not pay them myself, I was too shabbily dressed for that! I told him when I got home I hoped to paint a picture of the Organ Mountains, and to sell it for so much money that it would pay all my expenses; then at last he understood what I travelled for, for is not money the end of all things?

A few more hours of swamp and a most roundabout road brought us to the foot of the Petropolis sierra, up which I rode, though in time for the train of passengers from Rio. It was such a glorious evening; and while the poor animals were resting after their thirty long dusty scorching miles of road, I sat near some running water in the shade of a grand tree and enjoyed a rest also, where the mason brought me a tray of good coffee and bread, without any orders; and for this one kind thought alone deserves to have his name recorded—"Jose Luis Correa." He was as good a guide as could be wished for on such a journey, and had more than a common knowledge of plants and other things of the country, and I regretted much that I did not better understand his language to benefit by his information.

In three days more I was steaming towards England. I gave the steward a commission to buy me little singing-birds at Bahia, and he bought nine; they were all kept on the spare berth in my cabin, which went by the name of Bird-Cage Walk among the servants and children, with whom it was a favourite lounge. I gave most of the birds to my nieces at Clifton; and when one died soon after, it was buried in a lozenge-box, and half a lozenge was put in the box "in case it should wake in the night and feel hungry." A little girl came on board at Pernambuco who had a great talent for taming all living things. She brought some large locusts which were devoted to her; they came when she called, sat on her head or hand, and she made them pretend to be dead on her hand till she counted three, when they hopped away. It was very wonderful, considering the short lives of such creatures, how she had tamed them; of course they died from want of proper food on board. A little boy had a marmoset which he kept in a cocoanut shell, and fed on milk; poor little thing, it got very cold and shivered before we landed at Southampton on the 14th of September.

# CHAPTER VI

## TENERIFFE—CALIFORNIA—JAPAN—SINGAPORE

1875-77

THE winter after my return from Brazil I devoted to learning to etch on copper, Mr. Edwin Edwardes, who had illustrated the old inns of England, kindly giving me a few lessons. Friends seemed always accumulating round me and making life very enjoyable. I was called down to Netley to help to nurse my cousin, Dudley North, who had returned from Ashantee with three wounds and much fever, though he always maintained the savage who shot him was a gentleman, for he gave a yell first to warn him of the danger.[62] He lived through it, thanks to the care of the doctors and the nursing of Mrs. Deeble, the lady-superintendent there, who sat up with him for fourteen nights. She was a wonderful woman, looking always as if she had nothing to do, though she seldom slept more than one hour of the twenty-four at that busy time. Every bed in the hospital was full, the field in front also covered with tents full of invalids, and more were constantly coming home. I stayed about six weeks, partly at Colonel Gordon's (the Governor), partly at a lodging over the post-office. When my work there was over I returned home and paid visits among my friends and relations in England, Ireland, and Scotland, going to the meeting of the British Association at Belfast among other things; and after hearing Tyndall's wonderful opening address, I heard a sermon preached on it in a country church on the text, "It were better that man had never been born."[63]

The winter was an unusually cold one. After the experiences of the last two in Jamaica and Brazil I found it quite unbearable, so at last I determined to follow the sun to Teneriffe. M. E. and I started on New Year's day, 1875, in hard frost and snow, steaming from Liverpool in a wretched little steamer in unpleasant squally weather.[64]

On the 11th we landed for a few hours in sunny Madeira. I had a cousin there with a sick husband, and in spite of the marvellous beauty of all the surroundings I pitied her for having such a number of hopeless invalids all round her. I heard coughs and groans on every side, and saw poor bloodless faces carried about in hammocks on men's shoulders covered with white drapery, and looking like corpses. The other mode of locomotion was in clumsy bullock-carts, with a driver hanging on downhill and pushing uphill, continually greasing the great wheels with bits of rag. The place was full of colour, the gardens full of bananas and many of the bright flowers I had seen in Jamaica, the sea deliciously clear and marvellously varied in tints, with rich brown lava rocks in all sorts of grotesque forms sprinkled in and out of it, hanging creepers festooning the cliffs from innumerable pretty villa-gardens on their tops, and splendidly-formed hills rising behind. At sunset that same evening we saw the top of the Peak on the golden horizon, and on the morning of the 13th we landed at Santa Cruz.

We drove on the same day to Villa de Orotava, creeping slowly up the long zigzags leading to Laguna, where every one (who is anybody) goes to spend the hot summer months; in the New Year's time it was quite deserted, and looked as if every other house was a defunct convent. All had a most magnificent yellow stone-crop on their roofs, just then in full beauty; ferns too were on all the walls, with euphorbias and other prickly things. After passing Laguna, we came on a richer country, and soon to the famous view of the Peak, described so exquisitely by Humboldt; but, alas, the palms and other trees had been cleared away to make room for the ugly terraces of cacti, grown for the cochineal insect to feed on, and which did not like the shade of other trees. Some of the terraces were apparently yielding crops of white paper bun-bags. On investigating I found they were white rags, which had been first spread over the trays of cochineal eggs, when the newly-hatched insect had crawled out and adhered to them; they are pinned over the cactus leaves by means of the spines of another sort of cactus grown for the purpose. After a few days of sunshine the little insect gets hungry and fixes itself on the fleshy leaf; then the rags are pulled off, washed, and put over another set of trays. The real cochineal cactus has had its spines so constantly pulled off by angry natives who object to having their clothes torn, that it sees no use in growing them any longer, and has hardly any. When I was in Teneriffe people were beginning to say that the gas-colours had taken all their trade away, and had begun to root the cactus up and plant tobacco instead; but they could not re-grow the fine trees. These cactus crops had done another injury to the island besides that of causing it to lose its native trees. The lazy cultivators when replanting it, left the old plants to rot on the walls instead of burning them, thereby causing fever to rage in places where fever had never been before; they were now planting eucalyptus-trees with a notion of driving it out.

The roads were very bare, and the much-talked-of Peak with its snow cap was spoiled for beauty by the ugly straight line of the Hog's Back on this southern side. Nevertheless the long slant down to the deep blue sea was exceedingly beautiful, and a certain number of date-palms and dragon-trees, as well as the euphorbia and other fleshy plants, gave a peculiar character to the scene I have not seen elsewhere.

We found there was a hotel (and not a very bad one either, in its own Spanish fashion), and we got possession of its huge ballroom, which was full of crockery and looking-glasses, and some hundred chairs all piled up on the top of one another. This room had glass doors, besides other rooms opening into it, but served to sleep in well enough; and I determined to stay and make the best of it, for the climate and views were quite perfect. I did stay more than a month. M. stayed a fortnight, then went to the Smiths at Puerto, and home to England. The people at Orotava were most friendly, the gardens lovely. The nobles who owned them were of the very bluest blood of old Spain; but not rich—they seldom went out of the island, and had kept all their old habits and fashions. The ladies walked about in mantillas, flirting their fans, and wore no other costume even at their evening receptions, merely adding some jewels, and flowers stuck most becomingly behind their ears. They had no education beyond what they got in some convent, but were thorough ladies. One old lady seemed to reign supreme amongst them—the Marchesa della Florida. She was good enough to take me under her protection, and even asked me to come and stop in her house; but I valued my time too much to try such an experiment. Dr. Hooker had given me a letter to the Swiss manager of the Botanic Gardens, who also kept a grocer's shop. He was very kind in taking me to see all the most lovely gardens. The famous Dragon Tree, which Humboldt said was 4000 years old, had tumbled into a mere dust-heap, nothing but a few bits of bark remaining; but it had some very fine successors about the island, and some of them had curious air roots hanging from the upper branches near the trunk, which spread themselves gradually round the surface, till they recoated the poor tree, which had been continually bled to procure the dye called Dragon's Blood. When the good people found my hobby for painting strange plants, they sent me all kinds of beautiful specimens.

After M. E. left, the landlady gave me a smaller room opening into the big room with a good view into the street, where I could live in peace and quiet, without fear of interruption, and they fed me there very kindly too. Any one who likes bread and chocolate can live well all over Spain; I did not care if I got nothing else. My friend the gardener arranged with the farmer at the Barenca da Castro to take me in for three days; so I took some bread and a pillow, mounted my donkey, and rode thither through lovely lanes, mounting over the high cliffs till I came to my destination—an old manor-house on the edge of one of those curious lava cracks which run down to the edge of the sea, filled with large oaks, sweet bay-trees, and heath-trees thirty feet high. Half-way down was a stratum of limestone, from which a most delicious spring burst out. People came from all the dry hills round to fetch the water, and to wash and water their cattle. The ground was covered with sweet violets. There were green beds of water-cresses all about the sweet clear pools on the little theatre of green at the mouth of the cave, and then some pretty falls to the lava rocks on the beach some thousand feet below. People and animals were always coming and going, and were very picturesque. The men wore high top-boots, blankets gathered in round their necks, and huge Rubens hats. The women had bright-coloured shawls draped gracefully over their heads and shoulders, with red and black petticoats; sometimes hats on the top of their shawl-covered heads. They were all most friendly.

My quarters at the old house above were very primitive. A great barn-like room was given up to me, with heaps of potatoes and corn swept up into the corners of it. I had a stretcher-bed at one end, on which I got a very large allowance of good sleep. The cocks and hens roosted on the beams overhead and I heard my donkey and other beasts munching their food and snoring below. From the unglazed window I had a magnificent view of the Peak, which I could paint at my leisure at sunrise without disturbing any one. The family much enjoyed seeing me cook my supper and breakfast in my little etna morning and evening—coffee, eggs, and soup; soup, eggs, and coffee, alternately. I returned by a lower road, close to the edge of the sea, under cliffs covered with sedums, cinerarias, and other plants peculiar to the Canary Islands.

I stopped a while at the Rambla da Castra, on the sea-shore, standing almost in the sea, surrounded by palms, bamboos, and great *Caladium esculentum.* It was a lovely spot, but too glaring. After this little excursion I remained quietly working in or about Orotava till the 17th of February, when I moved down to Mr. S.'s comfortable home at Puerto di Orotava. Mr. S. when I stayed with him had a second wife, a most lovable Scotchwoman. He was seventy years old, and talked quite calmly of taking me up the Peak, not minding fifteen hours on horseback; but the weather fortunately remained too cool for such an attempt. I believe he knew every stone on the way, and had shown it to Piazzi Smyth and all the travellers one after the other. The latter gave me a letter to him.

I had a room on the roof with a separate staircase down to the lovely garden, and learned to know every plant in that exquisite collection. There were myrtle-trees ten or twelve feet high, bougainvilleas running up cypress-trees (Mrs. S. used to complain of their untidiness), great white lancifolium lilies (or something like them), growing high as myself. The ground was white with fallen orange and lemon petals; and the huge white cherokee roses covered a great arbour and tool-house with their magnificent flowers. I never smelt roses so sweet as those in that garden. Over all peeped the snowy point of the Peak, at sunrise and sunset most gorgeous, but even more dazzling in the moonlight. From the garden I could stroll up some wild hills of lava, where Mr. S. had allowed the natural vegetation of the island to have all its own way. Magnificent aloes, cactus, euphorbias, arums, cinerarias, sedums, heaths, and other peculiar plants were to be seen in their fullest beauty. Eucalyptus-trees had been planted on the top, and were doing well, with their bark hanging in rags and tatters about them. I scarcely ever went out without finding some new wonder to paint, lived a life of the most perfect peace and happiness, and got strength every day with my kind friends.

The town of Puerto was just below the house, and had once been a thriving place, some English merchants having settled there. Now only a few half-bred children remained, entirely Spanish in education and ways, though they talked their fathers' tongue after their fashion. I went off with a donkey-boy and a couple of donkeys for a week to Echod, all along the coast, sometimes high, sometimes low, with fresh views of the Peak up every crack. At Echod there is the best view of all; and a few miles above that place are forests of the Canary pine, which is something like the Weymouth, with very fine needles, but drawn up into slender trees of one hundred or more feet high. Echod is a lovely old place, full of fine big houses, with exquisite views up and down; but it rained most of the time. The Marchesa de la Florida had written to her cousin the Count of Sta. Lucia, who took me to see some fine coast-views, and insisted on walking arm-in-arm over ploughed fields and slippery pavements at an angle of forty-five degrees, much to my embarrassment. He was a regular Sir Charles Grandison of politeness.[65] Some other grandees, with terribly long strings of names, were most hospitable, showed me their beautiful villas and gardens at Corronel and Gorachico, and even pressed me to stay. The latter place is built on a glacier of black lava, and the next eruption will probably send the whole town into the sea. It was one of the most frightful bits of volcanic scenery I ever saw. The day I was there was wintry and dark with storm-clouds; the white waves ran in between the dark rocks, and sent up great jets of foam with an awful crashing and roaring.

Santa Cruz, to which I at first took a dislike, I found full of beauty. Its gardens were lovely, and its merchants most hospitable. I stayed there till the *Ethiopia* picked me up, on the 29th of April, with my friend Major Lanyon on board returning from the Gold Coast, where he had been filling the place of Colonial Secretary.

I landed again at Madeira. There was a Mr. C. on board with some dozen strange birds and beasts, including great ostriches and marabouts who made nothing of swallowing padlocks and door-keys. Major Lanyon also had a human curiosity in his charge—the son of King Coffee of Ashantee, whom our Government was to educate. He was a good-natured, nicely-mannered boy. The missionaries had already taught him how to eat with knife and fork, etc. There was nothing savage about him.

I got home on the 8th of May, and was soon in the full enjoyment of a London season among good friends, exhibitions, and concerts. On the 17th of July I went down to the most agreeable country house I know—that of Mr. Higford Burr, at Aldermaston. Some people I had never met before, Mr. and Mrs. S., asked me where I was going next, and I said vaguely, "Japan." They said, "You had better start with us, for we are going there also, on the 5th of August"; and, to their surprise, I said I would. All my friends said it was so nice that I was not going alone this time, particularly for that long Pacific voyage! What a pleasant time I had at Aldermaston! Mrs. Higford Burr was the very most charming hostess in the world, so alive and interested in every one's particular hobby, often knowing more about it than they did themselves, with that gentle, sympathetic manner which made even the dullest think they were themselves agreeable. She made every one feel at home. Naturally such a hostess was always sure of the most pleasant company in her house. Many a delightful walk I have had there under the great oaks and bracken (the latter nearly as tall as myself) with some of the best talkers in England. That time I was fool enough to slip on the polished oak floor, when running in for a cloak, and to sprain my ankle and knee. When I returned home I could scarcely move. Every one said I must "see a doctor." I was not in the habit of doing such a thing. I knew nobody in particular to see; but as my head had been full of etching lately, I thought I might as well consult the great etcher, Seymour Haden, as any one else. I went to him. He had gone away for a holiday; but his young son gave me a bandage, and told me I had had a bad twist, that was all. (I thought I could say that as well as he.) Then another friend came, and insisted on taking me off to see a famous quack in Mayfair, who came in in his shirt-sleeves, and got a big skin of wash-leather with some sticky stuff on the soft side, which he stretched cleverly all over the calf of my leg, from knee to ankle, told me to leave it three days, then pull it off, wash it with sea-water, and put another on. He gave me a bundle of skins, and told me if I wanted more, I was to ask Mrs. C. in Japan.

So on the 4th of August 1875 I went down to stay the night at Leasom Castle with Sir Edward and Lady Cust. The next day I went on board the *Sarmatian* at Liverpool, and found the S.s in the next cabin to myself; and he very kindly handed me in a cup of tea every morning when he made his own; for they carried every possible luxury, including canteen and box of books, and had made more journeys in less hours than any people living. We passed one or two hundred ice-bergs; some of them were said to be as big as the rock of Gibraltar; some of the smaller ones came too near to be agreeable. One night we had to tack about in a hurry, finally dropping anchor in the fog close to a huge cliff of ice. We had a most narrow escape, and how cold it was! The view of those great ice-islands at sunset was very striking, some in deep shade, others lit up and sparkling in the sun's pink rays. Some had bridges and arches from one to the other, while others stood up alone like giant Memnons or steeples. We also saw many whales playing near the ship, not half so gracefully as porpoises.

The lion of the ship was Lord Houghton, who was very good company. Samuel Butler, the writer of *Erewhon*, was also a passenger. He was talkative and agreeable by fits and starts, and did not believe in anything orthodox; but was inveigled into playing the hymn tunes on Sunday (Lord Houghton standing up beside him and singing most devoutly). There was also a mighty deal of heavy leaven among the passengers — men who looked like rich butchers and wool-collectors; the women with odd rings on their forefingers.

Fogs delayed us at the mouth of the St. Lawrence; we ran aground and stayed there all night till the returning tide set us free, and brought us safely up to the shore opposite Quebec. We had a cold troublesome journey through the custom-house, where my travelling companions' luggage gave them considerable occupation; the officers as usual did not even condescend to open mine. "You're a-going to paint pictures of Japan, are you? Wall! I wish you success; I should like to be going along too," the head-man said.

At Chicago we left the train for a night and lodged in a marble palace full of contrarieties — mirrors, chandeliers, whole regiments of black waiters, scaffolds, paint-pots, blue velvet sofas, and general higgledy-piggledy. We saw all the usual sights of that gorgeously-slovenly, machine-made, and inflammable city, then rolled on to an older city which interested me far more—that of the prairie-dogs. The pretty creatures were so accustomed to the trains then, that they did not even get up to see them pass, but basked in the sun on the tops of their houses by hundreds and winked at them; some of the younger and more silly ones sat up like hares and shook their paws at us. All that long prairie country was fine; there were hundreds of miles of sunflowers over it, and continual dust. After a day or two we went through a particular kind of alkaline dust which rendered one's skin like sand-paper; the natives never attempt to wash it off, and suffer less, we were told; but it was difficult either to breathe or see unless one attempted it (in the moderate way Pullman cars allowed). The accommodation of those much-vaunted carriages was still open to improvement. The ventilation at night was most ill provided for. I slept on a shelf under Marie (Mrs. S.'s Swiss maid). If I opened the scrap of window next my face, I was blown away and smothered with dust; if I shut it, I was stifled. I used to get up before the rest, and get my washing over as I best could in the airless little room at the end, with a stove almost red-hot and the guard asleep on its sofa. Except once, when we had food "on board" and a kitchen on wheels for twenty-four hours, we used to stop twice a day for a regular feed, every person having a dozen little hot dishes put round his plate in a semi-circle; and one must have been very dainty indeed if one could not find something to like amongst them. We also stopped long enough at the other stations to pick a few flowers; and the train always started again slowly, so that any stragglers could catch it up. Books, newspapers, and "goodies" were sold "on board."

At Ongar we turned aside by a branch railway and went to Salt Lake, and had the luxury of baths and real beds for three nights. It was to me a most unattractive and unpicturesque place. Mr. S. had a letter to Brigham Young, and took us to interview him; horrid old wretch! my hand felt dirty for a week after shaking hands with him.[66]

We passed along the sides of the great Salt Lake at sunset; its white edges looked really fine, backed by the purple hills, all so bare and dreary at other times. The next stage was in a horrible springless machine, holding an unlimited number inside and out, trusting to tight packing for keeping the passengers' bones unbroken by the jolting; for the roads were never mended, and all springs had long since become paralysed. Mrs. S.'s Marie was soon sea-sick. "She would be set down at the first stopping-place and go back; she would not endure it. Could I imagine any man who had ever been over such a road taking a lady there!" etc., with awful looks at poor jolted Mr. S. opposite. Mr. S. made a parliamentary speech to the honourable gentlemen outside, persuading them to change places; so they got in while the two ladies got out, leaving me to shake in peace the rest of the way. I had fourteen hours of it, combined with dust an inch thick all over everything. The next morning I got an old miner "guard" and a horse, left Clarks at six for the "Big Trees" of the Mariposa Grove, and had a long day's work among them.

The whole road was beautiful, through the biggest trees of the fir kind I ever saw, till I saw "The Trees." All the world now knows their dimensions, so I need not repeat them; but only those who have seen them know their rich red plush bark and the light green eclipse of feathery foliage above, and the giant trunks which swell enormously at the base, having no branches up to a third of their whole height. The little trees with wide base and tops made by shaving and narrowing the stem, which are to be found in every child's Noah's Ark, are exact models of the sequoia proportions. There were about seven hundred in that one grove of Mariposa alone, and three other groves within a day or two of them. They stood out grandly against the other trees, which in themselves would be worth a journey to see—sugar-pines, yellow-pines, and *arbor vitae*, hung with golden lichen. The forest was full of strange trails of big bears and other wild animals. I was told that the bear-steps were probably those of "old Joe," who had been known "just about there" for the last twenty years, and was a kind of Mrs. 'Arris to travellers. I was shown many of those funny little perforated larders the woodpeckers made for the squirrels to put their acorns in.

The descent into the Yosemite gave perhaps the very best general view of the valley; so I got our driver, after he had rested his horses and dined, to give me a lift up the hill again as far as that view, and leave me to paint it. He told Colonel and Mrs. M., who were going on with him, that "I was one of the right sort. I neither cared for bears nor yet for Ingins," and he absolutely refused to take a dollar from me when I offered it. But I had only two or three hours before dark. I could do nothing satisfactorily. The view was "very big,' but to my taste that was its chief merit. It was like a magnified Swiss valley, the gray granite cliffs looking as hard and inharmonious as Dolomites; they were shaped like them or like the Organ Mountains of Brazil, and even their great height (3000 feet of sheer precipice) was dwarfed by the enormous size of the pines on and about them. All the waterfalls were dried up, and there was dust instead of flowers. The whole was as disagreeable as nature could be at that time of year. It was most tantalising to pass acres of azalea plants, and I made a vow to return in May in some future year, and stay a while there. The next day my friends were too tired to go beyond the verandah of the hotel; so Marie and I mounted two very "sorry nags" and accompanied a large party of tourists all round the valley to the Mirror Lake (which might have been a bit of the Tyrol), then up ladders to "Snows," a kind of "Bel Alp" hotel, which must be quite divine in spring from the quantity of flowers and clear water. It was a hard day's work, and the S.s "did" the Yosemite far more comfortably, and perhaps as profitably, and decided they had had enough of it, and would go back to Clarks the next day.

The same driver drove us, a most villainous-looking bandit; but he was a real good fellow, and had taken a liking for me because "I cared for neither bears nor Ingins," and he gave me some rattlesnakes' tails and a great lump of bark from the big trees, looking like a brick of solid plush. His carriage broke down with the weight of Mrs. S.'s luggage (mere necessaries! the rest having gone on to 'Frisco and £20 to pay for extra). How the driver swore (and swearing was not of a mild sort in California), then he turned round quite gently to me: "Now don't you go for to take any of them lazy cattle of guides to 'The Trees' again; you are going a long journey, and it's the dollars you want; don't you waste them on such brutes. I'll tell Moore to give you a good old 'orse as I knows the ways of, and show you how to loose his girths, and you just stay and draw till you're tired, and tie 'im up and loosen 'im, and then tighten 'im again, and come 'ome quiet; and if you don't say nothing to nobody, nobody won't say nothing to you; you'll save your dollars, and that's what you want." So I did say nothing to nobody, because I never saw anybody to say anything to all day after the S.s went. I had a long day's work in that lovely forest painting the huge tree called the Great Grisly, whose first side branch is as big as any trunk in Europe. My old horse was very quiet; but as there was little for him to eat besides dust, I divided my luncheon with him, and came home rather hungry. Even the scraggy pines round "Clarks" were 170 feet high, and it was nice wholesome quarters to rest in and work. After that I went down to 'Frisco and became No. 794 in the Occidental Hotel.

An old Norfolk play-fellow, R. Brereton, now an engineer with an American wife and child, had made the whole journey with us off and on from England; he now looked me up, and was most kind in showing me the lions of the big new city—the Liverpool of the West. There was a local exhibition going on, full of Californian works and products. The cabinet-work was neat, also the buttons and other small things cut from the great ear-shells of the Pacific, and in the market which we visited at the fashionable hour (9 P.M.) we saw magnificent grapes, apples, pears, peaches, tomatoes, egg-plants, and all sorts of vegetables in great abundance, as well as clams, oysters, crabs, and lobsters. There was a tearing wind, and the streets were not agreeable; and as all the trees had been destroyed on the hills around, they had become scorched-up dust-heaps. The climate was most unpleasant. The city itself is a strange mixture of new Paris streets and Irish hovels, with its still stranger Chinese town in one corner, always amusing to fresh travellers from Europe. The hotel was admirably managed, with lifts to every storey, as well as grand staircases.

In the afternoon the consul called for me with Colonel and Mrs. M. (whom I had just seen in the Yosemite), and he drove us on the top of a pair of spidery wheels to Cliff House, to see the Islands of Sea-lions, or seals. Those rocky islands were some hundred yards from the balcony of the hotel, which had been built for the purpose of feeding and sheltering the cockneys of 'Frisco, who often spend a "happy day" in watching the crowds of sea-beasts through various telescopes which are fixed for the purpose. It is easy work and most enjoyable in the heat of the day, with the cool sea-breezes all round, coming across the great Pacific, with no land westward nearer than Japan. I thought if ever I had days to spare I should like to go and lodge at that hotel and draw there. The sea-lions came quite black and dripping out of the water, and climbed up the rock with a series of waddles and jerks to the very top, where they played or slept in the sun till they became dry and coloured like real lions. They kept up a perpetual roaring and happy murmuring sounds of different sorts; but on the island near them few ventured to land. It was possessed by a variety of sea-birds of different long-legged sorts, as well as gulls. The American Government protects these creatures, and no boat is allowed to go near them, or any shooting practised. A kind of park has been laid out on the road to Cliff House, and there is a huge race-course for trotting gigs on the way, which is the fashionable amusement amongst the young men. Tamarisk, euphorbias, and aloes were the chief decorations of the park.

The next day I returned and spent the day painting at Cliff House, and the day after that I started back to the "Summit Station," Colonel and Mrs. M. going with me as far as Sacramento, where there was a fair at which he hoped to see fine horses and cattle, but was disappointed. I continued in the train, which slowly climbed its 8000 feet and landed me at midnight at the top of the pass, in the midst of the Nevada Mountains, and I settled for a week in a very comfortable railway-hotel. One could go ten miles on either side under cover of one long snow-shed, east and west. The trains only went through in the middle of the night, except a few wood-trains for short distances; there was no village, so it was a most quiet locality. My other window looked over the bright rocks and trees and mountain-tops, with a few small lakes here and there, like the top of some Swiss pass. The house was still well filled with San Francisco people doing "Vileggiatura." The food was excellent, popped corn and cream being the thing for breakfast. Half-an-hour's climb took me to the highest point near, from which was a most magnificent view of the Donner Lake below, and all its surroundings. Of this I made two large sketches, taking out my luncheon, and spending the whole day on those wild beautiful hills, among the twisted old *arbor vitae*, larch, and pine trees, with the little chipmunks (squirrels) for company, often not bigger than large mice. The sunshine was magnificent; I could trace the long snow-galleries and tunnels of the railway, high along the projecting spurs of the mountains, into the far horizon. It was a most quiet enjoyable life, with few adventures beyond my cold tea being put into an unwashed Hervey-sauce bottle one morning. I took a good drink before discovering it, and did not like it, then sat down and laughed till the tears ran out of my eyes again—that air made one feel so happy.

My landlord drove a drag, four-in-hand, down to Lake Tahoo most days, and at the end of the week took me on there, driving down the steep descent to Lake Donner. We went along the whole length of its clear shore to Truckee, then followed the lovely clear river to its source in the great Lake Tahoo, a most lovely spot with noble forests fringing its sides. There was another capital wooden hotel there, where I could work again in peace. Behind the house were noble trees, fast yielding to the woodman's axe; huge logs were being dragged by enormous teams of oxen, all smothered in clouds of dust. They made fine foregrounds for the noble yellow pines and cypress-trees, with their golden lichen. The M.s picked me up there again, and after going round the lake in the little steamer we disembarked on the east side, and took a carriage with a driver who has been made famous by Mark Twain. We followed one long shoot of floating wood-logs for a mile or more, all tumbling over one another on the rushing water till one felt one must go too; it would be impossible to stand over it and watch the moving mass without throwing oneself in.

Two hours' rail at the end of this drive took us up the hills to Virginia City. The last half-hour of the ascent was through and over a continual succession of human beehives, surrounding all kinds of extraordinary machinery and gigantic mole-hills. All the hills were entirely bare of tree or verdure; nothing but salmon-coloured dust below and smoke above. Virginia City itself is just the surroundings of one big mine. There was gambling going on in every house; only one man who did not gamble was to be found there, they said—the canny Scotch manager of the mine. He showed us everything the next day, from the rough ore as it came up at the head of the mine to the great bricks and bars of pure silver taken out of the red-hot furnace by men whose work meant certain death to them, but who were never difficult to procure, from the enormous wages they got. The Colonel went down the mine while I sketched above; then we returned down the hills to Carson City, supped, and walked about the streets, looking in at the windows and watching the eager faces of the gamblers till midnight, when my friends went on east, I west, back to the Summit Hotel, which I reached at four in the morning. There were rough people in the train, but they were always good and civil to me, and gave me a couple of seats to myself. The landlord's little daughter took me the next day to see her lake, a lake that no one could find unless she showed the way, she said. She had a swing there, between two trees; and I tried to paint her, for she was a rare child, very beautiful, and not more than six years old. She knew all the birds' notes, and imitated them so well that the birds answered her, and she called up all kinds of pretty echoes for my entertainment.

After a few days I left at four in the morning, descended to Stockton, and by another line to Milton, thence by stage to Murphy, which I did not reach till nine at night in the dark. My driver was a very peculiar character. Every one called him "the Colonel," and chaffed him, but he never said a word. I asked him some questions, but I only got grunts in return. He dropped letters into all sorts of odd post-receptacles—hollow trees and baskets slung to branches in lonely places; and so we went on to our journey's end. When, after washing some of the dust off my face and hands, I came down to have some supper, behold, "the Colonel" also appeared in a white waistcoat and dress coat, the very essence of conversational politeness! The landlord and his daughter, who waited, treated him with every respect. He ordered them about as if he were a very great man, told me all sorts of interesting things about the country, and volunteered to get me a giant trap-door spider's nest before I returned.

The next morning I drove on to Calaveras Grove, found myself the last guest of the season in the comfortable hotel under the big trees, and stayed there a week. That was indeed luxury, to be able to stroll under them at sunrise and sunset without any delay or trouble. A stag with great branching horns was my only companion; he had a bell round his neck, and used generally to live in front of the house, but liked human company; and when I appeared with my painting things he would get up and conduct me gravely to my point and see me well settled at work, then scamper off, coming back every now and then to sniff at my colours. One of my first subjects was the great ghost of a tree which had had a third of its bark stripped off and set up in the Crystal Palace; the scaffolds were still hanging to its bleached sides, and it looked very odd between the living trunks of red plush on either side. The sugar-pines were almost as large, and even more beautiful than the sequoias, their cones often a foot long, and so heavy that they weighed down the ends of the branches, making the trees look like Chinese pagodas in shape. They are called sugar-pines from the white sweet gum which exudes from the bark, and drops on the ground like lumps of brown sugar; it is much eaten by the Indians. The cones of the "Big Trees" were small in proportion. About six miles from the Calaveras Grove was another with 1300 big trees in it. I rode there one day on an old cart-horse, and found that one hollow tree was used as a house by an old trapper. He was out, but his dogs strongly protested against any entrance in his absence. On a tree near were a quantity of rat and other skins. I was told that he had probably eaten the animals, and was not over-particular as to how he lived. He had been there three years, cut sticks and nick-nacks of "Big Tree" wood to sell, then, when he had made a little money, he would have a regular drinking-bout and drink it all up. My guide to those trees was an Alsatian who had left his country to avoid the Prussian conscription; he said many of his friends had run away for the same purpose to America, and never meant to go home to do soldiering for the Germans.

My time was up, and I had to go back to civilised life. At Murphy I heard a thump on my door at four o'clock, and "the Colonel's" voice shouted out the hour; and while I was swallowing my coffee downstairs, I heard his voice outside in the street: "What, you there, Jim?"—"Yes, I heard the lady was a-going down the valley; I thought I should like to come and see her off comfortable." Though this was said by the most ragged specimen of a live scarecrow I ever saw, I felt flattered by being particularised by the definite article and so bracketed with "the trees" and "the valley," the two greatest things Jim knew. The Colonel said nothing, but he took his carriage and two horses short cuts over the country, so as to avoid the road as much as possible, driving between two trees with not an inch to spare on either side, and making his horses go on each side of some tree stump only an inch lower than the floor of the carriage, then turning to me with a grunt of satisfaction. So I thought I would talk though he wouldn't, and told him how the Stockton railway-people had objected to giving me checks for my baggage because, they said, "it was too small and too heavy to hold wearing-apparel, and they only checked wearing-apparel." I asked him if he could check them straight to 'Frisco. At this he threw off silence and became excited: "Oh, they was nasty, was they? Like 'em, damn'm. You just give me your check, don't you say nothing; I'll settle 'em, I will; they was nasty, was they?—like 'em;" and he continued this at intervals till we reached Milton, when he put me into a carriage, then stalked off after the luggage, handed me in the checks to San Francisco just as the train was going, stalked off again, and I heard the same refrain fading in the far distance: "They was nasty, was they?—like 'em, d—'em."

I saw a good many Indians about the country at different times and in different places. They were the lowest of low types of humanity. The Republic allows them just money enough to drink themselves to death easily on, and they do that, and nothing else. They collected acorns in the summer, and buried them in the ground by way of storing them; then they made stones hot in the fire, dropped the acorns on them, and covered them over with earth, which took the bitter out of them and made them not bad. This was, I believe, the principal food that these poor wrecks lived on.

I had met at Lake Tahoo an old lady, Mrs. R., and her daughter, who invited me to go and see them near San Rafael, on the north of the harbour of San Francisco, which is like an inland sea, the Golden Gates at its mouth not being half a mile broad. Ferry-boats go across in all directions, some of them floating palaces, some humble market steamers. The city itself is built on a sandbank close to the sea and the Golden Gate. The little ferry for San Rafael gave me fine views of the harbour as it went along close to the north-west shore, where I could see the sea-lions playing on the rocks in abundance. Then I took the railroad and went northward, till I reached the station I had been told to make for. It was a perfectly isolated building, and I asked the guard for Mrs. R.'s. "Wall, do you mean the young or the old un? The old un? Then you just be spry, run and catch up the train again, and ask the guard to set you down at the old un's gate." Which I did, and in ten minutes more was "set down" at an avenue of deodaras, walked up to a pretty little country-house, and had a warm welcome from a very dear old Scotch lady and her daughter. Her husband had been one of the first settlers in California. She had no wish to return to her native land, and lived much the same sort of life old ladies live in the country at home.

I was very anxious to see some of the red-wood forests. They had been so destroyed that it was not easy to get to them, but the village doctor gave me a letter of introduction to the head woodman of an estate, some two or three hours up the northern line of rail, who took me to his house to sleep. It was only a small hut of logs, but they had a spare room, and made me very welcome. The wife was a capital woman. She gave us a wonderful supper of eggs, ham, cakes, apple-tart and cheese (together), and good tea. Her children were pictures of health. She came from New England, and complained of the dulness, but otherwise was well off. The red-wood trees are all about those hills, and are more like silver-fir than the other sequoias. My host took me some miles up a side valley to see some which were fifteen feet in diameter, and nearly 300 feet high. They were gradually sawing them up for firewood, and the tree would soon be extinct. Its timber is so hard that it sinks in water, and no worm can eat it there.

It is invaluable for many purposes, and it broke one's heart to think of man, the civiliser, wasting treasures in a few years to which savages and animals had done no harm for centuries.[67] I settled myself to sketch near a "bear's bath," hoping to see the big beast come and wash himself, but he didn't. I saw two pretty little deer and numbers of squirrels and birds, then walked back, and, after more apple-tart and cheese, was put on the engine of a wood train, as the passenger train had gone by some hours before. My engine was driven by a very intelligent young man, who had gone on an exploring expedition once over the Yellowstone country, and told me much about it. I had a very good time on that fire-eating beast, the engine. What a lot of wood it consumed while pulling up that steep ascent! After that we let the fire go out, and descended by our own weight alone. We stopped very often, and it was late before I got back to Mrs. R.'s. When I tried to slip a couple of dollars into the engineer's hands, he coolly opened my bag and put them inside. "Just you keep them things till you want 'em, and shake hands again to show you don't mind my saying so," he said; "the talk he had had with me had done him real good, and he didn't want pay."

On the 16th of October I took possession of a splendid, large, airy cabin in the *Oceanic*, one of the finest steamers afloat, fitted up in the most luxurious way, with an open fireplace in a corner of the great saloon, which we were very glad of after the first week, as we went by the northern route, which was too cool for pleasure. We also had a superabundance of head-winds, and did not get on as fast as our captain wished. I used occasionally to think that we had more dead Chinamen on board than was altogether agreeable to our noses, every ship being obliged to take a certain number of these strange people's bodies back to their beloved fatherland. All the waiters belonged to the same nation, and everything was well managed. Quite a cosy party gathered round the English fire-place, one lively little lady, in the tightest of dresses and highest of heeled boots, being the life and pet of every one. She made toffee and sang songs, took fits of hysterics, and was continually entertaining the party in some way or other. At the end of the voyage a huge bouquet was cut out by the cook, of turnips and carrots framed in the leaves of a large cabbage, with its stalk tied up in white frilled paper; and the captain presented it on his knees with a speech learned by heart, as a testimony of gratitude from all of us. Three weeks without seeing land at all is a long time, and latterly I suffered much from an attack of my old pain, brought on by the cold.

We jumped in one day from the 28th to the 30th of October, and at daylight on the 7th of November found ourselves within sight of Fujiyama. I watched the sun rise out of the sea and redden its top, as I have seen so well represented on so many hand-screens and tea-trays. The mountain is a much steeper cone than Teneriffe or Etna, but has about the same quantity of snow on it. The coast is beautifully varied with ins and outs, islands and rocks, the cliffs everywhere fringed with trees and higher than I expected to see them, the water of the clearest aquamarine colour. It was a real sight to see the boats which surrounded us from all sides filled with tiny men in the oddest dresses, some looking like the straw umbrellas they put over beehives, some in strange stripes and checks, some in no clothes at all, or next to none, but all good-humoured and sensible, with their funny tufts of back-hair turned over their bald crowns, like clowns in pantomimes, and all their ways of doing things so unlike the ways of the rest of the world. A boat's crew rowing or pushing, not together, but one man forward and the next back, with a jerky yet graceful movement, and curious subdued puffs and grunts which are not disagreeable or inharmonious to hear, though reminding one of small steam-engines—hupp, hupp, hupp, hupp, I can hear it still going in my head. Some of my ship friends landed with me. We drove out into the country, and took funny cups of yellow tea in a bamboo tea-house, with five pretty girls rather over four feet high, in chignons with huge pins, blackened teeth, and no eyelashes, laughing at us all the while. We saw the sunset on the white cliffs of Mississippi Bay, and all the funny little people manuring and watering their tea-gardens and cabbages.

The next morning I saw my friends off. The big ship departed, and I returned to the hotel at Yokohama—a sort of mongrel establishment, with neither the cleanliness of Japanese nor the comforts of English life. Mrs. C. soon found me out, and instead of my wanting more of my quack's clever plasters, I gave her the rest of my own supply, as my sprain had long ago recovered itself. As Sir Harry and Lady Parkes were said to be soon going away on an expedition round the coast, I started to pay my respects to them at eight in the morning.[68] The railway went alongside of the famous Tokado road much of the way to Yedo, and was always full of interest. The rice and millet harvest was then going on, and the tiny sheaves were a sight to see. They piled them up against the trees and fences in the most neat and clever way, some of the small fan-leaved palm-trees looking as if they had straw petticoats on. There was much variety in the foliage; many of the trees were turning the richest colours, deep purple maples and lemon-coloured maiden-hair trees (*Salisburia*), with trunks a yard in diameter. The small kind of Virginian creeper (*Ampelopsis*) was running up all the trees. These seemed generally dwarfed, except round the temples, which were marked all over the country by fine groves of camphor, cryptomeria, cedars, and pine-trees, as well as a small variety of bamboo. The little houses were excessively neat, and had beds of lilies growing on their roofs. Every single dwelling was a picture, exquisitely finished and ornamented, though all on such a miniature scale. Many of the town houses were built of black mud, which was fireproof, and looked like polished black marble, the shutters and doors being made to fit close with the greatest precision and security.

At the last station one of the Japanese ministers got into our carriage in the costume of a perfect English gentleman, chimney-pot hat included. He invited me to come and see his wife at his country-house, and at Yedo packed Miss C. and myself into two jinrickshas, a kind of grown-up perambulator, the outside painted all over with marvellous histories and dragons (like scenes out of the Revelation). They had men to drag them with all sorts of devices stamped on their backs, and long hanging sleeves. They went at a trot, far faster than English cabs, and answered to the hansoms of London, but were cheaper. So we trotted off to the Tombs of the Shoguns, most picturesque temples, highly coloured and gilded, half buried in noble trees, under a long low ridge or cliff. We left our cabs, and wandered about amongst them attended by a priest, a wretched mortal who would have sold even Buddha himself for a few cents if he dared run the risk of being found out. We then mounted the ridge above, and went to a famous tea-garden on the site of an old temple, with grand views over the city and sea, where we had tiny cups (without handles) full of yellow sugarless tea, ate all sorts of delicate cakes made out of rice and bean flour, finishing up with cherry-flower tea, which is made by pouring boiling water on dried blossoms and buds of the cherry-tree. The smell was delicious, the taste only fit for fairies, and very hard for big mortal tongues to discover. The tiny girls who served us were very pretty, and merry over our gigantic and clumsy ways. I felt quite Brobdingnagian in Japan.

We descended by a hundred steps, and were trotted on again and entered the Mikado's domain, round the outer wall of which ran a moat full of lotus lilies (*Nelumbium*), not what we in England call lotus, but the real Indian lily with its tulip-like pink flowers and flat, high-stemmed leaves. They were then in seed, and the seed is eaten by all the Pacific natives. Thousands of wild-fowl were swimming among the plants. I just missed seeing the Mikado by three minutes, his English brougham passing out of the gates just before we reached them, and though my biped took to galloping, we could not catch him up. He was surrounded by a company of cavalry in semi-European dress.

The English Legation was very new and very ugly, with many rare and beautiful Japanese and Chinese things in it, but the master and mistress of the house so genuine in their kindness and hospitalities, that one forgot the ugly shell. They kindly offered to take me with them in the Government steamers to inspect lighthouses all round the coast, thus giving me opportunities of seeing parts of the islands never visited by Europeans, taking a month or more to do it in. It was a great chance, but alas! that same night, on my return, I had a terrible attack of pain, so fearful that I sent for an American doctor, who injected morphia into my arm, and put me to sleep for twenty-four hours. The people in the hotel thought I was dead, and when I woke I was too weak to think of starting on any expedition for some time. My room was sunny, with a window from which I could see the river and bridge close by, continually crowded with people, who looked as if they had walked out of a fairy-tale, and a beautiful hill of trees and quaint houses on the other side. The climate in November was colder than suited me, out of doors. The camellia-trees were covered with bloom some twelve or twenty feet high, and chrysanthemums were in abundance in all the gardens. Mrs. C. sent down her "boy" with another about sixty who "speakit English a leetle," and wanted to be my "boy." His name was Tungake. As I heard no ill of him I engaged him, and whenever I looked at him he put his hands on his knees and slid them down to his ankles, and grinned. He was very useless, and anxious to make little percentages out of every bargain; but the language was so impossible to make anything of in a short time, that I could not have done without some such attendant.

After a few days' quiet, I started in the steamer for Kobé,

another of the European settlements of Japan—a pretty place on a quiet bay of the sea, with high hills behind it, and an interesting temple, at the entrance of which was a shed with a white horse in it of a peculiar breed, with blue eyes and pink nose, and hoofs turned up from want of exercise. This horse was kept in case God came down and wanted a ride. Plates of beans are put on a table near, with which pious people feed the horse as they pass in, dropping some money into a box at the same time to pay for them. A stuffed horse is kept in another shed close by, to be ready, in case the holy beast should die, to fill his place, and not disappoint the equestrian Deity. The entrance-arch of that temple was festooned with wistaria, the whole being shaded by a monstrous camphor-tree and cryptomerias. There was also a candle-tree loaded with yellow berries, with tufts of scarlet-leaved sumach grafted on it for ornament. Further on there was a winding road leading high into the hills, with some beautiful cascades and temples, and plenty of tempting little tea-houses at every beautiful point of view. Many had miniature gardens on tables in front, with dwarfed pine-trees under a foot high, perhaps fifty years old, rockwork, bridges, lakes, fountains, and rivers—everything in proportion, and the whole covering a space of not more than a yard square! The people seemed pleased to have them admired, and brought out their tiny cups of yellow tea, without seeming to expect pay for them.

Kobé was a very sociable place. Lady Parkes was not sorry to make me an excuse for escaping its heavy luncheons and dinners, and we started by rail for Osaka, where we took jinrickshas, with a tandem running in front, and trotted about ten miles to the valley of Minbo, famous for its maples. The hills were perfectly on fire with its different tints of red, crimson, scarlet, and every shade of carnation, even the different purples. We left our jinrickshas at the beginning of the valley, walked up by a winding path through the trees, with little chapels on all the most picturesque points of the road (as they have in Roman Catholic countries of Europe). Picturesque gateways, temples, and steps were hidden among these gorgeous trees, while peeps of the hills and distant plain were seen through them at every turn. Our luncheon was spread out in the priests' parlour. I spent a vast quantity of madder and carmine in trying to imitate that which could not be imitated, after which we all returned to Osaka.

Lady Parkes and her two A.D.C.s went back to endure a state dinner at Kobé, while I made my way to the inn (kept by a Frenchman), appointing to meet Sir Harry and his party at the railway at nine the next morning. I started first from end to end of that vast city in order to pay a visit to Mr. Frank Dillon, the artist, who was staying with his son at the Mint in a regular English-looking house. I had no idea of the distances, and had but little time to spend on my visit; only a most bewildering rush through the streets, my frantic biped howling and shouting all the way, and flinging off his drapery till he appeared in a complete suit of tattooing and nothing else, one great serpent winding round his right leg, round his body, and down his left arm to the hand, on the back of which its head was painted. The Osaka streets were crowded with people, and the shops were full of things I longed to examine. We passed over many bridges, and the rivers below seemed as full of life as the land was.

We started with our luggage in fifteen jinrickshas, with two men in each, one in the shafts and one running tandem in front. They trotted over thirty miles that day. As they got heated they peeled off their draperies and flung them into our carriages, leaving nothing on but a bit of rag from the waist, and a very decent allowance of tattooing all over. They never got in the least tired, but did the last part of the way up the High Street of Kioto at a gallop after nearly seven hours of hard running. The road was generally very narrow; the bridges, placed at right angles to it, rather steep up and down and without parapets, were very disturbing to one's nerves, as the men never broke their pace, but swung one on and across and round again in one even jog. First went the ubiquitous landlord of the Kioto hotel, then Lady P., then myself, Sir Harry, young L., Mr. G., and Mr. A. (the great Japanese scholar), then the Chinese valet and Tungake, and six more jinrickshas of luggage. Wild yells were given at every sharp turn and bad bit of road or steep bridge by No. 1, and echoed by the rest of the men. One of them, with highly-illuminated legs in the rarest mediæval style, ran backwards and forwards like a dog, keeping the line straight. We passed through the richest cultivation—rice, tea, buckwheat, cotton, mulberries, bamboos, camellias twenty feet high, full of single pink and white blossoms. Oranges, persimmons, and Japan medlars seemed the common fruits. The little houses were models of neatness, with their bamboo frames, paper windows, and little stacks of rice straw piled round them as extra padding to keep out cold. Everything was arranged daintily and prettily. Mats were spread on the ground in front of the houses, with rice drying on them; and every now and then a group of noble trees showed us where some temple was hidden. The place where we stopped for our half-hour of rest was on a bend of the great river, and the rooms of the tea-house were supported on poles over the water, so that we could watch the loaded barges going up and down it.

Before dusk we were in the long suburbs of the old capital of the Tycoon. Our men went faster and faster, till they nearly galloped us up the long High Street and steep ascent to the hotel; and soon after the Governor of Kioto (in corduroys and shooting-jacket, and about four and a half feet high) appeared to pay his respects to Sir Harry and beg us all to go and dine with him. His Excellency begged to be excused, but promised to have luncheon with him the next day at a tea-house the other side of the valley, for the Governor was starting on an official journey, and that would be on his way. His official interpreter also came to pay his respects—a gentleman who spoke good English and was our guide during the days the Ambassador stayed. He promised to be my "Protector and Pass" when Sir Harry left, which nearly set me laughing most uncivilly, the little man being quite Liliputian, and much embarrassed with a European beard and moustache. It was also "the thing" not only to speak indistinctly but to put the hand before your mouth while doing it. The bows of ceremony were endless. Whenever Sir Harry looked at a Japanese he bent double, with the hands sliding down from knees to ankles, like machines. We worked hard next day, and saw many wonderful temples and palaces all of wood, with a beautiful concave curve in their overhanging roofs. Inside they were richly gilded and painted, with pine-trees, storks, flowers, and people, on a gold ground. The temples were full of exquisite bronzes, china, and fresh flowers.

We drove out to the Governor's luncheon party at the tea-house, which had one side of the room quite open towards a pretty garden and a clear view. On the table were vases of chrysanthemums, tied on all the way up sticks a yard high, so as to show all the flowers and hide the stalks. The ornaments were of rare old Satsuma porcelain; the food which came from our hotel, being of the knife-and-fork order, not interesting. After luncheon we took leave of the Governor and pulled up the river, getting out where the valley narrowed to walk along its banks. I saw the leaves of *Primula sinensis* and ferns, but there were few flowers at that season. We also saw many lovely kingfishers.

We all went in a string of jinrickshas to the lake of Biwa, going through a long street full of china-shops, where the modern cream-coloured porcelain was exhibited which is sold in Europe as Satsuma. The paintings of birds, insects, and flowers on it are exquisite, though it is extremely cheap. We met fishermen trotting along with great bundles of fish slung on the ends of bamboos over their shoulders, and fruit-carriers with brightly-polished orange persimmons, making the real oranges near them look quite dull. We climbed up a steep ascent and through a mountain gorge to visit several groups of fine temples in magnificent groves of cryptomerias, one of them being built like a Chinese pagoda. All had delicious views of the great blue lake of Biwa, with the town of Otsu spread out like a map on its shores. Close to the water's edge was a huge pine-tree, its branches trained from childhood so much in the way it should go, that the top of it resembled a bed of well-clipped turf. This tree shaded a quarter of an acre of ground.

We went on along the edge of the lake to its southern corner, where the river Yodo flows out, crossed by two really magnificent bridges divided by an island, over which the great Tokaido road is taken, turned up the hill-side between strange gates and figures, then up magnificent stone steps to the great temple of Isbyama, one of the most holy in Japan. The buildings, though less ornamented, were more elegant than most of the temples I had seen; they were supposed to be of unknown age. Prosy, realistic travellers put them down as having been built in the twelfth century, which is no great age after all. The rocks were piled about and planted, as were the flowers and shrubs, with infinite taste and care, the views everywhere most lovely. Below the temples were the houses for pilgrims. The next day Sir Harry and Lady Parkes and their suite departed, leaving me in sole possession, with a special order from the Mikado to sketch for three months as much as I liked in Kioto, provided I did not scribble on the public monuments or try to convert the people; for it was still a closed place to Europeans.

Sir Harry himself had been nearly murdered on his last visit there, and Sir Rutherford Alcock was never even allowed to enter.[69] But I was perfectly safe all alone, and comfortable too, in the old temple building some centuries old, which had been turned into an hotel for Europeans, with the addition of a few chairs and tables. It was kept by a Japanese named

Julei, who was always dressing himself up in native or European costumes, the latter being of a monstrous plaid pattern, with a prodigious watch-chain and breast-pin. He spoke a little English and kept a French cook. My room was made of paper, with sliding-panels all round, two sides opening to the frosty air and balcony, the other two only going up about seven feet, leaving abundant ventilation between them and the one great roof of the whole house, with the advantage of hearing all my neighbours' conversation beyond. I had a pan of lighted charcoal on a chair to warm me, and two quilted cotton counterpanes on the bare floor to sleep between. When I complained of cold, they brought me in gorgeous folding-screens, and made quite a labyrinth around me, all painted with storks, cherry-blossom, bamboo, and all sorts of lovely things, to keep the draughts out. The worst of my quarters was that I could not see through the paper windows to paint without opening them and letting in the half-frozen air or damp rain; but I much preferred my quiet life in Kioto among the purely Japanese people and picturesque buildings, to that in one of the European settlements.

One of the screens in my room was especially beautiful. It had a gold ground with red and white pinks, and pink and white acacia painted in the most lovely curves on it, as well as two kingfishers and a stork. There was also most fascinating crockery. One large creamy and crackled vase of modern Satsuma had beetles, grasshoppers, mantis, and moths carrying flowers, drawing a coach, holding mushrooms as umbrellas, etc., as well as lovely borderings of flowers and leaves. In that vase were chrysanthemums of different colours, tied on an invisible bamboo stick, so that the bouquet was a yard and a half high. From my windows, when I pushed back the paper sliding shutter, I saw a most exquisite view (for the house was perched up high on the side of the hill, with the most lovely groves and temples all over it), and below the great city of over 200,000 inhabitants. Nearly all the houses were one-storeyed, and great high temple-roofs rose among them like Gulliver amongst the Liliputians. Beyond the city were beautiful purple hills with tops sprinkled with fresh snow. A most eccentric garden was in the immediate foreground and all round the house. The top of one of the favourite trained pine-trees came up like a terrace of flat turf to the level of the balcony; it looked so solid that I could almost have walked over it. Groups of gray boulders, and small clipped azaleas, heaths, and camellias, with many other flowers and small tufts of pampas-grass and bamboo, filled the rest of the garden, varied by little miniature lakes and canals. The drawback to me was the cold, which was intense at night. The charcoal pans (most classically shaped) were a poor substitute for fires, but the ventilation and draughts of the rooms were so great that one was in no risk of suffocation from the fumes.

There were six Europeans in Kioto including myself—a German engineer and his sub, a clever Prussian doctor, and a lady who was paid by the Mikado to teach forty Japanese girls Lindley Murray and the English language (!), with her husband, who was a sporting character.

The great temples of Nishihongwangi belonged to a set of reformed Buddhists. One of them, who called himself the "Canon of the Cathedral," had been two years in England, and spoke our language remarkably well. He was really a most charming person, and gave me much interesting talk about his religion while I sketched. He called it the Protestantism of Japan, and it seemed as pure and simple as a religion could be. He said he believed in an invisible and powerful God, the Giver of good, but in nothing else—not even the sun or the moon, they were both made by that same God. His priests (including himself) married, and drank wine. He had been to hear all sects preach in England, and thought the Unitarian most like his own. There were no sort of idols in his temple. He introduced me to many old priests in gorgeous robes, who did not look as full of brains as he did. He had a table brought out beside me with tea and cake, and a pan of charcoal to warm my hands over, and took the greatest interest in my work. He gave me two of the usual conventional drawings of a woman and a piece of bamboo, done by his little daughter. The mile of street which led me home was one succession of fascinating shops. I never passed through any of those streets without picking up some beautiful little "curios." The hill behind the hotel was covered with temples, tombs, and bells, some of them very large. The great bell of the Cheone was eighteen feet in height, and made a fine subject, surrounded as it was by trees dressed in their autumn colours. Whenever any one felt devout, he used to go and strike one of those bells, by means of a kind of weighted battering-ram fastened to the scaffold which supported the bell (for they are generally hung in buildings by themselves). Through all the dark hours of the night these devout fits seemed to seize people, and did not improve the sleep of others on that hill of temples.

Horses were rare sights in Kioto. I saw only one man riding. His horse had its tail in a blue bag, tied up with red tassels, its mane tied with the same colour. He went on at an ambling jog. The post-boxes were in the corners of the streets, and stamps were put on the letters in the usual European way; but they were carried afterwards in square boxes hung on the two ends of bamboos, and balanced on men's shoulders. These men ran day and night over their appointed number of miles, finding relays waiting to shoulder the bamboo and continue running without losing a single moment, or breaking the sing-song chant which it is the exclusive privilege of postmen to sing. On our journey up, our jinricksha men had begun singing it, and had been stopped by some officials, who said it was against the laws for any one but the postman to sing that song.

The Japanese are like little children, so merry and full of pretty ways, and very quick at taking in fresh ideas; but they don't think or reason much, and have scarcely any natural affection towards one another. Everybody who has lived long among them seems to get disgusted with their falseness and superficiality. One never sees a mother kiss or caress her baby. The poor little thing is tied on to the back of a small sister in the morning in a well-padded bundle, and tumbles about with her all day, roaring piteously. People only laugh if one pities it.

As I sat at work, plenty of people came to look at me; but they never got in my way or between me and the place I was sketching. They always seemed to understand what I was about. The women were very merry, but pretty and lady-like in their ways. The young men, with their attempts at European clothing and manners, were comical with their great-coats and wideawakes over petticoats and pattens. Like Tungake, when they spoke to people they considered inferior, they used a most guttural tone of voice, and I used to fancy strangulation must ensue after much of it. His delight was to get the "boy" of the hotel to carry his lantern or go errands for him; he never did anything himself if he could help it. Then he would put on his coarse, patronising tone, and spin yarns to the "boy," who made bows and laughed at his jokes, and jerked out "heh" continually.

The Doctor and Mrs. W. came for me one evening in the full moonlight, and took me in a third jinricksha to the Kyrinitza Temple, upon a height in a lovely nook of the hills, backed by fine trees. All the quaint gates and porticoes stood out grandly, as well as the city below, almost as clearly seen as in the daytime. And the effect of the white light among the crowded pillars of the temple ought to have been very fine, only the moon did not look that way at that hour—a fact the Doctor was much vexed at having forgotten. He proposed to wait; but we shivered, and begged to be excused. There is something magnificent in the simplicity of those wooden temples, entirely without paint outside, the great round pillars showing all the colour and grain of the wood. Inside they are gilt, full of rich things and colour, with quite a Byzantine look. I made a study of the great Cheone Temple in its almost too dark interior. The priests delighted in watching me, and were most eager over my progress. Their ways were funny. At twelve every day they carried little boxes of lacquer, containing cups of hot tea and rice, to the different altars, one of them beating a gong or other musical (?) instrument. Tungake said, "Him tell God tiffin ready." They hardly left the food long enough to cool, but took it away and ate it themselves. The Shinta temples were red, and full of all sorts of idols. There are many different degrees of Buddhism, the highest of which appears a very reasonable religion. After our moonlight expedition that night, we returned to dine with the Doctor, whose little dinners would have been thought extra nice in Paris or London. Mrs. W. lived quite in the country. After sketching all day amongst the dead leaves, and morning white frosts, I used to be scarcely able to stand from stiffness and coming rheumatism, and had to hold on by a tree at first, till I could use my feet. Then I often tramped off through the temple gardens and fields, to pay her a visit, and restore circulation if I could. She also would keep me to dinner, and send me home in her own jinricksha.

The upper classes seem to have melted away in Japan since the new state of things there, though it is supposed that some of the Princes of the Tycoon still sulk in the country amongst the hills.[70] One day Mr. W. sent me his own jinricksha man and two others, to run double tandem, and they hauled me up to the top of the highest mountain in the neighbourhood, from whence I had magnificent views all over the lake of Biwa as well as the city. It was white with snow and frost at the top, and too cold to stay long. At the foot of the hill were many pilgrimage-temples, with those curious gates, their tops like inverted bows, whose origin I had never succeeded in making out. Some said they were copied from the top of a tent, and the droop of its canvas between two poles. Some said they were intended for birds to rest on. They were never wider than one stone, which often rested simply on two other upright ones, but it was always hollowed out in the same curve. We passed through hundreds of stone lanterns also, a peculiarity of Japanese temples. We saw people washing vegetables in the streams, with a small wooden tub for each foot to stand in, in the water, having loops in them for the great toe to go into. Thus they could patter about without wetting their feet in the shallow streams. They usually walked on pattens, and had many falls, which were highly amusing to all the spectators, who went into roars of laughter at their misfortunes. That very day my three bipeds got into such a state of delight when the steep hill-road was left behind them that they started into the city at full gallop, tearing round the corners and yelling like wild things, and finally fell down like a pack of cards, upsetting me at a street corner. I heard my skull go crack against the wall of the house. When at last I picked myself up again, and put up my hand to feel if my head was still there, I saw a crowd round me, holding their sides and roaring with laughter. All my three men were more or less bruised, but grinned also, so I followed fashion and did as they did; but no one attempted to help us in any way. They looked on the whole, apparently, as a little scene got up for their amusement.

Kioto was a terrible place for emptying purses. While Lady Parkes was there we had a perfect bazaar every night of wonderful embroideries, china, bronzes, and enamels, the latter being expensive but very lovely, with porcelain linings. I went to see them made. First a fine scroll-work of wire was stuck on, then the crevices were filled with clay, after which the whole was baked, coloured, baked again, filed down, and polished—a slow process, taking many hands to do the different details.

I went to a place where they sell live pets, and saw the

most beautiful gold and silver pheasants, mandarin ducks, monkeys, and gazelles, and hideous brown salamanders from Lake Biwa, two feet long; also tortoises in a tank. The tortoise with a green tail Japanese are so fond of embroidering, is merely one with green algæ growing on its shell. The Doctor had one in an aquarium, besides gold fish with ruffs round their necks and fringes on their tails. He showed me also a leaf from a tree on which one could scratch some writing with a pin, which became black like ink, and would last so for years. Japan was most attractive. There was always something new and interesting to meet me every day. I had hoped to stay over the winter, and to go to the hills and Nikko in the summer, but I got stiffer and stiffer, and at last could scarcely crawl; so on the 19th of December I ordered a boat to Osaka, and set myself to pack as well as I could, with a fool to help me, crippled hands, and bones full of pain.

We started in jinrickshas, at 8 P.M., for a two hours' rattle through the suburbs to the river-side. In the boat-house the men were roasting their bare legs and dripping garments over the great pots of burning charcoal, and the tiny neat little women were offering them thimblefuls of the hot stuff they called tea, while cooking was going on in another corner, and a bright glaring light was on the people's faces. I shivered and ached, and could barely crawl out to the "house-boat" prepared for me, and in through one of its windows on to the heap of quilted coverlets, on which I passed the night; with abundant ventilation, but not more than in my late quarters at Kioto. The sailors ran round and round over my head, pushing, pulling, and shouting, till morning, when they landed me near the Osaka railway-station, in which I found a good fire and boiling water for my tea, to say nothing of a comfortable dressing-room (European comforts are nice sometimes). I took the first train to Kobé, and good Mr. B. opened the door to me himself, telling me his wife had said I would come that morning, when she saw so much snow on the hills. They were kind, and gave me extra warm clothing, packing me off in the steamer the next day for Yokohama, where the C.s again received me.

I was in the doctor's hands for ten days with rheumatic fever. I could not even feed myself during part of the time. I sent off Tungake, and hired a small nurse of about four feet high, who tyrannised over me like a genuine Gamp, perpetually running in and out at night with a horrid lantern, whose tallow candle she used to blow out close under my nose and leave to smoulder.[71] Then she curled herself up in the hearthrug, putting a wooden pillow under the angle of her jawbone, so as not to disturb her beautifully arranged hair and chignon (which was only dressed once or twice in the week). She had no idea of keeping up a fire, and used to pour water on the coals to make them last, she said, and I suspect she intercepted and carried off a good deal of the food my kind hostess ordered for me, till I was half starved on one roasted lark. I lived in the house in the garden, and out of reach of the family care. Mrs. C. was extremely kind, went on board the steamer with me, and secured me a good cabin to myself.

The Messagerie boats are certainly the very best in the world. That one was so beautifully warmed and sweet, that it seemed like a change of climate when I entered it. I got better every day, all was so clean and the cooking so good. I sat next the kind old captain; and his devotion to a small puppy "bull-dog-mastiff" was most amusing. For the next day or two we were always within sight of some island or another. One of them was very striking, with smoking sulphur springs on its side, and a great natural arch in one of its buttresses over the sea. We had beautiful calm weather, and entered the harbour of Hong Kong about eight in the evening, when its semicircle of lights were bright as the stars above. The number of fine ships and odd junks (looking like ill-tied bundles of bamboo) was very striking.

In the next morning's sunrise, the colours of the rocky hills which surround the bay or inland sea were perfectly marvellous—pink, rosy red, and salmon-colour, with scarcely any vegetation except just round and above the town itself. We all moved our things out of the nice little ship we had come in to a larger one of the same company on the other side of the harbour, and passed on our way the steam-launch Commodore Parish had sent for me; but it soon followed and took me on board the old Hospital Ship, the *Victor Emmanuel*, which had brought my Cousin Dudley and so many other wounded from Ashantee, and was now a man-of-war again, and anchored off Hong Kong. The Commodore had turned its great saloon into a perfect museum of Chinese and Japanese curiosities, and he now established me and my lame feet on a comfortable sofa. He then took me on shore, having his own chairs and bearers waiting at the landing-place, with the very longest of bamboos to sling the chairs on, hoisted on their shoulders. There must have been at least twenty feet between the bearers, and they went, like the Japanese, at a continual trot, but were huge men, very different in their whole ways and appearance. It is difficult to believe that they can both be descendants of one race. The Chinese have far more originality and power of thought, as well as bodily strength and endurance. We did not do more than we could help, but saw enough to give me an idea of how pretty those hills might become in a few years by irrigation and good management. After luncheon we went and had tea at Government House, and Miss K. gave me a drive in her pony-carriage along the shore, with a Sikh outrider, of whom, she told me, the Chinamen had the greatest horror, as they said the Sikhs were not men but devils. The Chinaman's tail is a great help to the police; it gives them such a handle to catch him by. They often tied several men by their tails and drove them on in front of them. The clever thieves had learned two dodges to escape this — one was to have false tails which came off in their hands; the other to plait them full of needles and pins, which did their captors "grievous bodily injury."

I slept at Government House in a real good bed for once, with a roaring English fire close to it—this was no small treat—waking next morning to look out on that wonderfully-coloured circle of mountains, and the blue bay with a foreground of exquisite garden shrubs and flowers. If I had only been well I could have stayed on there, and gone the next week in the Commodore's steam-launch up to Canton, one of the wonders of the world; but it was wiser to get nearer the equator, and four days more took me into heat enough at the French settlement of Saigon, and the mouths of the river which leads to that wonderful old forest full of ruined palaces and temples, Cambodia, about which so little is known.

Two more days brought us to Singapore, where I landed on the 19th of January 1876. I could barely hobble from the office of the hotel to my rooms at the other end of the building, through its lovely garden; but how delicious that still warm air was, with exquisite blue sky, lilac shadows, and white lights! The figures which squatted under the verandah and portico had a grace about them which I had never seen before, and their rich dark complexions were the real thing, and not white turned brown or yellow by fading or scorching. Their turbans, sashes, and draperies of pure colour, and the sprinkling of gold and silver, were in such perfect harmony with their skins. Many of them were simple bundles of white calico. It was such a pleasure to look at these figures that I could not even scold when they persecuted me to buy a hundred things I did not want. I found a lemon-tree close to my room, covered with tailor-ants which had sewn up the leaves into most ingenious nests, the pretty flowers opening their sweet petals close to them.

One of my windows was quite blocked up by a great bread-fruit tree covered with fruit as big as melons, with leaves two feet in length, gloriously glossy, and I set myself at once to make a study of it. While at this work Mrs. S., the banker's wife, and her father, Major MacN., came to see me. The former insisted on my moving at once to her comfortable house outside the town. Like all the houses of Singapore, it stood on its own little hill, none of these hills being more than two hundred feet above the sea; but they were just high enough to catch the sea-breezes at night, and one could sleep with perfect comfort, though only three degrees from the equator. This house had belonged to the House of Guthrie for two generations, and was surrounded by every sort of fruit-tree. Of these there were perhaps more in Singapore than in all the rest of the world. The lovely Mangosteen was just becoming ripe, and the great Durian, which I soon learnt to like, under the teaching of the pretty little English children, who called it "Darling Durian."[72]

No garden could have been more delighted in than that one was by me. Every day I was sure to find some new fruit or gorgeous flower to paint, Mrs. S. working beside me all the hot day through in her deliciously airy upper rooms. Then we drove out among the neighbouring gardens in the late afternoon and evening, and went to bed soon after dinner; for Mr. S. and the other gentlemen came home far too tired after their days in the hot bank in the town for playing at company or sitting up at night. They had a delightful little monkey called Jacko, who considered that all dogs were made to be teased; and it was strange to see how the big creatures submitted and humoured it. This monkey was a bad sitter, and seemed to have a malicious pleasure in throwing itself upside-down whenever I looked at it. Then if I scolded, it held out its paw to shake and be friends. There was also an ourang-outang who used to be led in by the hand by the Malay butler. They were exactly alike. Both had the most depressed expression, which the small one never varied. The big one grinned sometimes,—when he looked even more like a monkey than he was before.

The Botanical Garden at Singapore was beautiful. Behind it was a jungle of real untouched forest, which added much to its charm. In the jungle I found real pitcher-plants (*Nepenthes*) winding themselves amongst the tropical bracken. It was the first time I had seen them growing wild, and I screamed with delight. One day we drove out to have luncheon with the Doctor and his family, who had a country house about five miles off, near the coast, in the midst of plantations of cocoa-nuts. The Doctor showed me all the process of crushing and clearing the oil—not a particularly agreeable one. But the pictures of people at work, the glorious trees, and certain plants under them, were very interesting. One wild plant I saw there for the first time, the *Wormia excelsa*, which abounds in the different islands of "Malaysia," and is often planted as a hedge. It has a glossy five-petalled flower of the brightest yellow, and as large as a single camellia, with large leaves like those of the chestnut, also glossy, and separate seed-carpels which, when the scarlet seeds are ripe, open wide and afford a most gorgeous contrast of colour with its waxy green and scarlet buds. I know few handsomer plants. All the tribe of Jamboa fruits (magnified myrtles), too, were magnificent in their colours. There were said to be three hundred varieties of them. Some of them had lovely rose-coloured and pink young leaves and shoots. The nyum-nyum was another curious fruit coming out from the trunk and branches like the blimbing, with tiny red flowers and pink young leaves which looked like blossoms in the distance. That cocoa-nut plantation was a most enjoyable place. A narrow path led in five minutes through its shady groves to a quiet sandy bay, where one might bathe all day long without fear of interruption.

I was taken to see a neighbouring lady one day, who told me that a few years ago she was sitting in her central room at work when the roof suddenly came down on her. The white ants had eaten the wood of the supporting beams all round the nails, till the whole gave way without warning and crushed the house; but she and her children were all dug out unhurt, having been under the gable or cone of the roof, which kept its shape. Houses at Singapore are generally built with a high central cone and a windowless room under it lighted from the other rooms, with no ceiling between it and the high-peaked roof, under which are more lights and ventilation. Those rooms were always too dark for painting in, and too airless for my taste. I prefer hot air to none, and could never get into the tropical habit of sleeping while the precious daylight lasted.

Mrs. S.'s verandah was full of rare plants; orchids, caladiums, and other exquisite things. On one, *Ficus Benjamina*, were planted some score of phalœnopsis in full flower, like strings of white butterflies hovering in the air with every breath of wind. One day we went to have tea with Mr. Wampoa, the famous Chinaman, whose hospitality and cordiality to the English have been so well known for half a century in the Straits.[73] He showed us all his curiosities; but his garden was to me the great attraction, rare orchids hanging to every tree, and the great *Victoria regia* in full bloom in his ponds, as well as the pink and white lotus, and blue and red nymphæas. Many of his plants were cut into absurd imitations of human figures and animals, to me highly objectionable, but amusing to the children. He had also several live creatures and birds of great beauty. He showed us a tortoise from Siam with six legs. The hinder ones it only used when trying to get up a very steep bank or steps, propping itself up with them, while it struggled on with the four front ones. The Siamese cat was a remarkable little creature, coloured like a fox, and might possibly have been a mixture of the two animals originally. It had sky-blue eyes, and was very shy. Mrs. S. had one which followed her about and slept in her room.

The Maharajah of Johore, a near neighbour and great friend of her husband's, came to dine and play at billiards with him in a black velvet coat with diamond buttons, worn over the usual Malay petticoat or Sarong. He wore a rich turban on his head, and spoke good English. After a fortnight I went to stay at Government House with Sir William and Lady Jervois. It was a huge building with fine halls and reception-rooms, but very little bedroom accommodation. It stood on the highest hill of the district, and overlooked all the town of Singapore, its bay and islands, and miles of the richest country covered with woods and cocoa-nuts. Close under my window was a great india-rubber tree with large shiny leaves and fantastic hanging roots. In the front of the garden was a gorgeous tree of *Poinciana regia* blazing with scarlet blooms. I immediately begged a branch and hung it up to paint, but made a most absurd mistake. I placed it the wrong way up. It was stupid, but I was consoled afterwards when I found that that clever Dutch lady, Madame van Nooten, had actually published a painting of the poinciana growing in the same topsy-turvy way! Nothing approaches this tree for gorgeousness; the peculiar tender green of the acacia-like leaves enhances the brilliancy of its vermilion tints. The amherstia was also in great beauty in the same royal garden, with scarlet pods and delicate rosy-lilac young leaves. The beaumontia creeper was there too, with its white waxy bells and beautifully embossed leaves. It was curious to see how little the English people cared for these glories around them. Lawn-tennis and croquet were reigning supreme in Singapore, and little else was thought of after business was over. Lady Jervois and her daughters were exceedingly kind to me, and played Mozart deliciously of an evening, while Sir William was a most genial host.

# CHAPTER VII

## BORNEO AND JAVA

1876

AFTER a fortnight at Government House, Sir William wrote me letters to the Rajah and Rani of Sarawak, and I went on board the little steamer which goes there every week from Singapore.[74] After a couple of pleasant days with good old Captain Kirk, we steamed up the broad river to Kuching, the capital, for some four hours through low country, with nipa, areca, and cocoa-nut palms, as well as mangroves and other swampy plants bordering the water's edge. At the mouth of the river are some high rocks and apparent mountain-tops isolated above the jungle level, covered entirely by forests of large trees. The last mile of the river has higher banks. A large population lives in wooden houses raised on stilts, almost hidden in trees of the most luxuriant and exquisite forms of foliage. The water was alive with boats, and so deep in its mid-channel that a man-of-war could anchor close to the house of the Rajah even at low tide, which rose and fell thirty feet at that part. On the left bank of the river was the long street of Chinese houses with the Malay huts behind, which formed the town of Kuching, many of whose houses are ornamented richly on the outside with curious devices made in porcelain and tiles. On the right bank a flight of steps led up to the terrace and lovely garden in which the palace of the Rajah had been placed (the original hero, Sir James Brooke, had lived in what was now the cowhouse).[75] I sent in my letter,

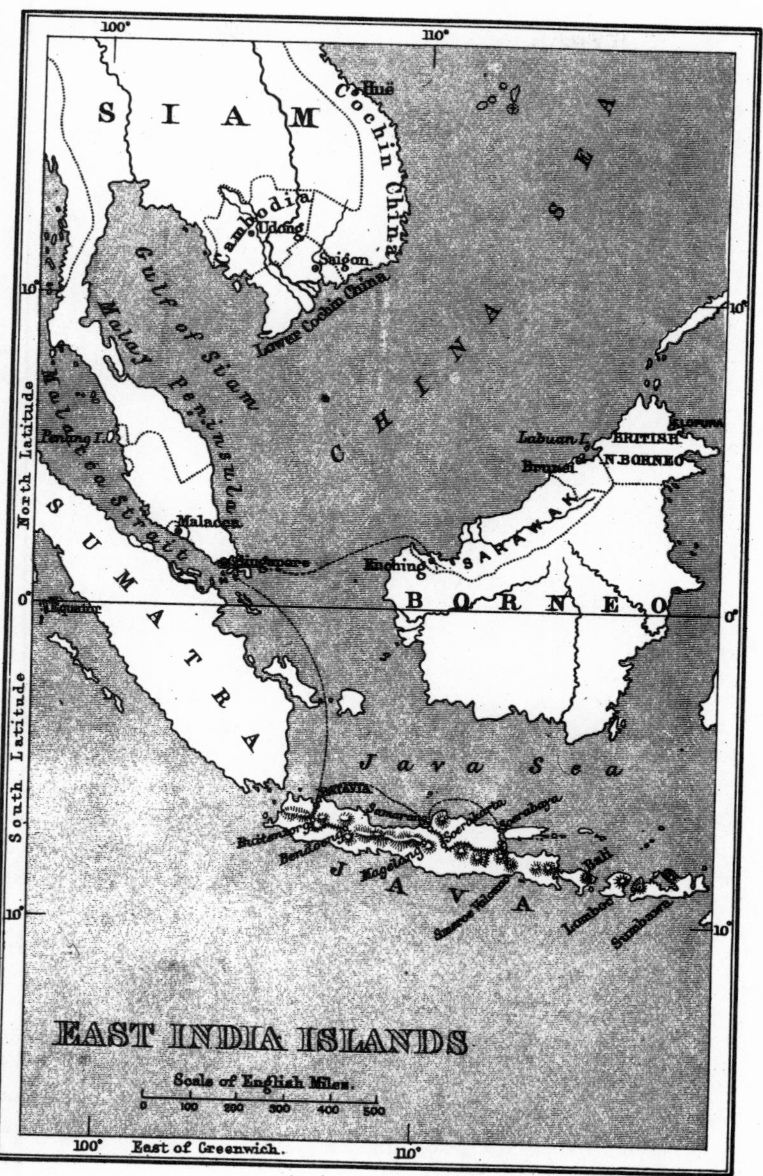


London: Macmillan & Co.

and the Secretary soon came on board and fetched me on shore, where I was most kindly welcomed by the Rani, a very handsome English lady, and put in a most luxurious room, from which I could escape by a back staircase into the lovely garden whenever I felt in the humour or wanted flowers.

The Rajah, who had gone up one of the rivers in his gun-boat yacht, did not come back for ten days, and his wife was not sorry to have the rare chance of a countrywoman to talk to.[76] She had lost three fine children on a homeward voyage from drinking a tin of poisoned milk, but one small tyrant of eighteen months remained, who was amusing to watch at his games, and in his despotism over a small Chinese boy in a pig-tail, and his pretty little Malay ayah. The Rajah was a shy quiet man, with much determination of character. He was entirely respected by all sorts of people, and his word (when it did come) was law, always just and well chosen. A fine mastiff dog he had been very fond of, bit a Malay one day. The man being a Muhammadan, thought it an unclean animal, so the Rajah had it tried and shot on the public place by soldiers with as much ceremony as if it had been a political conspirator, and never kept any more dogs. He did not wish to hurt his people's prejudices, he said, for the mere selfish pleasure of possessing a pet.

He had one hundred soldiers, a band which played every night when we dined (on the other side of the river), and about twenty young men from Cornwall and Devonshire called "The Officers," who bore different grand titles,—H. Highness, Treasurer, Postmaster-General, etc.,—and who used to come up every Tuesday to play at croquet before the house. Some of them lived far away at different out-stations on the various rivers, and had terribly lonely lives, seldom seeing any civil-ised person to speak to, but settling disputes among strange tribes of Dyaks, Chinese, and Malay settlers.

The Rajah coined copper coins, and printed postage-stamps with his portrait on them. The house was most comfortable, full of books, newspapers, and every European luxury. The views from the verandah and lovely gardens, of the broad river, distant isolated mountains, and glorious vegetation, quite dazzled me with their magnificence. What was I to paint first? But my kind hostess made me feel I need not hurry, and that it was truly a comfort and pleasure to her to have me there; so I did not hurry, and soon lost every scrap of Japanese rheumatism, the last ache being in the thumb which held my palette—it is usually the limb that does most work which suffers from that disease. Every one collected for me as usual. Orchids and pitcher-plants were pulled for me most ruthlessly, the latter being of several varieties, from the tiny little plants which grew in the meadow near, and whose pitchers were not half the size of thimbles, to trailing plants of six or eight feet long. The common pepper-plant, too, was much cultivated and very elegant, as well as gambier and other dyes, sago, and gutta-percha, the former growing thirty feet high, with grand terminal bunches of flowers from the centre of the crown (very unlike the small cycads people had called the sago palm in other countries). It takes fifteen years before it flowers; then, before the fruit has time to ripen, the whole tree is cut down and the pith taken out and washed. Wallace says one tree could supply a man with food for a whole year.[77] The gutta-percha trees were fast disappearing. They ought to have been protected by law, and the people compelled to bleed them as in other countries, not to sacrifice the great trees for one crop—trees which had been a hundred years growing, and could not be quickly replaced.

Nearly every evening I used to go for a row up and down the river with the Rani. It was quite alive with canoes and other picturesque boats, from good-sized merchant vessels to mere hollowed logs of wood, so small that the paddlers seemed to sit on the water, and might easily be snapped up by alligators; but they did not often come so high up the river. When they did there was an immediate crusade; traps were baited with monkeys or cats, and the beast was caught. The Rajah gave a large reward for one, and a still larger sum if, after a *post-mortem* examination, the brute was proved to be a man-eater. It was always buried under one of the garden trees, to the great improvement and delight of the latter.

The little town was full of life and civilisation, the bazaars and houses gay with colour, porcelain panels with raised flowers and griffins being let into the walls. At night the lights got so magnified in reflection that one could fancy oneself almost at Cologne or Mayence. Above and below for miles the semi-amphibious Malays had built their basket-like dwellings on stakes in the mud or on the banks above—thatched, walled, and floored with the leaf-stalks of the nipa palm, which delights in growing in brackish water, being almost drowned at high tide and almost dry at low. The Malays get wine, salt, and sugar from its juice, and oil from the nuts, which are contained in a cone as big as a cannon-ball. The sunsets were superb on the river. When the tide was very high we used to go up some of the small side-streams, and push our way under arches of green tangle, which broke off bits of our boat's roofs, as well as the rotten branches over our heads. We watched troops of monkeys gambolling in the trees, chattering and disputing with one another as to who we were, and what we came for. One day we were overtaken by darkness in one of these expeditions, and made a short cut home overland, with a native to guide us by an almost invisible path through the bush, very suggestive of snakes, but we saw none. The wild jungle came close up to the garden on three sides, and none but native eyes could discover paths beyond or through it.

There were acres of pine-apples, many of them having the most exquisite pink and salmon tints, and deep blue flowers. These grew like weeds. They were merely thinned out, and the ground was never manured. They had been growing on that same patch of ground for nine years. They were wonderfully good to eat. We used to cut the top off with a knife and scoop out the fruit with a spoon, the truest way of enjoying them. The mangosteen, custard-apple, and granadilla were also in abundance. The mangosteen was one of the curious trees people told me never had a flower. But I watched and hunted day by day till I found one, afterwards seeing whole trees full of blossoms, with rich crimson bracts and yellow petals, quite as pretty as the lovely fruit. This last is purple, and about the size of an orange, with a pink skin inside, divided into segments, six or more, which look like lumps of snow, melting in the mouth like it with a grape-like sweetness. The duca was a still finer fruit of the same order, growing in bunches, with an outer skin or shell like wash-leather, and a peculiar nutty flavour in addition to its juiciness. The custard-apple was well named, for it is a union of both words. Its outside is embossed with lozenges of dark green on an almost creamy ground, and over the whole a plum-like bloom, very difficult to paint, and indescribably beautiful.

My dresses were becoming very ragged, so I sent for a bit of undyed China silk and a tailor to make it. He appeared in the morning in a most dignified and gorgeous turban and other garments, and squatted himself in the passage outside my door at his work; but when I passed him on my way to our midday breakfast, all these fine garments, even the turban, were neatly folded in a pile beside him, and he was almost in the dress nature made him. Every one peeled more or less in the middle of the day, many going regularly to bed in dark rooms. I never did, but worked on quietly till the day cooled into evening, and I could go out again. The Rani gave me entire liberty, and did not even make me go with her for her somewhat monotonous constitutional walk every afternoon, crossing the river to the one carriageable road, tramping nearly to its end and back, always dressed to perfection, and escorted by the Rajah or some of the "officers." She used to time those walks so as to take me for a row before the splendid sunsets were over, and I never minded how long I sat in the boat waiting for her, watching the wonderful colours and the life on the river.

Now and then the wives of some of the rich Malays used to come and pay her a visit, dressed in all the brightest silks of China or Japan. They wore many ornaments of gold, much worked, and coloured rose or lilac, with ill-cut diamonds and other stones set in them. They had exquisite embroidery on their jackets, but were most proud of their heads of long hair, and delighted in letting it down to show us. The Rani took me one day to return a visit of ceremony from the family of the principal shipbuilder, a member of the Rajah's council. He and his son received us at the landing-place, and we mounted a high ladder (over the stilts) to his house, and were taken into the great barn-like room, where fifty Malay ladies had been invited to meet us in their gaudiest dresses, covered with gold bangles and dangling ornaments. They all sat round against the walls of the room, on the floor; we were conducted by our elbows to some chairs round a table in the middle, on which were two wax candles lighted in our honour, while coffee, with two large trays of curious cakes, was brought. At the end of the room were five big drums, some singing women, and many babies in and out of clothes. A most frightful noise began. Once our host got up, went and spoke to the orchestra, and returned to tell us he had told them "to play louder; they were not making half noise enough!" (the more noise the more honour being his maxim). Many of the women were pretty, and their manners very sweet and gentle. On our return another boat followed us with the two candles and trays of cakes as presents. The latter were made of rice-flour, gum arabic, and sugar, in different proportions, flavoured with almonds and spices. They all had a great family likeness to one another.

One night we found about fifty Sea-Dyaks all squatted round the luxuriously furnished English drawing-room when we came out from dinner. They had very little dress except tattooing, long wild hair, and coloured pocket-handkerchiefs round their necks, and sat perfectly silent, only giving a gratified grunt if the Rajah made an observation or relit his cigar, till the Rani got up to say good-night; then they also departed, apparently contented. They had come down the river in a long canoe from a great distance, to ask leave to take the heads of another tribe which had insulted them, and had been told they must not have that pleasure. They seemed to submit without a murmur to the prejudices of civilisation. Of course these people were full of superstition, and we used often to see small canoes and cocoa-nut shells full of burning oil floating down the stream with the receding tide, having been started from some house where there was fever, to scare away the malaria, and save doctors' bills. They used also to beat drums for the same purpose, which was much more disturbing to the neighbourhood.

There was a magnificent specimen of the Madagascar ravenala or travellers' tree, close to the house on the other side of a small bend of the river, and the Rajah had had the good taste to leave all its younger off-sets round it uncut. I spent some afternoons in drawing that view, and used to see numbers of graceful water-snakes swim up the creek with their heads curved well out of the water. Iguanas I also saw, and monkeys which used to come down to the edge of the garden and laugh at us. Sweet singing-birds were very plentiful. There was a bush under my window to which a pair of honeysuckers came regularly every night and morning. They were no bigger than humming-birds, but did not hover like them; clinging on tightly with their feet while they plunged their long curved bills into each flower in turn. The cock had a head and back of bright metallic blue and a yellow waistcoat; his wife was greenish.

One day a letter came, announcing that Captain Buller, R.N., was going to bring the new Consul of Labuan in his war-ship to pay a visit to the Rajah; so, as his spare rooms were only two, I persuaded him to send me off out of the way to his mountain-farm at Mattange.[78] The Rajah lent me a cook, a soldier, and a boy, gave me a lot of bread, a coopful of chickens, and packed us all into a canoe, in which we pulled through small canals and forest nearly all day; then landed at a village, and walked up 700 feet of beautiful zigzag road, to the clearing in the forest where the farm and châlet were. The view was wonderful from it, with the great swamp stretched out beneath like a ruffled blue sea, the real sea with its islands beyond, and tall giant trees as foreground round the clearing, which was also full of stumps and fallen trees grown over with parasites—the most exquisite velvety and metallic leaves, creeping plants, "foliage plants," caladiums, alpinias, and the lovely cissus discolor of all manner of colours, creeping over everything.

Great parasitic trees were standing there with their stalks all plaited together. They had strangled the original tree on which they had lived the first years of their treacherous lives, and were now left like tall chimneys of lattice-work, their victim having rotted away from the centre. There were masses of tree-ferns; one group round a little trickling spring which supplied the house with water, I could not help painting. Life was very delicious up there. I stayed till I had eaten all the chickens, and the last remains of my bread had turned blue; then, having seen the smoke of the parting salutes through my telescope in the swamp far below, I came down again, my soldier using his fine long sword to decapitate the leeches which stuck to me by the way. When I got to the end of my walk, I found I was expected to go back in a big boat so covered with matting that I should have seen nothing of the beauties outside, and to sit on a flat pillow and not get the cramp, in the company of a native Malay family of high caste in gorgeous clothing, who were arranging each other's heads in the Italian fashion. I could not endure that, so insisted on getting into my old canoe with my soldier and the lads, and letting the old cook have the honour and glory of a voyage with his native grandees. I had a most enjoyable day; for we hunted up all sorts of orchids, pulling under the thick overhanging trees, while the boys ran up the branches like monkeys, cut them through with the soldier's silver-mounted sword, and let the tangled masses tumble down into the water below with a great splash and a flop, nearly swamping my small nut-shell of a canoe. We picked off all the treasures, and soon had a perfect haycock of greenery in the middle of the boat, to carry home and hang up to the Rajah's trees in the garden. One great tassel, like rats' tails, two yards long, of every delicate shade through blue, green, yellow, and red, was as much as one man could lift. Beautiful orange rhododendrons were also growing on those branches, but dropped their flowers at the slightest touch.

The Rajah was very glad of all the things I brought; but hanging on a dry tree over a well-mown grass lawn is a very different thing from living in a swamp over the water, and I fear few of my treasures lived long. He and the Rani went one expedition with me in the yacht, first going down the river to the sea, then up in a boat over the sand-bar, and up another big river, past groves of casuarina-trees, winding in and out, almost back to the sea again, as far as Loon Doon, where we found a very nice house, and the magistrate, Mr. N., a most hospitable host. The forests behind his house were really magnificent. *Clerodendron fallax*, whose blooms used to be employed by the Dyaks to dress the heads of their enemies taken in battle, and the large kinds of mussænda, were particularly striking, the white bracts of the latter catching one's eyes at every turn. The blue crabs all over the mud at the edge of the river were very pretty too. Our host had a most cheerful character, and said he never felt dull, though the only European within reach was a wretched old drunken missionary with a Malay wife, and he saw no other from year's end to year's end. He had found and seen the great rafflesia in bloom on the hill-jungle near, a sight I would have given much to see.

The Rajah had planned taking me to some other stations, but his wife was suffering. We went back instead, and soon after I started in the small steam-launch up the river, with Mr. B., the good-natured Scotch-manager of the Borneo Company's mines in Sarawak, and a young Devonshire giant, rejoicing in the title of "His Highness the Rajah's Honourable Treasurer." The banks of the river were a continual wonder all the way up, with creeping palms or rattans binding all the rest of the greenery together with their long wiry arms and fish-hook spines. I traced this plant far up into the high trees. No growing thing is more graceful or more spiteful. We slept at some antimony-mines near the river. I found that the manager and his wife came from Hastings, and their belongings lived in its old High Street, and knew myself and my dear old father perfectly well by sight. How the poor sickly little woman enjoyed a gossip! She had had a constant struggle for life ever since she came out, but contrived to make her house most comfortable, and to make good cakes and bread and butter for her husband, in spite of the fever-giving climate. We continued our journey next morning on a springless tram-cart, with our feet hanging down behind, and considered that a rest for three hours; after which an excellent little pony carried me, while the men walked, through a marvellous forest for fifteen miles, except when we came to the broken bridges and I had to balance myself on cranky poles while the pony scrambled through below.

Some of the way was under limestone cliffs. The rock ferneries round the springs, caves, and masses of standing stalactites were exquisite, and the character of the vegetation was different from any other in the country. The rattans or creeping palms were everywhere, and the tall tree-stems were plastered over with exquisite coloured leaves, so exactly regular in their arrangement that they looked like some

French artificial trimming, mounting up fifty feet or more without a branch till they suddenly changed their character, and turned into great bushes, their ladders of leaves rotting off, the stem spreading as a root, which grasps the victim-tree, gradually encircling it, and strangling it to death. No one misses it; for the parasite has taken its place in the forest, and becomes itself a tall tree with a crown of branches and leaves at the top like the others. It seemed difficult to believe that those delicate velvet leaves and crimson stalks which ornament the tree so kindly at first should start with the express intention of murdering it and taking its place! But there are plenty of other murderers which start as parasites from a seed dropped in the branches by birds or wind, and throw their roots downwards. Some of those roots were full of fresh water. The Dyaks used to cut them in two and drink from them when no other good water was near.

That forest was a perfect world of wonders. The lycopodiums were in great beauty there, particularly those tinted with metallic blue or copper colour; and there were great metallic arums with leaves two feet long, graceful trees over the streams with scarlet bark all hanging in tatters, and such huge black apes! One of these watched and followed us a long while, seeming to be as curious about us as we were about him. When we stopped he stopped, staring with all his might at us from behind some branch or tree-trunk; but I had the best of that game, for I possessed an opera-glass and he did'nt, so could not probably realise the whole of our white ugliness. At last we reached a deep stream with a broken bridge, too bad even for the pony to scramble over or under, and he was sent back. Four Dyaks were waiting with a chair, but I was too anxious to examine the plants to get in, and walked on till near the end of the journey, when, in order not to disappoint them, I got in, and they started at a run and carried me up the steep little hill, and then up the steps of the verandah with a grunt of satisfaction and triumph.

It was an enchanting place that bungalow at Tegoro, entirely surrounded by virgin forest and grand mountains. Just opposite rose a small isolated mountain, full of quicksilver, with a deep ravine between us and it, and huge trees standing upon its edge, festooned with leaves, their branches adorned with wild pines and orchids—for foreground. I was taken to the top of the mountain the next morning, where I saw all the process of collecting and purifying the quicksilver. On that hill one might hear Scotch, Malay, two kinds of Dyak, and seven kinds of Chinese spoken. The Chinese were the only really efficient workmen. They were clever and handy, but not lovable. Nobody liked the "heathen Chinee." The poor, simple, lazy Dyak used to bury his wages in the earth. The Chinamen dug them up again, and gambled them away.

The Dyak has a sweet expression and much nobleness of figure, which he does not hide with superfluous clothing. His voice is gentle, and if asked a direct question he will give a truthful answer, and is almost the only savage who does. Every one told me the same story about them. At five o'clock a gong used to sound. We heard a great thud and shout, and the Dyak threw down his load, wherever he might happen to be, and ran home rejoicing—work was over. Sometimes they would come up at that time, sit on the steps of our verandah, and gossip with Mr. B., who joked with them, treated them like good children, and then sent them off with a present—an empty bottle, or a bit of toast off the tea-table; and they went away quite happy. They just stayed long enough to make sufficient money to buy a gun or a blanket, then returned to their homes. I felt quite sorry to think that fine old mountain was steadily being blown to pieces with gunpowder. Every bit of it was said to be impregnated with quicksilver or cinnabar, and one could pick up lumps of pure vermilion as one walked over it. It was a cruel process too, sweeping out the flues; and though eleven out of the twelve men employed twice a year on it lost their health or died, fresh hands were always to be found for the work, being tempted by the high rate of pay. I plunged my arm into an iron bath of the pure metal up to the elbow, and found it very hard work to get it in. Much more is said to be in other hills around.

I never saw anything finer than the afterglow at Tegoro. The great trees used to stand out like flaming corallines against the crimson hills. It was lovely in the full moon, too, with the clouds wreathing themselves in and out of the same giant trees around us. We had our morning tea at half-past six on the verandah, and a plum-pudding in a tin case from Fortnum and Mason was always brought out for the benefit of our young Cornishman, who was always ready for it.

Mr. B. walked him off after it, and I had all the day in perfect quiet to work in the wild forest or the verandah on different curious plants. One creeper with pink waxy berries like bunches of grapes was particularly lovely; and the scarlet velvet sterculia seed-case, with its grape-like berries, most magnificent in colour. Mr. B. soon started for some other mines, and I was left to the care of his assistant, Mr. E., who had been sent out originally as a naturalist by Sir Charles Lyell, in search of the "missing link," or men with tails; and after searching the caves in vain, kept himself alive by "collecting" for different people at home.[79] Mr. B. found him out and sent him to Tegoro. He was full of wit and information about the country. I found him a most delightful companion, as good as a book to talk to, and he was delighted to find one who was interested in his hobbies. One day he came up with a native carrying a toucan on a stick over his shoulder. The creature was tame, but bit and poked with the sharp end of his long bill. I wanted to see his wonderful tongue, and we opened his mouth, but could see none; he had curled it all back on its curious spring head. He had beautiful black eyelashes standing straight out, and his full-face view was very funny, the two eyes looking so separately at one, on each side of the big red nose. I gave the toucan a large berry from the bunch I was painting, and he swallowed it whole. I saw it roll slowly down his long throat and rest in the curved part. He had that berry up again several times during the time I was sketching him, playing with it at the end of his long beak, then letting it roll down again. When I offered him bread he gave his head a jerk of disdain. "White people's rubbish," he croaked out with indignation. They were most odd birds, with their huge beaks and crests, having the queer habit of plastering their wives up in a hollow tree when they began to lay, with a hole left for ventilation just large enough for her beak to go through, to receive the food the cock brought her. This he continued to bring till the young birds were fledged and able to fly away and feed themselves. Mr. B. told me he used to pass one of these nests constantly, and see the beak of the old hen sticking out of its prison, and could not resist giving it a flip with a liane hanging in front of it, when the beak disappeared with a croak of disgust. The hollow crest is thought to be a sounding-board or drum, which helps them to make their odd trumpet-like call.

Mr. E. was a cousin of Millais, and his sketches and illustrations of his different adventures in pen and ink were most excellent. I tried hard to make him publish them. Once he was bitten by a lemur he had caught and given to a friend. Walking back over the mountains, he felt as if his shirt was throttling him, but found his fingers so swollen he could not unbutton it, and soon fell down in a swoon, remaining there till the morning, when some Dyaks found him all puffed up and unable to move or speak. He said he did not suffer, but was utterly powerless. They poured half a bottle of gin down his throat, and carried him home, and he got well in time. It was all the effect of that small cat's poisonous bite.

One day Mr. E. took me into the great forest by a regular Dyak path, which means a number of round poles laid one in front of the other over the bogs and mud. It requires some practice to keep one's balance and not occasionally to step on one side of the pole, in which case one probably sinks over the tops of one's boots in the wet sop, lucky if one goes no deeper! We crossed the river several times on the round trunks of fallen trees, which, when rendered slippery by recent rains, are not altogether a pleasant mode of proceeding, particularly when there is a noisy rushing deep river a few feet below. Now and then there was a bamboo rail; but as they were generally insecurely fastened and rotten, one was as well without them. We passed one or two large gutta-percha trees which had escaped the usual reckless felling, and had the scars of present bleeding; and I was taken inside the trunk of a splendid parasitic tree, a gigantic chimney of lace-work, the victim-tree having entirely rotted away and disappeared—I could look straight up and see the blue sky at the top through its head of spreading green. The lace-like shell was not two inches thick, and it must have been over 100 feet high.

We went back another way along the banks of the stream, under rocks more in than out of the water—such clear cool water with grand ferns and rattans dipping into it from the banks above. Mr. E. found me a Green Stick insect, which curled its long tail over its head like a scorpion and looked most vicious, but was perfectly harmless. It had gorgeous scarlet wings to fly with, but on the ground was invisible as a blade of grass.

At last I had to leave Tegoro. Mr. E. walked down two miles with me; then we got into a canoe and shot the rapids for many more miles, with the great trees arching over the small river we followed, and wonderful parasites, including the scarlet æschynanthus, hanging from the branches in all the impossible places to stop. We sat on the floor of the canoe, held on tightly, and went at a terrible pace, the men cleverly guiding us with their paddles and sticks. Sometimes we stuck, then they went into the water, pushed, lifted, and started us again. We met other canoes returning, and being dragged up by the men. Some were going down like us with three stone jars of quicksilver in each, very small things, but as much as two men could lift.

At last we got out and walked again through the wonderful limestone forest and out to the common or clearing round Jambusam, where there was a long-forsaken antimony-mine. Mr. B. had kindly arranged for me to stay there, and had sent food and furniture to meet me.

Mr. E. went up a mountain near and brought me down some grand trailing specimens of the largest of all pitcher-plants, which I festooned round the balcony by its yards of trailing stems. I painted a portrait of the largest, and my picture afterwards induced Mr. Veitch to send a traveller to seek the seeds, from which he raised plants and Sir Joseph Hooker named the species *Nepenthes Northiana*. These pitchers are often over a foot long, and richly covered with crimson blotches.

Mr. E. took me to the entrance of a limestone cavern, and cleared a path for me; but as we had few lights and it was very slippery, I stuck to my old rule of not going willingly anywhere where I could not see my feet. The ferns and mosses were in the greatest variety about there. One of the sterculia trees was loaded with orange bells, but without leaves, looking like a solid mass of colour against the green hillside. Then I said good-bye to Mr. E., who did not think he should find another person to talk to about the wonders of the forest for years perhaps, no one caring there for the things around them; and at Tegoro there was only one European besides himself. Poor little Mrs. R., at the antimony-mines, really cried at losing sight of the one white woman she had seen for so long, and I lingered till the last day before returning to the Rajah's at Kuching.

The next day I said good-bye to His Highness, who came on board to see me off at seven in the morning, like the real English gentleman he is, a quality which no amount of sentinels presenting arms or yellow umbrellas can knock out of him. Mr. B. also went across to Singapore with old Captain

Kirk, and we were a pleasant little party of three on deck. The weather was so calm and warm that we had our meals under the awning. Those two would not let me land in the ordinary way, but made me wait till the Company's boat fetched me, with a grand native in a gorgeous turban to look after my luggage, and put me into somebody's smart open carriage, which conveyed me with great dignity to Government House.

Lady Jervois had sent to meet me by the last mail, and this one was before its time; but she made me very welcome, and I stayed there till the Java steamer started—a most comfortable Messagerie boat with few passengers, but a most entertaining monkey belonging to the captain. It was entirely gentle, with an amazing amount of curiosity. Every man who would submit had all his pockets searched and the contents examined, tasted, and smelt one by one. My thimble puzzled him much. He could not get it off. He went from it to the middle finger of my other hand and found no thimble there. At last he gave the puzzle up in despair, and made up his mind it was a particular deformity of mine, having a silver tip to one of my fingers only. He used to take pinches of snuff out of a snuff-box, sneezing with great enjoyment afterwards; and when a glass of water was given to him he would dip his hands in, then rub them over his poor wrinkled old face to cool it. I got quite fond of Jacko. He used to cross his arms, put his head on one side, and look as sentimental as a young Oxford Don.

There was one Englishman only on board. He remarked that he thought it was very hard that the little beast should have the luxury of re-enjoying his dinner whenever he chose to take it out of that great cheek-pouch, thus having one pleasure more than human beings. He also contradicted me flatly when I talked of the *Amherstia nobilis* as a sacred plant of the Hindus. I said I thought Sir W. Hooker told me it was so, and he said Sir William had been a great botanist, but was not a Hindu scholar. I had made a mistake, and I began to look at the little man with respect, and found he was Dr. Burnell, the famous Indian scholar and Judge of Tanjore, making a pilgrimage to Boro-Bodo during his short spring holiday; so we became friends, and continued so till he died. I like a real contradiction when it has a reason behind it, and there were plenty of reasons in Dr. Burnell.[80]

When we reached the roads off Batavia we were transferred to a small steam-launch, which took us for a couple of miles through a long walled canal with sea on each side beyond the walls.[81] It was said to be almost impassable in bad weather, and looked very Dutch and straight, full of barges with the sort of sails and rigging I had often seen in old Dutch pictures. At the Custom House my friends handed me over to the care of its Head, who would not look at my luggage, but told me to wait a little till the train started for Buitenzorg. After an hour, during which time I sat still on my trunk sketching boats and banana-trees, he returned to tell me the train had gone an hour ago, and there was no other till the next day; so he packed me and my trunks into the smallest of dog-carts, with a mite of a pony to draw it, which I expected to see lifted off its legs by the weight behind.

It took some time to start the poor little beast off, but being once set going, he dashed at a furious pace all the way to the hotel, which consisted of a straggling collection of ground-floor rooms, with verandahs and sleeping men on rocking-chairs all round them in the lightest possible clothing. The landlady came out quite composedly in her night-gown, her hair down her back, and was very efficient and kind. I did not think so much of her husband, and suspected that "schnaps" would soon be the death of him. About five o'clock I put on my best dress and took my letter to the President of the Council, M. van Rees, a most courteous and agreeable man.[82] His wife was in the hills, where he said I must go and see her, and he handed me back to the carriage as if I were a princess, and told the driver where to take me so as to have an idea of the outside of Batavia. The fashionable part in which I was is like one huge garden, with toy houses dotted about in it, half hidden amongst the trees, most of them being of one storey with Grecian porticoes as large as themselves, and verandahs all round. It is divided by many canals and roads crossing at right angles to one another. All "The World" was walking or driving about in the cool evening air, without cumbering themselves with hats or gloves. The ladies looked very nice, with natural flowers stuck into their hair. Everywhere in Java this habit prevails, and it is really sensible. What is the use of heating heads and hands with protection from the sun when it has gone out of sight? Visits too are paid at that hour. If a family is "at home," lights are lit in the portico or verandah; if not, nobody thinks of going. They never think of staying to dinner, which is the family supper, when all the small children of the family reign supreme. Dutch children are awful! Being much left to native nurses who give them their own way in everything, their manners are not improved by the constant society of the nurse's children, over whom they domineer much as they used to do in the old days of slavery.

The roads are watered most systematically by natives, with two watering-pots suspended from the two ends of a bamboo on their shoulders; they run about as fast as they can go, guiding the pots with their hands. Every man in Java is obliged by law to water the ground in front of his own domain.[83] The common Javanese men all wear a painted dish-cover on their heads. They have fine figures, but hideous faces, from the habit of stuffing tobacco and betel-nut between their lower lip and teeth, causing the former to project in a horrible way; but they are honest good people.

Batavia was a most unpleasant place to sleep in, full of heat, smells, noise, and mosquitoes. I started as soon as possible the next morning in the train for Buitenzorg, which, though only a few hundred feet above the sea, has pure cool air at night.[84] Every one (who is anybody) has a villa there, and merely goes to the city on business and as seldom as possible. The old French landlady said she had been expecting me a long while, and gave me a cheerful little room with a lovely garden on each side, with such cocoa-nut, breadfruit, and bananas that it was a real joy to sit still and look at them; and I resolved to stay quiet for a month or more, and learn a little Malay before I went anywhere else. Mr. and Mrs. F., who lived close by in the most exquisite little garden that ever was seen, promised to make all easy for me both at Buitenzorg and on my future travels, and they abundantly fulfilled their promises.

The order of everything in Java is marvellous; and, in spite of the strong rule of the Dutch, the natives have a happy, independent look one does not see in India. Java is one magnificent garden of luxuriance, surpassing Brazil, Jamaica, and Sarawak all combined, with the grandest volcanoes rising out of it. These are covered with the richest forests, and have a peculiar alpine vegetation on their summits. One can ride up to the very tops, and traverse the whole island on good roads by an excellent system of posting arranged by Government. There are good rest-houses at the end of every day's journey, where you are taken in and fed at a fixed tariff of prices. Moreover, travellers are entirely safe in Java, which is no small blessing. Mrs. F. used to drive me about in the very early morning, and show me lovely views and forest scenes, with tidy little native houses hidden among the trees and gardens, made of the neatest matting of rattan or bamboo, with patterns woven in black, white, and red, and slight bamboo frames hung round with bird-cages. These houses generally have sago-palms, bananas, cocoa-nuts, as well as coco, coffee plants, and breadfruit trees, belonging to them. The sago-palms were just then in full flower, with great bunches of pinkish coral branches coming out of the centre of their crowns. The fruit when ripe is like green satin balls quilted with red silk

The famous Botanic Garden was only a quarter of an hour's walk from the hotel, and I worked there every day, but soon found it was of no use going there after noon, as it rained regularly every day after one o'clock, coming down in sheets and torrents all in a moment. On one occasion I was creeping home with my load, without the slightest idea of a coming storm, when down it came, and in five minutes the road was a river. I had to wade through some places a foot deep in water, when a kind lady saw me from her window, and sent her servant running after me with an umbrella.

The Governor-General asked me to dinner in his grand palace in the midst of the garden. There were several people there, and some great men with fine orders on their coats; and when a little dry shy-mannered man offered me his arm to take me in to dinner, I held back, expecting to see the Governor-General go first; but he persisted in preceding the others, and I made up my mind that Dutch etiquette sent the biggest people in last, only taking in slowly that my man *was* his Excellency after all. We ought not to be led by appearances, for he was very intelligent, and talked excellent English. But as Madame de Lonsdale (a Spanish lady) only understood French, the conversation was mostly carried on in that language, and I floundered about in my usual "nervous Continental."

There was another hotel in the place, with a most magnificent view from its terrace, which I painted, looking over miles of splendid plantations of cocoa-nut and every kind of fruit-tree, with patches of rice and other grain between, leading up through grand forests to the most stately volcano, with a wide river winding underneath, full of people wading, washing, and fishing. Those amphibious people always prefer to go through the water rather than over it on bridges, and they go in, clothes and all, in the most decent way. Men and women dress almost alike, in all the brightest colours, with rich Indian scarves thrown round them, and always the inseparable umbrella. One sees perfectly naked children going along with an umbrella, or sometimes balancing a banana leaf as a substitute on their heads. They delight in flying kites of different kinds. Quite old men used to come and practise that amusement on an open space near my window every day, with their gray beards thrown up in the air, and their respectable turbans falling off their heads. The studies from that window were endless in their variety of colour, everything seeming brighter than in other places, even the fruits. One rather mawkish variety of the jamboa (myrtle) was pear-shaped, and of the brightest pink and scarlet colour as well as white. These used to be threaded on bits of cane, tied in bunches, and sold with bananas and oranges, in baskets slung from the ends of a bamboo over a native's shoulders, the native wearing a grass-green jacket, scarlet sash and turban, and crimson sarong or petticoat. No colours were too bright in the north of Java. In the south, indigo-blue was the prevailing tint, a fashion which had probably come over from South India with the Hindus who settled there many centuries ago.

No Malay uses his hands if he can help it; the smallest weight is put on the head, or slung to the bamboo on his shoulder, which wears quite a deep groove in the flesh. The nurses are also very gay, with lovely Indian scarves thrown over their heads and shoulders, often interwoven with gold, and the children are carried on their shoulders or hips very easily and gracefully.

The Dutch food in Java was peculiar, but good, the principal meal being at twelve o'clock, when one found on every plate a mountain of well-boiled loose rice. To that one added chicken bones, rissoles, sausages, cutlets, poached eggs, salt fish, curry sauce, stewed bananas, and a dozen other incongruous things, and ate them all together with a spoon. They had also beef-steaks and potatoes in some places, and dessert; but the former was often of buffalo flesh, which is blue and black, not tempting, and one had little desire for more food after the first mixture. The evening meal was a much lighter one. In the afternoon at four o'clock cups of tea were taken into every room, and the world dressed itself up to pay visits and walk or drive. The baths, too, in those regions are taken in an odd but very agreeable manner, in marble baths with the water coming from a spout overhead, and running out at the bottom, merely splashing one all over, with a bit of perforated wood to stand on. It refreshes one much more than soaking in water. There were abundance of baths in the hotel, and they could be taken at any hour.

The Botanic Garden was a world of wonders. Such a variety of the different species was there! The plants had been there so long that they grew as if in their native woods—every kind of rattan, palm, pine, or arum. The latter are most curious in their habits and singular power of emitting heat. All the gorgeous water-lilies of the world were collected in a lake in front of the palace. The Director was most kind in letting me have specimens of all the grand things I wanted to paint. The palms alone, in flower and fruit, would have easily employed a lifetime. The blue thunbergia and other creepers ran to the tops of the highest trees, sending down sheets of greenery and lovely flowers.

The view from the bridge in the very High Street of Buitenzorg was the richest scene I ever saw. A rushing river running deep down between high banks, covered with a tangle of huge bamboos, palms, tree-ferns, breadfruit, bananas, and papaw-trees, matted together with creepers, every individual plant seeming finer and fresher than other specimens of the same sort, and the larger such plants were, the grander their curves. Then they had the most exquisite little basket-work dwellings hidden away amongst them, and in the distance was a bamboo bridge—a sort of magnified human spider's web. Looking straight along the street from the bridge was another pretty view—little shops full of gaily coloured things, such as scarlet jamboa fruit, yellow bananas, pomelas, melons, pines, and hot peppers of the brightest reds and greens. Pretty birds in bamboo cages, people in every shade of purple, scarlet, pink, torquoise-blue, emerald-green, and lemon-yellow; small copper-coloured children carrying all their garments on the tops of their heads, grass-cutters carrying inverted cones of green fastened to their bamboos and almost hiding them. Long avenues of huge banyan trees bordered the principal drive to the palace, with large bird's-nest ferns growing on their branches, each tree forming a small plantation of itself, with its hanging roots and offsets from the branches. Herds of spotted deer used to rest in the shade under these trees, and parties of the great crested ground-pigeon, as big as turkeys, were always to be found there. It was a delightful place to work in, even in the heat of the day.

The market was a very busy one, full of odd groups and queer things, and if one bought anything there it was done up in a bit of banana-leaf, pinned together with a spine of the wild palm, and tied with a strip of its leaf. I watched some common coolies getting their breakfast at a Chinaman's stall, out of fifty little saucers full of odds and ends, taking a pinch from each, with a rice cake to put the morsels on. Fingers were their only tools, and by the end of the day the saucers must have had a strong fingerish flavour, I should think. Chinamen in Java tucked up their tails and wore gray wide-awakes an inch too big for their heads, and did not look picturesque. The Malays often wore hats as big as targets, and coloured like them with all the colours of the rainbow. More dandified characters wore highly polished dish-covers, gaudily painted, over their turbans or head-handkerchiefs. The servants of great people stuck glazed chimney-pot hats with cockades on the tops of their bright-coloured turbans, which had a very ridiculous effect.

After more than a month at Buitenzorg I left my heaviest trunk and started for Batavia, with a big letter in my pocket from the Governor-General to all officials, native and Dutch, asking them to feed and lodge me, and pass me on wherever I wished to go. I found the hotel at Batavia quite full, but the landlord and landlady made me live in their rooms and eat with them, most kindly putting me up for the night "somehow," and charging nothing. I made a pilgrimage to see the Dutch flower-painter, Madame van Nooten. She was very poor, and the Government had helped her to publish a large volume of prints, oddly and badly selected and not over-well done, but she was an interesting and most enthusiastic person, and she pressed me to come and stay with her. I bought a copy of her book and sent it home to Kew (it being far too big to be kept in a flat), but the ship was wrecked and it never reached its destination.

I also went down to the business part of Batavia to Bryce's shop, where I was told I could "buy anything," but found they only sold things wholesale, so I was reduced to making little purchases from the Chinaman pedlar as usual. I had a letter to Mr. P., who scolded me for not giving it before, and made me promise to go and stay up at his country-house with Mrs. P. on my return. I found there was a good deal of division between "sets" of people in Java, and that one set was very jealous of another—thus I saw only the set to which Mr. and Mrs. F. belonged.

There were three kinds of public carriage in Batavia—a comfortable open one, which was expensive and too heavy for the small ponies which dragged it; secondly, a kind of dog-cart, which was light and uncomfortable, without any rest for the back, but it was the fashionable cab of the place; thirdly, a very light and comfortable car, with a seat behind the driver facing the horses, and a back to it, but that conveyance was unfashionable, and considered not "the thing" to go in. Mrs. F. had kindly arranged with the captain of my ship to call for me at six the next morning and take me with him on board. He was a great big laughing young fellow, and rolled about the deck in loose white trousers, a shirt, and a meerschaum in his mouth, but was thoroughly efficient. The ship was full of great people, three Residents (Lord-Lieutenants of Counties) and a Colonel of Engineers going with his men to make a railway. Most of these people talked English, and most had large families of disorderly children and servants. One perfectly round old gentleman used to sing songs and tell stories to the children, and got nearly torn to pieces by them. Most of the people on board were of the same shape and of the Pickwick type of countenance. Their loose trousers were made out of the national sarongs, whose enormous patterns and gaudy colours looked strangely out of character round their short legs, the upper half of their dress being a flowing white shirt (not tucked in at the waist). They had no stockings, and heelless slippers which went flop-flop about the deck. The ladies dressed in the same fashion, only the sarong was put on like a petticoat. They were generally very fat.

At Samarang every one but myself and the captain went on shore, but as I was to return that way I preferred staying quiet and painting the glorious view of its harbour and the five volcanoes from the deck. There was no snow, but they were all about 10,000 feet high, and their slope was steeper than that of Etna or Teneriffe. One smaller one was still smoking, the others were quiet. It was a wonderful scene, for those mountains are not the mere satellites of a great volcano, but each a perfectly separate one of great size. The country at their feet seemed magnificently cultivated and peopled. Soerabaja is much more fitted to be the capital than Batavia.[85] It is a very busy place, with a lovely landlocked harbour in which the biggest ships could anchor close to the shore. Both it and the streets were full of traffic and movement.

I only stayed a night at the hotel, then Mrs. F.'s nephew, the town-clerk of the place, took me off to his house in the suburbs (where everybody had their villas), and gave me a delicious room in his garden. His wife was most hospitable.

They were the first people who had really shown that virtue, though many talked of it. The Dutch are generally so taken up with the idea of money-making that it does not occur to them to entertain strangers, though they would always be willing to help one if told how to do it; but Mrs. S. H. made me quite at home. I promised to return straight to her house after my expedition to the mountains. I drove from their house "post" with four horses, which went full gallop and were changed every three or four miles without a moment's loss of time (as well as their coachman and groom, to whom I always gave a fee of twopence each), through an almost continuous avenue of tamarind trees, which met overhead, shading the long straight road most deliciously. This was mended and swept as smooth as a carpet, the bullock-carts and heavy traffic being forced to go on a parallel road outside the trees.

I stayed three days in an excellent hotel at Pasoeroean, which had a civil landlord (who charged it in the bill). From hence I made an expedition to Blauwe Water, the site of an old Hindu temple, where there were some hundreds of tame monkeys in the trees, protected by Government. Everybody who visited the place took fruit for them, and when they heard a carriage coming they came down to receive the new-comer, and with much chattering and disputing divided the spoil, then swarmed up into the branches again. It was an odd sight, and the stillness which reigned between the arrivals of the carriages was very curious. I began a sketch of the old Hindu temple ruins and tank. After an hour or two, feeling hungry, I took a biscuit out of my pocket, which I began to eat leisurely as I went on with my work. I was disturbed by a pull at my dress, and found a huge monkey sitting close beside me, looking reproachfully at me with the expression of "How can you be so greedy? Why don't you give me a bit?" Of course he did get it, and then departed and hid himself in the leaves overhead. They had one old king among them—a very big monkey, who always helped himself first, and allowed none of the others to interfere with him.

My polite landlord would not let the banker do anything for me. Oh no! He would himself drive me to the end of the road and see me mounted on horseback. So we drove to Paserpan, past magnificent crops of corn, rice, millet, mandioca beans, tobacco, and sugar-cane; and after sitting awhile on rocking-chairs under the chief's verandah my little horse was brought out (a real beauty), but with a man's saddle covered with velvet, brass-headed nails, and embroidery. I do not enjoy that sort of seat, but the pony carried me beautifully, and the way was all interesting, though the poor coolies who carried my trunks found paper and paints very heavy, and some sat down and declared themselves "sakit." We wasted a couple of hours trying to find others, and at last succeeded; they were all nice good-humoured fellows, with whom I felt quite safe. After a while we mounted up to the region of coffee, and finally to that of cinchona and tea, and all manner of European vegetables, which were sent down every morning for the poor gasping people on the hot plains to eat.[86] My land-lord at Tosari was very angry with the other for not having taken the trouble to telegraph to him, to send down his horse with a side-saddle and strong hill-coolies for the luggage. He was a very nice person, originally a civil engineer. His jolly fat wife and children were all most friendly and kind to me, and after the other guests left I took my meals in their private rooms.

Tosari is 6000 feet above the sea. Its season was over, and it was cold at night, and generally wrapped up in clouds. The scenery is very curious, the steep volcanic hillside ploughed up into great furrows from top to bottom, often 1000 feet deep, and the tops a few yards across. One could talk to people on the opposite hill-slope, though it would take hours of hard scramble or roundabout paths to reach them. Those steep slopes were cultivated in the most marvellous way. I never met such an industrious people, and where other crops are impossible, the gaps are filled by tree-ferns and almost alpine flowers—marigolds, nasturtiums, balsams, guelder-roses, raspberries, great forget-me-nots, violets, sorrel, etc. The country was splendid beyond the casuarina trees, which are tall and transparent like poplars, and invaluable for foregrounds, as they cut the long horizon-line of sea, plain, and sky, and do not hide the landscape. The people (like all other mountaineers) were honest and friendly, and every spur had a village perched on its sharpest point. The politeness of every one was overwhelming. As I was mooning along collecting flowers one day, the chief of the district rode up, with half-a-dozen wild men in attendance on bare-backed ponies. They all dismounted and made bows while I passed on, then remounted and disappeared.

My good landlord himself accompanied me on the great expedition of the place, to see the Bromo volcano and Sand Sea. He said there were wild horses feeding on the Sand Sea which might be troublesome to my little horse, so he put his gun over his shoulder on the chance of shooting some small birds for me. We started over the white frost in the early morning, and the only bird he shot at was a peacock—an enormous one, which flew across the road with a great yell and fluster, and I hope was none the worse for my landlord's small-shot. The Sand Sea was the original crater of the Bromo, which fell in and sent up that flat plain of sand, like a moat round the present crater, surrounded on all sides by high rock walls 800 or 1000 feet high, down which we walked or slid with our ponies following us. Then we crossed the sand for some miles, and climbing to the edge of the present Bromo crater, looked down on the sulphur and smoke within. It is considered very holy by the 8000 Hindus who still exist in that southern end of the island, and who go on a particular day every year and throw chickens in, which generally fly out again, and are caught and eaten. In early times human sacrifices were made there, then animals ; now these rites are next to nothing. We mounted up on the other side of the Sand Sea cliffs, and got a view of the cone and smoke of the Smeroe volcano over the other mountains, the only really active one, and the highest, it being 13,000 feet above the sea.

We rode on to the first village, and had a fine view of the sea and all the eastern end of the island. My landlord was a most entertaining companion, speaking perfect English, and knowing the whole place well. He and his wife were both musicians. They had a piano and harmonium, and sang really well. He also did a good deal of doctoring, giving ten grains of quinine or an electric shock from his machine in exchange for a fat chicken. This the natives considered a fair exchange. The place was constantly in the clouds, but it seldom rained, and the air was dry. The native houses were all made of bamboo, which was first soaked for some days in water, in order to drown the small weevil which lives in the wood. Then it was used in a thousand ways. The roofs, floors, indeed the whole house and frame were made of it. It was split, flattened, and plaited into mats, which formed the walls of the houses, and it always looked clean, neat, and polished. The three children of the house were real beauties, the girl of thirteen like the fairest woman Rubens ever painted, with golden wavy hair and an exquisite complexion, which was taken no care of. She was out all day long without a hat. It was real genuine beauty, and wanted no dressing. She and her brother employed themselves in making bamboo cages and catching birds to put in them. They knew a little English, and the youngest child went by the name of Klein-baby. He was a real pickle, and used to perch himself on the edge of my window, sit on his heels, and chatter Dutch at me for an hour at a time, and never bored me. He wanted no answers, and never would believe I did not understand him. The cinchona is a tiresome crop to grow, as it takes seven years before it is fit to bark. Some people had tried taking half off for some years running, but it did not answer so well. At Tosari they cut the whole down, when each tree was worth about ten shillings. They were planted at about a yard and a half from one another, with tea underneath to take their place when cut.

It seemed like leaving home again to come away from those kind people. I had been told in the plain the road I wished to go was full of difficulties and dangers, but I found none. First we mounted up to the hoar-frost and view of the smoking Smeroe, then down a valley of ferns and rushing water, through miles of old coffee plantations all left to grow their own way, in the Java fashion, with erythrina trees covered with their coral-like flowers to shade them, and numbers of natives creeping about under them collecting the berries, singing and talking to one another. Once a great toucan came floundering along, almost knocking against me with its awkward wings and huge crest. Beautiful blue birds and butterflies came constantly up the path. We stopped half an hour for luncheon, and the good little horse carried me all day without any food most merrily, till we got on the plain and a straight hard road, when we all became weary of the last five miles over it.

At last we got to Pakis, and rode to the house of the chief, with a letter from my Tosari landlord asking him to send me on to Malang. He was a model Javanese, and I felt quite safe with him; paid my men and horses, and sent them back. I sat myself calmly down in the universal rocking-chair under the verandah, or rather the steeple-shaped roof which covered the open Court House. The chief informed me I should be sent on soon, and a good deal more in a more than unknown tongue; for in that part of Java they talked Javanese, not Malay, and the former language I had not even attempted to learn. Then he went and stood in the road, looking up and down, as if he expected a pumpkin drawn by six griffins to come round the corner. He also sent men galloping off in different directions on their toy ponies. A big cock came and sat down beside me, and shared the bit of bread I was eating, giving contented clucks at every crumb I threw him, allowing me to stroke his broad back as if he were a cat. People in Java delight in taming birds. Every house is hung with cages, and one sees the children walking about with Java sparrows sitting on their heads and shoulders, a string tied round their leg. Cages of doves are hung up among the feathery foliage of the tamarind trees, with cords and pulleys to get them up and down by, and are supposed to attract wild birds to build and perch near them.

Presently they brought me a delicious cup of tea, with a tortoiseshell cover over it, and a bottle of antique biscuits from Reading; and after a while the lady of the house returned, and I was put with my trunks into her carriage—a sort of big wheelbarrow with a roof over it and no seats, the driver sitting on the shafts. It was lined with red flannel, and I stretched myself at full length, and rather enjoyed its hard floor after my long day's ride. A loose horse trotted on in front of us during the first post, and then was put in the place of the original, which went home. The trunks were packed on another horse. All this was done for nothing, the chief writing to my landlord at Malang that as there was no post-carriage he had sent me in his own. Mr. MacL. received me most kindly. His father had left the Highlands in 1804, and he called himself a Javanese; but in spite of his untidy, disreputable exterior, was a true Scotch gentleman. He was once a millionaire, but had become very poor, and had many trades, amongst others that of keeping a boarding-house, which he did not in the least understand. It was full of business men, a quiet, depressed set, with wives and untidy families, who got their food from the kitchen and ate it anywhere. Mr. MacL. took me himself to see the Resident with my Governor-General's letter.

He sent me about in his own great open carriage and four

horses, first to Singosari, where I saw some huge and hideous old Hindu idols, half human, half animal, carved elaborately out of a stone which is not found in that end of the island, and sitting among palms, ferns, and frangipani trees. The whole neighbourhood of Malang abounds in Hindu ruins, the richest tropical vegetation, running water, and fevers. Once while painting in my room I was called out to see the young "Controller" of Batoe, a very limp young man in spectacles, who said the Resident had told him I wanted to go to Ngantang. He was going to conduct the Regent (native prince) there in a day or two, and would take me also if I liked. Could I start at once? Of course I could. I knew Ngantang was called the gem of all Java. So I bundled all my things into the wardrobe, gave the key to Mr. MacL., and started with next to no luggage in the carriage and four, with my new friend, who was very much like Lord Dundreary, minus the whiskers. His English was well-intentioned but peculiar. He "feared there would be *no eat* in Batoe," but I found his house and food both perfection; he only wanted a good wife to make his home a model. He had masses of roses in his garden, and beautiful hills beyond; and, to his great delight, I began sketching at once, while he tumbled in and out of the verandah, watching, no easy task, as he was so short-sighted; he never saw anything three inches beyond his nose. He was thoroughly happy, having put himself into a light and flowing attire peculiar to Dutch officials in Java—loose pink cotton trousers, white flowing shirt, collarless and cuffless, no stockings, and heelless slippers, rather startling to a stranger at first, but I had got quite accustomed to it. The ladies were even more untidy, and literally wore a nightdress, with their hair hanging down their back, till sunset.

The Controller took some hours to write a long letter on two sides of a slate (sucking his pencil between every few words), to beg a particularly good horse from a native chief for me to ride the next day. He had brought a lady's saddle from Malang with him. The next morning he gave a great gasp of relief as I jumped into it from the ground, for he had a sort of horrible dread that he should have to lift me on, and did not know how to do it. Poor fellow! He was in the highest degree of nervous anxiety about the Regent's reception and his responsibility for it, for that personage was very great indeed, descended from many generations of native princes, and a "knighted Sir of Holland," my host told me. So I thought it kind to take myself out of his way early, and rode on ahead, with a mounted official before and behind to guard me. They wore big "uglies" over their turbans, red jackets and waistcoats *au naturel*, with daggers stuck into their sashes behind, hitching up the red jackets, and their great toes stuck in through the stirrup, like dark people all over the world. The whole road was crowded with people expecting the Regent, in holiday dress, gay with colour. When they saw my red-jackets they thought he was coming, dismounted, and forced their animals into the bushes and ditches, squatting down themselves, and even pretending to pray at me with their two lifted hands. They were rather disgusted at seeing only a woman.

I went through miles and miles of coffee covered with white bloom. In that one district alone 1,300,000 lbs. of coffee had been sold to Government that year, so that every one was in great spirits. Then we came to beautiful winding valleys, with the river far below like a torrent, often quite hidden by the jungle of bamboos and other green things over it. I never saw anything more lovely. At last the valley widened, and we reached Ngantang and the Chief's house, where there was a great gathering to greet the Regent, and a wonderful orchestra to entertain me while I sat in the open verandah waiting for him. Two large frames like bed-stretchers held each sixteen tin kettles, forming two very imperfect octaves. There was also a kind of viol of metal plates. Both these were struck with bamboo sticks, and there was also an odd collection of drums. They gave me tea, and the usual bottle of biscuits, and a nice airy room—the floor of mud, the walls and ceiling of bamboo-matting, so closely plaited that I could not find a chink of light through it. There were four such rooms, with clean beds, snow-white linen, and mosquito-curtains. In all provincial capitals of Java there is such a house for official people to lodge at, and with my letter I had a right to use it also.

After about two hours the great people arrived, the old Regent the very picture of good-nature. He wore a black wide-awake over his turban and big gold spectacles. He was a thorough gentleman in manner, and very popular, insisting on putting me into the place of honour at dinner and supper. We had the usual rice, with so many curious little dishes to eat with it that they seemed to require numbering and a catalogue, but I did as the others did—I took a pinch of each, and found the food very entertaining. The thing I missed in Java was bread. It is seldom eaten, very costly, and only brought when called for, so I tried to learn to do without it also.

The forests round Ngantang were full of curious things—parasitic trees with extraordinary outside-roots, buttresses, and leaves looped over the branches. There were clearings full of coffee and tobacco, and above all, the Kloet volcano and other hills clothed up to their tops with rich woods. I spent a delightful morning and evening wandering amongst it all, perfectly alone, while the great people held their court in the verandah. Most of my return ride was made between them, the Regent on a pretty little cream-coloured horse looking the image of Pickwick, and bringing constantly to the tip of my tongue the sentiment, "Bless his old gaiters," for he also wore those appendages on horseback. The young Controller on my other side rode a miniature Javanese pony, and almost touched the ground on each side with his feet. We had a great train of attendant chiefs behind us, while the people who had collected in the road to see us pass fell down like packs of cards as we came up. The lower they went, the higher their great man rose above them and was magnified; so many of them went flat down in the ditch. About half-way they turned off by another road to visit a different district, and I went on with my two red-jackets, the Regent's Head Man, as an especial honour, in front; and after a mile or two I met an Englishman!—an engineer who was mapping out the country. He said he had heard of me, and it was a real treat to have a talk with a countrywoman, so he turned and rode a while with me, and let his tongue go, then said he felt better, and good-bye. It was a pleasant meeting for us both.

I stayed two more nights at Batoe, and left my host a portrait of his house. The Regent, who was staying with his son-in-law, a chief near, came with his principal wife and a daughter to pay us a visit in state, drawn by four horses, with betel-nut, umbrella-carriers, etc. How the old man laughed, cried, chewed, and slobbered, all at the same time! The wife was a daughter of the people, who had once sold fruit at the road-side. She was very small, but her manners were quite easy, and she talked sensibly, my host said. She had no children, but the old gentleman had a moderate supply of other wives, and thirty-four children, and was said to be rich enough to provide for a great many more. The lady wore a Paris hat and chignon, but the rest of her costume was like that of the ordinary native. Such visiting seemed very liberal on the part of a Muhammadan, but the Dutch officials in Java had made themselves generally trusted and respected by the upper-class natives, who seemed everywhere on the best terms with them. Then my host took me back to Malang, and the Resident, M. de Vogel, came and called on me, and looked over my work. He was just like an English gentleman, knew every place and plant, and arranged to send me the next day in his carriage to Djampang with four post-horses, which went like the wind (for which I paid nothing), changing every three or four miles.

At the first post I found my old friend the Chief of Pakis waiting for me, who made me signs I was to follow him, pretending to pull off his jacket and uttering the magic word "water." He took me down to see a beautiful blue lake and bathing place, and showed me the spring bubbling up through it; then conducted me back to the carriage, mounted his mare, and cantered on beside me to Pakis, where horses were changed; then brought me on to Mr. Netcher's, the Controller of Djampang. He was a descendant of the famous painter of white satin petticoats, and a very nice and intelligent man of the world, a great contrast to my late kind host. His wife was a victim to fever. He borrowed a spirited little horse from a neighbouring chief (which went well after the usual commencement of standing on its hind-legs and kicking out behind at starting), and sent me with five attendants to see and sketch the beautiful little temple of Kidal. The luxuriance of the bamboos and palms was even greater than any I had seen elsewhere. The rivers and waterfalls were surrounded with ideal ferneries and huge-leaved plants. Both the temple of Kidal and that of Djampang (close to the Controller's house) are small, but perfectly covered with the richest and finest carvings, quite like cameo-work of pure Indian designs. They are almost smothered in foliage, grown over with ferns and lycopodiums, and have small tanks and springs of water near them.

My poor fever-stricken hostess appeared in the evening,—a mere hopeless skeleton-woman, who took cod-liver oil, cream, and enormous quantities of food, beer, wine, etc., but never got fatter, she said. The next day she was pretty well, and her husband took me for a glorious ride and walk up and down the steep spurs of the Smeroe, followed by about twenty chiefs. They had an idea that if anything were to happen to a white person in their district some dreadful misfortune would happen to them, so they always followed him about like dogs, watching every step he took. My little horse stood twice on its hind-legs, so it was led all the way by a chief on either side, and when I got off to walk they kept close to me, ready to pick me up if I tripped, and seemed to expect me to throw myself down every steep bank, though it would have been difficult, with the stout bamboo-railings and good roads, to come to any harm. On the top of every ridge was a village and chief's house, where we sometimes rested and took tea all round, including our numerous followers, who squatted in a semicircle in front of us. If we stopped on a hill to breathe, they all squatted on their heels round us, as it is disrespectful to stand in the presence of Government officials, or to allow their heads to be on the same level as ours.

We went through glorious scenery—deep dells full of ferns of endless variety, anthurium leaves nearly a yard long, and higher than myself; then through endless plantations of coffee-trees, pollards, but growing naturally to the height of twenty feet, they were thirty years old. The best coffee was said to be picked out by a little wild cat or racoon, which eats the fleshy part and leaves the berries on the ground to be picked up and sold. At Djampang many of the fine old forest-trees had been left to shade the coffee; some varieties of banyan were very curious. We saw a crowd of monkeys in one tree. One of these creatures made a jump which might almost be called a flight. It was a land of jumping or flying creatures—lizards, frogs, foxes, and even spiders flying, or seeming to do so. I saw a huge spider turn and fly at a man who was trying to catch it. He was not frightened (though it was said to be poisonous), but got hold of all its legs in a bunch behind, so that it could do him no harm.

On my return to Malang, Mr. MacL. arranged that I should go back to Soerabaja by the direct road, fifty miles in a country cart for fifteen guelders; the post would have cost eighty-five, and I preferred this mode of travelling, as I should see more of the country. The carriage was a long covered machine on two wheels, with no seat but a hard board-flooring, on which I stretched my poor old bones, setting my back against my trunk. The cart had springs and two capital ponies to drag it, but these were so small that I wondered how they ever kept on their legs when I got in. The driver sat on his heels in front. I enjoyed going slowly and stopping to rest often, when I could sketch the people in the little wayside places; but after the first half of the way my driver got tired (not his horses), and tried to sell me to every carriage he passed, but none of them would take me on as cheaply as he wanted, so he had to go on, grumbling all the way. At sunset, when we were resting at a roadside tea-house, another cart came up also to rest, containing a young German, and as neither he nor I had met any other Europeans all day we had a grand talk, when he found I "could Deutsch," and he gave me some Rhine wine out of a long-necked genuine bottle which was most delicious. Those native tea-houses are very convenient; all sorts of nice rice- and corn-cakes with eggs and tea could always be bought there, and the people in them were friendly and kind.

When the moon rose the German and I went on our different ways, my driver grumbling so incessantly that at last I stopped him before a brightly-lighted verandah, went through the garden up to it, and asked a lady and gentleman sitting there what language they could speak. They were most kind and hospitable in French, and the gentleman came out and ordered my coachman roughly to go on to his destination, and not to bother me by his grumbling. He said the horses were all right; the man was only lazy and in want of a scolding. After that he went on quite gaily, singing (as they call the tremendous noise natives make in their throats when happy!). The moon was bright, and the bananas, palms and breadfruit trees looked lovely with all the neat little houses. The long suburbs of the great city were as light as day, and by about half-past eleven I had guided my coachman through all the right streets to the house of my friends, to find every one asleep and the lights all out! I wandered about after paying my man, knocking at doors and windows in vain, and prepared to sleep on my trunks under the verandah. I must do my driver the justice to say he would not leave me alone there, but determined to stay too till some one came, and squatted down by his horses to wait. I was tired of rubbing my edges off against that old portmanteau, so went to the great verandah to get a rocking-chair, and found the natives sleeping there. After some kicking and shouting the magic word "Ingus" they roused up and went in search of my key. At the same moment Mrs. S. H. drove home from a party and welcomed me most heartily. She had told her servants to keep a light in my room and expect me, but these town-people were not like the mountaineers, and did nothing when the eye of the mistress was off them.

It was nice to wake up in a comfortable room the next morning, and to find the little charcoal-heating machine on the table outside, with its excellent pot of coffee and milk on the top, and pretty china cups. I felt almost friendly even to the seven children with eyes all on their finger-ends, who swarmed in and out all day with their black attendants and dogs. Their poor father was oppressed with work, and used to come back half-dead every night from his hot office in the city, and must have been more bored by the weight of such a lot of unmannered babies than any one else; but he escaped the midday meal, which was like a scene at the Zoo, and did not improve one's appetite. My hostess lived in disorder all day, and never went out till it was nearly dark. She said there was no beauty in the place (which she never looked at by daylight). I found much to admire along the edge of the fine river, full of strangely coloured barges, shaded by palms and fine trees, with picturesque native as well as Dutch houses interspersed, and grand distant volcanoes peeping over them.

Of course every one drove along the same road of an evening.

One saw odd collections of people, big Chinamen riding small ponies with their feet barely off the ground, and gorgeous native princes with gilt umbrellas held high over their heads by servants squatting behind, not to shelter them from the sun (which had long sunk behind the horizon), but to show their rank. They had noble English horses and gorgeous liveried servants, with brightly-coloured sarongs, and hats like shields or dish-covers elaborately gilt and painted on their heads over their turbans, and tied with a cloth under their chins as if they had the toothache. They kept flocks of geese in all the gardens round the houses in Soerabaja, as the noise they made was said to drive off all snakes! But the weather was too hot for enjoyment down by the sea, so I took the next steamer and returned to Samarang, where the captain was good enough to land me and put me in a carriage, telling it to take me to M'Neil's. Instead of to the bank, it took me to the manager's house—a splendid villa with marble floors, and Japanese pots of roses and carnations all round the verandah. A nice English nurse came out and told me master was getting up and would soon come. I felt quite sorry to spoil his Sunday's rest. He was most kind, wrote me letters, and put me into a nice cool room to wait till it was time to catch the train, and sent me in a good breakfast after my bath. So I got out of Samarang before the mosquitoes even knew I was in it, and reached Solo or Soerakarte at sunset by a slow train, which took me through a rather desolate tract of country with burning forests, showing plainly we were out of Dutch rule and order.

I found quarters in a little mat inn close to the station, and the next morning had two hours' drive about the city, and satisfied myself I did not care to see more, the Emperor and his 999 wives included. I called on the Resident, who said, "Oh yes, Prambanan was well worth seeing," and he would give me a letter to the Assistant Resident at Klaten, who could easily take me there. Of course I thought it must be close to Klaten; so I started by the next train, and on arriving at a rather lonely station, got a boy to show me the way to the Assistant Resident's—a good half-mile through scorching sand, about three o'clock, and in the full heat. I found the whole household taking its siesta, and when at last a black servant appeared, he took my letter, pointed to a rocking-chair, and put his finger on his lips. I waited, fell into the way of the house myself, and took a sleep too. Then tea appeared, and the master, who said Prambanan was two stations farther off, and I must stop the night with them, then his brother should go there with me in the morning. The family was a charming one. They walked about with me, showed me all their curiosities, and took me for a delicious moonlight drive, when the fireflies gave almost as strong a light as the moon herself; and next morning I was taken to the old Hindu ruins. They were much scattered over the plain; and the chief, who considered it his duty to accompany us, would insist on having his gilt umbrella of state held over my head. It had a stick three yards long, with no end of fringes and ornamentation, and I felt the dignity almost too much for me.

The ruins were more curious than beautiful, with many colossal figures of the gods, the same as those I afterwards saw in India. Mr. Jan Bor had to leave before me, so I stayed and finished sketching, with the umbrella and the Resident's Head Man to take care of me; and then was driven in by the heat to the station, to wait three hours, to the great enjoyment of its poor Tyrolean master, who seldom got a chance of talking his native tongue to one who knew Meran, his beloved Vaterstadt. Poor fellow! he was a victim to fever, and had a decanter of carbolic acid and water ready mixed in the corner of the room, which he said suited him better than quinine. He was a good fellow, and I liked listening to his sixteen years' experience of Java. The country all round him was filled with rice and indigo (mostly under water), irrigated by a system of terraces in as perfect a way as could be shown in any part of the world. The great slate-coloured buffaloes were ploughing through the mud, with the driver raised on a high seat behind them, and often a small boy on their backs. Those beasts have been known to kill a tiger in defence of the children who take care of them.

While I was gossiping in the room of the Tyrolese, the chief came in with his umbrella and followers, one of them bringing a teapot, cup, and sugar under a cloth for me. They all squatted round us, and would not go till they had seen me into the train and off for Djocia—the biggest town of Java, the residence of its native Sultan, and a great stumblingblock to Dutch order in the island.[87] But every one said he would soon be bought off and pensioned. The great square in front of his palace is surrounded by big trees cut like umbrellas, the symbol of greatness in those parts; and a huge elephant is chained up at one corner. There was an excellent hotel, and a charming doctor living there who had written books about the Java volcanoes. He worshipped Darwin, and had his photograph, which he showed me. The Resident sent me in his carriage to the tombs of the Sultans, which were poor things, but curious in their way. There was a huge yellow turtle in a tank, which was fed on meat. I could not make out what he had to do with the tombs. He was probably a last remainder of Hinduism. There were also some fine carved gateways and banyan trees. The chief had had a most tempting breakfast spread out for me under one of them, of cakes, fruit, tea, and what he thought most of, bread and butter; and he insisted on my taking a large water-melon away with me, in case I got thirsty on the road. When I departed all the population clapped me as if I had been a successful comedy.

I left Djocia in a grand post-carriage and four, with two extra horses to drag me up the hills, and ten men waiting to haul and push me over the dried-up river beds and lava streams (for grand volcanoes were on all sides). We crossed a most primitive ferry on a great bamboo-mat floor, laid over two boats, with men in hats as big as targets, pulling the thing over by two ropes made of rattan of enormous length. The horses were taken out, the carriage taken down and dragged up from the ferry by men. It was a most lively spot, always full of people going and coming, and animals standing or swimming in the cool clear water. Soon after passing it, we came to a huge cotton-tree, which had nearly strangled and swallowed up an exquisite little temple. Two sides of it were hidden entirely by the roots, between which the poor, crushed, but finely-carved stones peeped out. It was the tallest tree in all the country round, and towered up twice as high as the cocoa-nut plantations near it. The stem must have been quite a hundred feet high before it developed any branches. Another sort of cotton-tree was planted along all the post-roads to act as telegraph posts, the peculiar way its branches were arranged at right angles to the stem being very convenient for isolating the wires. All the pillows and beds were stuffed with the contents of its pods.

About a mile beyond the giant tree and tiny temple we came to the great pyramid or monastery of Boro-Bodo, or Buddoer.[88] At its foot an avenue of tall kanari trees and statues of Buddha lead up to a pattern little mat rest-house, and the farmhouse of its manager. The house contained a central feeding-room and three small bedrooms. From the front verandah we had a good view of the magnificent pile of building, a perfect museum, containing the whole history of Buddha in a series of basso-relievos, lining seven terraces round the stone-covered hills, which, if stretched out consecutively, would cover three miles. There were four hundred sitting statues of the holy man, larger than life, the upper ones under dagobas or hencoops of stone. Many of them had been knocked over by earthquakes, which had cracked the whole, and thrown the walls so much out of the perpendicular that it was a marvel how it all held together without cement of any kind. The whole was surmounted by a dome. From the top terraces was the very finest view I ever saw: a vast plain, covered with the richest cultivation—rice, indigo, corn, mandioca, tea, and tobacco, with the one giant cotton-tree rising above everything else, and groves of cocoa-nuts dotted all over it, under which the great population hid their neat little villages of small thatched baskets. Three magnificent volcanoes arose out of it, with grand sweeping curves and angles, besides many other ragged-edged mountains. Every turn gave one fresh pictures; and if Boro-Bodo were not there I should still think it one of the finest landscapes I ever saw.

The sun used to rise just behind the highest volcano, tipping the others with rose-colour, throwing a long shadow on the miles of cotton-wool mist below, through which the cocoa-nuts cut their way here and there. In half an hour all the clouds rose and hid the great mountain for the rest of the day. It is the second highest in Java, and only a few years ago buried forty villages during an eruption. At sunset the mountains were generally seen again. I never missed climbing the pyramids night or morning, and was always rewarded by some curious and beautiful effect of colour and cloud, and always found new stories in the great stone picture-book on my way up and down. I longed for Mr. Fergusson at my side to explain it all to me. Some of the carvings are very fine. The figures have often much beauty and expression in them, and are divided by exquisitely fanciful scroll-work, arabesques of flowers, birds, and mythological animals.

I had the place all to myself, and the good farmer and his son gave me all sorts of good things to eat, all on one plate. I never had any idea what they were made of. The house was surrounded by cows and goats, cocks and hens, and was a genuine farm. The landlord was very fat, and not elegant in the afternoon, when he dropped all his clothes but his sarong or petticoat; but he was a capital old fellow, and took the greatest care of me, walking up with an umbrella to fetch me home himself one day, when he thought I had sat too long out in the midday sun.

I wished there had been no watchman at night. He used to beat a drum incessantly, sometimes a mile off, coming closer and closer till all the dogs got mad with fury and ready to fly at him. He used to tell the people to "wake up and guard the house," then all the people who were sleeping on the different mats round it screamed at him. That took place three times every night. Sleep was impossible. What use could such a noisy guardian be? All the thieves could hear him and get out of his way. He carried a long fork of wood, and caught evil-doers by the neck with it, "they" said.

Four miles from Boro-Bodo was the other curious monument of Mendoot, only accidentally discovered a few years before I went there, under the mound of earth by means of which it was originally built up. It is said to be Hindu, and its carvings are worthy of the old Greeks, so polished and exquisitely designed and finished are they. The statue of Buddha inside, preaching, might have done for a Jupiter or a Memnon. It is of gray granite, quite colossal, as are his two friends on either side of him. The calm beauty of the great preacher haunts me still, and I was fool enough to waste two mornings in trying to paint it, of course failing utterly.

I took some bread in my pocket the second day, and told the old landlord I should walk home; but when I had got over the ferry I found him waiting for me, with his brisk little ponies and double-seated carriage like two arm-chairs, one person only being intended to sit on each, and the front had been well stretched to hold him. He laughed and shook half the way home at the idea of my being allowed to walk, and would charge nothing in the bill for fetching me, good old fellow!

All the animals were fond of him. His house was surrounded by bird-cages, and the great kanari trees were full of singing-birds' nests. Those trees grow very tall. The timber is invaluable, and the nut good to eat when once the shell is broken, but it is as hard as a stone. Wallace's description of the way the black cockatoo gets through it is one of the most interesting things in his book, which was a Bible to me in Java, all he says being thoroughly true.

I had nearly exhausted my purse when I got to Magelang, two posts farther, with four horses; but I had a letter of credit on the landlord of the hotel there. The bankers have a capital system in Java of giving one credit for small sums on private individuals, so that one has not to run the risk of taking much money about in one's pocket. Magelang is a large place, the capital of Kadoe, with the usual central square of banyan trees. Every one was most kind, and the Resident asked me to come and stay; but I did not wish to linger there. There was a grand view of the Soembing volcano from his garden, with the whole gently-rising plain covered with rice terraces and running water over them, trickling from one to the other. Deep below was all that was left of the river at that season, crossed by a bamboo bridge like a gigantic cobweb made of those great canes, which grow to a hundred feet high about Magelang. The garden was full of rare trees, but had been much neglected by the former Resident. Some of the statues from Boro-Bodo had been placed in it. A Chinese artist was "restoring" these, lengthening the eyes, flattening the nose, and turning them into regular Chinese ideals of Buddha.

Mr. van Baak wrote a letter for me to the Resident of Wonosobo when he found I was determined to visit the Diëng; and my landlord sent on a horse to the foot of the pass to which I drove. There the chief as usual made me welcome, introducing me to his principal wife, a nice sensible old lady, who took me into her rooms and introduced me to

Number Two, whom she visibly considered a "gay giddy thing," covered with gold ornaments, and did not care that I should admire her too much, but soon took me by the hand and led me away. I had a good old horse, but a man's saddle, which I never enjoy, and three coolies to start with, who hung the trunks on two crossed bamboos tied at the junction, running themselves at each corner of a triangle, one at each side of the trunks and one behind; but as soon as they were out of sight of the chief they declared themselves hungry, and would have stopped to sleep too if I had not started back to appeal to him, when they begged forgiveness and came on. It was nearly eleven before we were off; then one of them sat down and declared himself "sakit," but the two others took off the cross-bamboo and went on twice as well without him. It was a long drag, and nearly sunset before I mounted the hill of Wonosobo—a perfect marvel of richness, and a great contrast to the bare hills we had crossed. On them, however, I found one gem—a perfectly green orchid, and looked forward to a day's rest in a comfortable house, and time to paint it. But there was no hotel, and the Residency was being painted, the family in Europe. However, after some delay the Resident appeared, a singularly nervous man but very good, and before we parted next morning he had become quite hospitable in his offers that I should remain or return, and wrote me many elaborate directions for my future proceedings.

The only horse to be had, I was warned, was "peculiar"—a mild term, for he required three men to hold his head when I mounted the next morning, and two to lead him the first mile, after which he tossed me off and tried to macadamise me, while my foot was still caught in the stirrup and his heels close to my head. I felt sure death was coming, and felt quite comfortable, but thought he was a long while about it. I had no fear, only wished it over. Then the stirrup-leather broke, the brute got out of my way, and I got up none the worse. I never saw such a picture of fear as the face of the poor fellow who had led the horse. He was trembling all over; his eyes were starting from his head; he could not move for some moments. I only thought of restoring him to his senses: I stroked his poor hand which had been hurt by the horse crushing him against the rocky bank. Then I set to work to make a new stirrup with a bit of twisted bamboo rope he had in his hand, after which I remounted, and we went on placably and even sluggishly when the steep hills came, the only drawback being that the side-saddle had been made for a small child, not a woman, and was always turning round under me.

At Garoeng I gave my letter to a most practical and gentlemanly chief, who wore a very stiff stick-up collar and cuffs under his jacket. He proceeded to dictate a letter to some other chiefs about me, telling them to feed me and pass me on, and tightened up my saddle himself. His house had most picturesque high roofs with carved terminals, and was already in the clouds. He and his Head Man accompanied me on to the next chief, and we went round to see a lake black as ink (an old crater probably). Old carved stones were scattered on the banks. The next chief gave me more tea and biscuits, and a state umbrella carried by his Head Man to accompany me. Strawberries were flowering in his garden, and cinchona growing over them. I passed fields of tea full of flowers. The road got always steeper and my beast lazier, and I walked all the last part of the way. The scenery grew very wild, like the top of the St. Gothard; then the plants became like those of Europe (except the tree-ferns, ground orchids, and hollyhocks). At last I reached the rest-house and small village of the Diëng, 6000 feet above the sea, on a small filled-up crater, a pass between the tops of two mountains.

It was so cold that I was delighted to roast myself by a great wood-fire. My bed was against the other side of the chimney, and I was right glad of the blanket I brought up with me. The next morning the chief's cream-coloured pony was brought for me to mount, and absolutely refused to hear of such a thing, turning round and round, kicking, neighing, and snorting, so I sent it back and walked. The men all tried to get up, with the same success. It was a funny scene. We all laughed, including the pony! I had a most interesting walk among the scattered ruins of tombs, temples, aqueducts, and foundations of big buildings, whose very use and history are unknown. We passed lovely lakes of different colours, saw the mud springs boiling up, and the coils of smoke from them in all directions, with a strong smell of sulphur. Only certain narrow paths were safe to tread on, the rest being a mere treacherous and broiling crust, which would bear no human weight. It was all rather horrid, and the cold caused me such suffering that I determined to get down the shortest way to warmth again.

I had a most pleasant ride of eight hours on an excellent horse, which made the Wonosobo saddle also unobjectionable. The young chief, after running round and round the cream-coloured pony for a quarter of an hour, succeeded in mounting him, and rode with me, while my pretty horse went as quietly as a lamb. Java ponies have a habit of resisting their riders' getting on their backs, and showing fight at first, but are excellent and untirable after they are once started. The views were magnificent. We had a long mountain-pass to cross, and much bare moorland. At one place we passed acres of tea in flower with cinchona amongst it. I saw a white-coated Dutchman looking after his coolies, and to my surprise he came up and called me by my name, asking me if I had had a pleasant journey. He had seen me on board one of the steamers, and welcomed me like an old friend (though I had no remembrance of him). He offered me all kinds of hospitality. I had seen no white man for three days, and enjoyed a talk with him, he in Dutch, I in German; and he explained to my chief that I wanted a country carriage to take me on to Temanggoeng when I got down to the road.

The next thing I saw was the source of a great river and glorious springs, surrounded by another monkey colony, a remnant of the Hindus. There were great trees and rose-hedges all round. Then we descended again till we reached Nagaredge and went to the Controller's house, a half-caste hunchback, who scolded my chief for even suggesting a carriage, and ordered him to go on and not bother him. I produced my magic letter, when the manner changed instantly! I was bowed into a rocking-chair, and a cup of tea produced; but I did not like the man, and decided to go on over five miles of glaring road under the midday sun to the next place, where the chief lived who had started me over the pass to Wonosobo.

Our horses went on quite gaily. I was received like an old friend. My Diëng chief was treated like a gentleman and given a mat and tea too, and soon they packed me into a carriage which brought me to the Resident's big house at Temanggoeng, where two most dear ladies covered me with kindness. They were like female Cheeryble Brothers. One of them was constantly suggesting to the other some new thought for my comfort; the other thanked her most humbly, and blamed herself for not having thought of it earlier. The children were lovely and in perfect order, and the Resident one of the best specimens I had seen of a high-class Dutchman. I spent four delightful days there, and had a huge garden-room all to myself. The roses outside would have taken prizes in any show. They had my clothes mended, washed, and brushed, and took me lovely drives all round the country, seeming as if they could never do enough for me, speaking perfect English too. The crops on the plain about Temanggoeng and under its five volcanoes were enormous, water flowing everywhere from magnificent springs, economised with a marvellous system of terrace-irrigation—rice, sugar, coffee, tea, maize, indigo, and great groves of every kind of fruit-tree. The massive sugar-palm I had never seen in such quantity before. One tree would more than fill a good big cart with its fruit.

There were grand markets, and people used to walk miles with nothing but bundles of banana-leaves on their heads: these were used in a hundred ways—to cook on, eat on, as paper for doing up parcels (pinned with a thorn of the wild palm or cactus), to thatch houses and keep sun or rain off young plants or seeds, as well as to make mats and baskets of.

In the evening the two fair-haired little girls of eight and six used to sit on their little chairs at a tiny table and play at cards, with a negro servant in a gorgeous livery squatting between them, all three most intent on their work, with a background of roses in pots under the marble-floored portico, and the moonlit garden and mountain beyond. It always reminded me of Millais' famous study of Mr. Lehmann's little daughter. This would have been a still better subject for him. They had grand dogs of noble race, and all sorts of other pets. One night we went out to see a Chinese festival and fireworks. The latter were all fastened to tall poles, beginning at the base and lighting their way gradually to the top without any human hand touching them; they became in the end a perfect pyramid of different-coloured fires. The effect was very fine; also the curious crowd below, with the bright light on their upturned faces, was a sight worth coming across the world to see. There are some thousands of Chinese about that country. They come first into a district carrying a few tapes, buttons, or sugar-plums for sale on their own shoulders. Two years afterwards they have a man or two to do the carrying for them, then a horse for themselves to ride, then a shop, and finally they become rich men with horses, carriages, and liveried servants; but they always retain their pigtail and simple dress, and generally stick to the original district where they are known and respected. I got a great respect for them at last, and used to ask my way or other help of a Chinaman in preference to any other Asiatic. They were always so practical and quick to understand what I wanted.

My journey on to Amberawa was a difficult one, the road very bad and horses worse; one poor thing lay down five times, and at last had to be tied up to a tree and left behind. However, the whole day was before me, and I did not care to hurry; but as every hole was full at the hotel, I had again to claim hospitality of the Resident, who had three strangers already in his house, but made room for me somehow. I met another man who had met me in a steamer. Every one seemed to know all about me. My host was an old bachelor, with a clever brown housekeeper who kept everything in apple-pie order. He took me to see the camp near by and the great fortress—a piece of extrvagance the Dutch now bitterly repent of, as a few Armstrong guns could knock it all to pieces, and it is in a most feverish position.

My host said if I would stay longer he would show me many curious things, but I went on the next day to Samarang, thence by steamer back to Batavia, and thence up to my old quarters at Buitenzorg for a few days' rest; after which I took a country carriage with three horses, with extra men to push it when necessary up the very steepest hills, and walked myself up most of the splendid road over the Megamendoeng Pass. There were strings of people going and coming all the way, carrying heavy loads on their heads or from the ends of the bamboos on their shoulders. Near the top is a deep black lake in an old crater which I went down some steps to see. While there a shower of rain came on, and my guide picked two wild banana-leaves and covered me up with them instead of with an umbrella. The large-leaved ferns and arrow-headed leaves of different sorts were most magnificent. Then we descended considerably to Sindang Sari, where there was a kind of hotel and hospital for soldiers managed by an old and somewhat eccentric Dr. Plum. He was a philanthropist, and took up odd people who did not always turn out creditably.

His housekeeper was so drunk she could neither speak nor stand when I arrived. But when I had got rid of her over-anxiety to help me I soon made myself at home, with a good room and delicious wide verandah on the upper floor.

There were some nice invalid officers and their wives in the house, who were very friendly. The garden was full of foreground studies—ferns, aralias, daturas, and areca-palms growing in a half-wild and most picturesque way amongst rocks and running water, with delicious baths large enough to swim in, through which the water ran in and out continually; beyond all were the grand forests and great volcanoes of Gedé. Under it, about four miles off, was a branch of the Botanical Gardens about 5000 feet above the sea, where the director had a bungalow and spent the summer. His wife talked good English, and took pains to inform me she did not live there, but merely endured it for some months every summer, for the sake of her children. She little knew how I envied her position, within a few minutes' stroll of the wildest virgin forest. The aralias and pandanus were most elegant, and there were masses of a large cane-like plant with a creeping root, called the "patjuy," which produces great bulb-like shoots from the root, of the most beautiful carmine tint, having scarlet flowers and fruits hidden inside. These resemble miniature cobs of Indian corn, full of refreshing juice. They are quite treasures to thirsty travellers.

The doctor mounted me on a splendid piebald horse, which took me one day four miles further into the forest, to see a waterfall in a regular grove of tree-ferns. We went up and down perfect ladders, and my horse was so entirely sure-footed that I never thought of dismounting. Above the forest, on the Gedé volcano, many curious alpine plants are to be found, the most famous being the *Primula imperialis.* The doctor procured some plants for me; but they were all out of flower, which was a great disappointment.

Houses do not take long to build in Java, and eight men

can move a mat house in a few hours. They have generally some neat geometrical pattern woven in them; the terminals are carved into the shapes of lobster claws and goats' horns to ward off the evil eye. The natives used to sit before their houses, weaving a coarse kind of cotton dyed beforehand, and generally produced an ugly plaid pattern of large squares which they delighted in. The sarongs they printed in a peculiar way by painting the white part with some kind of wax, so that when soaked in the indigo or other dye, that part is protected and comes out untinted. Some of the patterns are like those on the Persian rugs.

From the doctor's, six hours in a country cart took me to Tanchur, where the Assistant Resident and his nice family took me in. I stayed there two days and painted the vanilla, with its lovely greenish-white orchidaceous flower-pods and fleshy leaves. I had not seen it in full beauty before. It was a real pleasure to look at my hostess and her sister, they were both so fair and amiable. The Resident of Bendoeng picked me up there, and took me on with him in a grand carriage with six horses. He had ruled the Preanger for eighteen years, the highest and largest province of Java, and was a very great man indeed.

We went like the wind. Buffaloes were waiting at all the hills, and coolies to push and pull at the steepest parts. The road was magnificent except at those places, and we went as fast as on an ordinary railway, with a train of mounted chiefs before and behind. At every district fresh ones joined us on the most frantic little horses, and when the great man deigned to speak to them they dismounted, and went down on their heels like frogs; every living soul did the same as the carriage came up, getting into the lowest ditch they could find. It looked very funny in the markets and crowded places we passed, to see every one suddenly lowered. Besides myself, the Resident had picked up an artist lately married, who was suffering from his lungs and going to try a month in higher air. We crossed the river twice, going down almost perpendicular roads and up again, and were dragged by men over a most picturesque ferry; but a new road and bridge were making, and in ten years there would probably be a railway too. We also crossed a range of chalk hills covered with woods in their autumn tints, which might have been in England. The people got more and more civilised, and the chiefs wore black alpaca suits like Europeans, all but their heads, which were still neatly turbaned over their knot of back hair and its comb. I was sorry to see the European dress creeping in; it never looks dignified on an Asiatic. Their horses were marvels of grace and activity, though very small.

We passed, at the top of a pass, a lake quite full of huge long-stalked pink lotus (nelumbium) in full flower, a glorious sight; but the weather was so uncertain that the Resident advised my going on the next day while it lasted tolerably fine, and staying with him on my return; so he started me himself and gave me breakfast at five the next morning, packing my trunks into the carriage with his own hands. He was a most wonderful man, and never spared himself. He could improvise on the piano most tastefully, and when he got tired and bothered with work or worry, had the habit of sitting down to the piano to refresh himself; he said he felt quite a new man after a few moments of "fantasieren."

My driver was an Indian with hair on his mouth and chin, very unlike the natives of Java. He had jokes for every one, and took the greatest care of me. His two tiny ponies only rested once in forty miles, but it was a hard pull; four times we had four men to drag and push us over the steep hills. The scenery was so exceedingly beautiful that it was worth coming, in spite of the almost incessant rain. I was glad to get out my blanket and put it over my knees and shoulders to keep out the cold and damp. The rain-clouds cast a bloom over the mountains I had never seen before. The greens seemed greener, and the colour of the nearly ripe rice was quite dazzling. We passed over a rich plain a little before five, and soon after arrived at the ferry of Garoet, to find it broken and impassable, the boats half-full of water, and the yellow river rushing like a mill-race. A motley crowd were waiting till it was mended, and I had a long conversation about it (in our respective tongues) with a Chinese merchant, who was travelling with a train of coolies, with piles of Lancashire prints hung to their bamboos. But my man said it was of no use waiting, and we turned back and begged hospitality of the "Assistant," as he called him, of Trogan, a young Dutchman with a pretty delicate wife and four babies. They were both too nervous to talk anything but Malay at first, but in time they found some words of French.

I was on my way to the house of Herr Hölle, who lived on the hills behind Garoet, and I had a letter to the native Prince or Regent there to send me on. I was taken the next morning to see the hot springs at the foot of the Goentoer volcano, whose lava-stream looked fearfully fresh and new. The hot water rushed out of a tangle of the richest hot-house vegetation I ever saw. A succession of tanks below the spring were divided by green banks covered with bananas, grasses, and huge caladium leaves, dark volcanic stones making delightful backgrounds to those green masses, while little bamboo-houses on stilts were reflected in the water. There were people in red sarongs bathing and fishing in the warm water, their fishing being done with hand-baskets like sieves. We had some of these brought for us to look at, and they were full of strange little green shrimps, beetles, and other nasty things, all of which they dried and ate with their everlasting rice.

The children had some strange pets in the house. One of them was a "fretful porcupine" which ran about loose, and delighted especially in hiding under my bed! She liked to have her nose tickled, and to nestle close to my feet. She was on the best of terms with the dogs, but rustled up all her spines if interfered with, and the dogs had the sense to leave her a wide margin. There were two little pumats, something between cats and ferrets, with very beautiful fur, but not good countenances. These were the small animals who picked out the best coffee, eating the outer part, and leaving the nibs for the humans to collect and sell. Cats in Java, like those in the Isle of Man, have hardly any tails.

The house was buried (like all the native houses) in a grove of cocoa-nut trees. It faced the high road, and every native who passed got off his horse and led it past the house of the white official, though my host was only a humble specimen of his class. They only pay the same respect to the Dutch they do to their own chiefs, and I still think we should have done more wisely in our Indian colonies if we had kept up the same old manners of the country. Ignorant people think very much of outside signs of respect, and take us at our own estimation.

I had almost made up my mind to wait no longer but return to Bandoeng, when Herr Hölle walked in, a grand man with a strong look that reminded me of Garibaldi, the same curious mixture of simplicity, power, and gentleness. The Governor-General called him "our great civiliser." He wore a fez on his head and sandals with wooden soles like the country people, but was otherwise more decently dressed than many of the Dutch country gentlemen. He talked all the dialects of the country, as well as Sanscrit and Arabic, and his English was excellent. He devoted his life to improving the condition of his people in the province, had written books in their languages, and established schools and other institutions for their enlightenment and comfort. He had had the ferry mended, and drove me, in his little single-seated carriage with two small spirited ponies, through Garoet and up a zigzag narrow road, 2000 feet above it, to his village and pretty little house, a model place in every way, ornamented with carved wood and terra-cotta mouldings, all made by natives under his directions.

No door or window was ever locked day or night. The people passed through the garden from cottage to cottage, and never stole a flower. Herr Hölle said he liked to see them moving about, and to know they were not afraid of him. They often came great distances too to beg him to doctor them or give them advice when in difficulties, and to work in his tea plantations, which covered miles of hillside. The winding paths were bordered with cedars, sheds being built at intervals to shelter the pickers from the rain. No scene could be more picturesque than those hills crowded with gaily-dressed people amongst the tea-bushes, the plain of golden rice and palm-groves below, with grand mountains beyond, two of them always smoking.

Eleven years before, all this small paradise was a swamp, the home of tigers and rhinoceros; now, the dear creatures were not to be seen. I went up to the edge of the impenetrable forest, where some said they were still hiding. Near it cinchona and coffee took the place of tea, while the ferns and wild bananas were growing on every scrap of uncultivated ground. I did plenty of painting, but my chief delight was in hearing my host talk, and seeing him among his people. One evening he took me to see the children shaking the trees to collect cockchafers, which they roasted and ate with their rice. They had a bit of burning wood on the ground, the insects flew to it, and were caught by the eager little creatures. So picturesque they looked in the firelight, the whole under the brightest moon I ever saw. The Government constantly sent Herr Hölle to mediate and arrange difficulties with natives all over the island. He knew all their peculiarities, proverbs, and idioms, and could always manage them. His great friend was the Mufti of Garoet: orthodox old ladies used to say he was a Muhammadan himself. He knew the Koran by heart, could convince the people by their own arguments from it, and met them half-way in most things; he allowed no pork on his table, no dogs in his house.

He had plenty of books and illustrations of the antiquities of Java, and showed me how both Boro-Bodo and Mendoet were raised, by covering up with earth as the builders went on, so as to form a long slanting road to the work, over which they could bring the stones to the very top. Mendoet had never been uncovered till a few years before, which accounted for its great smoothness and preservation.

I was taken to see the whole process of tea-culture—picking, drying, and packing, all so nicely and cleanly done. The boxes had English labels stuck on them, as they went chiefly to Australia. The great buffaloes brought their little carts up the hills and took them away. I saw the schoolhouse, with the maps and drawings for the pupils to copy. They were then away during the feast of Ramadan.[89] Herr Hölle had them taught to read and write Dutch; he said there had been a great outcry against it at first, but he thought it good to break down the boundary of races as much as possible, never hurting the feelings or rubbing against the prejudices of either, if he could help it. The Muhammadans in the Preanger were liberally inclined, never having more than one wife, and letting her go about with her face uncovered like other Javanese women. He took me to have breakfast with his friend the Mufti or Priest of Garoet, a most intelligent man, who sent his daughter to a mixed school for boys and girls. She was very clever, and had taught herself Dutch so well that she had made translations of some of Hans Andersen's stories, and published them in her own language (the Serbanese).

After leaving the Mufti's I was sent on alone, stopping to shake hands with my kind friends at Trogan,—to Bendoeng, where I found my magnificent room kept for me still, and a kind welcome from the energetic Resident, Mr. Pahut. His wife was still at Buitenzorg with her sick child, so I had all my days to myself, and painted a study of the rice-harvest, which was going on all over that rich high plain on which the city stands. It was a bright scene, with the golden stacks, sheds, and stubble, in which the gaily-clothed people and hideous buffaloes were buried up to their knees, with glorious sunshine over it all.

The nice little governess and her invalid artist brother were still in the house; the latter complained of the colouring of Java being "so monotonous," nothing but the *same* green! I never saw the same in any two trees; the lilacs and blues of the hills were delicious, the bamboos were just then quite yellow, and the rice-fields of every tint, from brown-gold to yellow and green, all full of variety; I longed to shake the stupid blind conceit out of the poor limp fellow! Bendoeng is a large thriving town, covering a very large space of ground, as every house stands separately in its own garden. The Chinese street or bazaar must have been quite a mile long, and the mosque looked very picturesque, amongst banyan- and mango-trees, cocoa-nuts and areca-palms.

I found a nice quiet place to paint in, from a raised terrace in front of a school close to the public road, so that the admiring crowd could not get within eye-rubbing distance of me. Once I saw them suddenly sink low in the dust, and found a beautifully dressed native squatting at my elbow; it was the Regent, or native Prince, with whom I had a long conversation in unknown tongues (I hope he was the wiser!). He watched my work a long time, then departed, and the population rose again. They were such a gentle people, never in one's way. I could not say as much for the Chinese school-children, who crowded round me in a most unpleasant throng when their school-hours were over. One boy stood at the back of my easel, staring at me, so I calmly raised my brush a little and put a dab of blue at the end of his nose, and the applause in the street below was uproarious.

The people of the Preanger were far merrier than in other parts of Java, and wore every shade of red in preference to the dull indigo blue, which is the favourite colour elsewhere. The Resident's carriage took me and fetched me from my sketching-place, with outriders and noble English horses. He himself was a remarkable young man to have done such an amount of work in so few years. His father had been Governor-General, and he had known the country and people all his life; he was a perfect king in his own province. Meals were very irregular, sometimes hours after the appointed time. I used to get painfully hungry, but could not be cross when the great man came and fetched me himself on his way to the dining-room so good-naturedly, talking, laughing, and entertaining us all while dinner lasted as if he had nothing else to do; after which he set to work again till the next meal came. Sometimes they had parties in the great portico, and three whist tables would be in use at once under the hanging baskets of exquisite ferns and orchids. All round were stands of splendid flowers—begonias and geraniums. Stag's-horn ferns were hung up like gigantic green brackets in every corner, with a perfect cascade of seed-leaves hanging underneath them. It looked very gay, but I did not find the society amusing. I was too sleepy after my hard day's work to sit up for a late supper, and was allowed to go off to bed.

The Resident arranged to send my trunks back by post, he stuck the labels on with his own hands, then packed me off in a great open carriage with the limp artist and Herr von Müller, Head of the "Woods and Forests" of Java. The latter was always flying about in that great carriage, which had been built for him in Manchester, with an awning added at Bendoeng. He had a table to screw into its middle, and a bed to pull out at night, and was a most genial good fellow, who liked to give a lift to his friends, and seldom went alone. He told me his carriage had already gone 16,000 miles, and had never had an accident or gone wrong in any way. The young controller of Bendoeng made a fourth to the party, and we flew over the ground with six horses all at full gallop, buffaloes and men to push when needed, ordered to be ready by telegraph everywhere. We passed two trees quite black with flying-foxes, taking their siesta with their heads downwards, wrapped up in their own wings instead of mackintosh cloaks, which these so much resembled. And near the top of the pass was a lake, full of the grand Indian lotus.

Travelling with great Javan officials almost takes one's breath away. We seemed perpetually trying to catch some phantom train; horses were waiting at every station, buffaloes at every hill, men running like furies beside the horses, shouting, whipping, pushing, and hauling; people and animals rushing into ditches to make way and show respect. The Assistant Resident and his pretty wife were not at Yandjor, so we ate our rice at a nice little hotel, quite smothered in greenery. The trees were loaded with fruit all round us, and it was very hot; but the road up to Sindang Sari is magnificent. The doctor gave me my old room and a most kind welcome.

The next day I rode back to the forest of Tchi Boelas, and the two officials went on a ride of inspection, while "our artist" and myself painted the same bit of forest scenery so differently, that no one would take our productions to have been painted in the same country. His might have been as well done at home in Holland, with some old Dutch pictures as his models, all discoloured by brown varnish. How odd it is that artistic people persist in seeing Nature everywhere alike and through smoked spectacles!

## CHAPTER VIII

### CEYLON AND HOME

1876-77

AFTER a few days I returned, to pack up and take leave of my friends at Buitenzorg. I called on Madame van L. and heard the billiard-balls in the next room cease to rattle; the Governor-General came in and talked for about ten minutes, then pleaded urgent and important business and disappeared: I heard the billiard-balls rattle again, and soon after took my leave. I was rather limp myself and wanted rest and home, so I gave up my idea of going to the Moluccas, and went back to Singapore in the same steamer which took me to Java.

As there was no first-class cabin to be had, the captain (who always lodged at my hotel at Buitenzorg when on shore) promised to keep quite as good a one among the second class for me. He made me sit next him at dinner, and on my other side I had a first-class Dutchman, who had the good taste to talk the captain's language, so we were a sociable little party. In the tug going down the long Batavian canal, a dreadful woman, with high heels to her boots, two parrots, and one baby, came and sat next to me. She put down the baby between the two birds, who deliberately bit the poor child till it roared with pain, when the woman cuffed the parrots and the baby too. I could not help remarking to my next neighbour the famous words of Dundreary, "Th' old 'ooman's a lunatic!" I made a friend for life of a most jolly young Scotch boy named K. Whenever I went on deck, he got me chairs, telling me all about himself and his belongings as if I were an old aunt. He said it made him feel at home to talk to a real Englishwoman again. When the pilot declared it was too late to take the ship into harbour, young K. and I went and spoke the old man fair, and made him land us in his boat. He saw me up safely to Mr. D.'s, the colonial secretary, where I had promised to stay for the few days I remained at Singapore that time, he himself going on to the house of his twin-brother.

The D.s had a cockatoo loose on a perch, which used to take restless fits and walk over all the furniture in the room. One morning it walked up on to my knee before I could remonstrate, and sat there perfectly contented while I stroked and rubbed it as I would have rubbed and stroked a cat—every now and then muttering, "Pretty cockatoo, cockatoo is a pretty creature."

After three days of gossip among my many friends, I started in the great French ship *Amazon*, with a good cabin but unpleasant people. The Dutch passengers sulked by themselves at one table, the Chinese at another. I was put among a mixed lot of Britishers, and never spoke a word for four days. There was a good deal of sea too off Sumatra. At last a wild Irishman, who had been wandering all over Australia and New Zealand with his eyes and ears open, took compassion on me and landed me and my trunks at Galle, after which he went on to pass the winter on the Nile and "see if there was anything to shoot there."[90] What a killing race the British are!

The Oriental Hotel at Galle is famous all over the world. I stayed there ten days, and saw it in its different aspects. On mail-days some hundred people thronged into it, and the street outside became a perfect bazaar. A crowd of ragamuffins of every sort and nation were to be seen there, amusing and cheating the Britishers in the verandah—the latter not being a choice collection of their kind, and much given to "brandy and soda." Young reckless boys, with hats on the back of their heads, sent out of England to make fortunes because they were incapable of doing anything at home, are not the class to succeed as emigrants. They dressed in all sorts of ridiculous head-coverings, and used strong language, because they thought it "manly" to do so. Then there were plenty of limp ladies, babies, and nurses, going home escorted about by poor used-up Anglo-Indians. There was a large old monkey which played tricks, and had done so for thirteen years, whenever the mails came in. His master found showing him off so lucrative that he had refused very large offers to buy him. The monkey looked horribly bored, and hated the sight of an Anglo-Indian. He had quite a different manner when I met him one day between the mails; he shook hands and seemed glad to see me, but could not abide mail-passengers. There were men with little parrakeets which sat on their fingers and were all drugged, and died as soon as they were bought from the ultimate effects of it; Jews with sham sapphires, for which they asked £30 and took a rupee (they were all glass), inlaid boxes, ivory elephants, bangles, sticks, crochet-work and lace from mission-schools; no end of clatter; also people who had sham fights with the two policemen, ran away, and came back on the other side.

On other days than the mail, Galle was quite dead and every one slept, not a soul moved in the streets. Mrs. Barker, the landlady, made me most comfortable, sending all my meals into my room, and I fixed on a "garry" driver I liked, and had him every morning to drive me out. I do not think I knew what cocoa-nuts were till I saw those at Ceylon; there they are the weed of weeds, and grow on the actual sea-sand. The sand was most golden, and the tropical crabs ran over it like express trains. There were also lovely rocks of rich red and golden tints scattered about in front of the sea, and the edge of the sand was bordered with the beautiful sea-grape (as it was in Jamaica), with masses of pandanus on their stilted roots. The sea-waves were exquisitely coloured and clear.

Galle itself stands out into the sea, and is almost an island, with old walls and forts round it like Cadiz. The Cingalese looked quite handsome after the Malay race, and were extremely neat and peacocky in their dress, which was almost always white; their hair was most carefully combed, oiled and knotted up, both men and women wearing the same combs. One man came to converse with me when out sketching, with a quart bottle in his hands which he made me smell. "It is medicated oil; I always use medicated oil for my hair. Do you use medicated oil for your hair?" he said, all spoken in the most priggish and precise way. I used to have the carriage put somewhere under the trees, the horses taken out, and so could sketch out of reach of the crowd, who were not so well disciplined as the people of Java, and came much in one's way. But the children were very pretty. "I'se a Christian," said a monkey when I asked it not to shake the wheels. My driver and his horse had long baths while I painted, and the former drove me home with the long shiny waves of black hair spread over his shoulders and bare back to dry, just as the elegantly dressed young ladies used to do in *Punch* a few years ago.

I screamed with delight at the sight of a bright green chameleon with a long tail and scarlet comb which ran over the rocks near. My driver made a noose out of a palm-leaf and caught it for me, but the creature's scarlet comb changed to green, and he wriggled so much that I let him go again. It was quite a different creature from those of the Mediterranean.

Bona Vista is a most lovely rocky point three miles from Galle, on which are a chapel and missionary school, and Mr. M. (the clergyman there) invited me in to breakfast on his verandah. They had the most delicious views. The house stood amongst granite rocks, which were covered with parasites and ferns and shaded by palms, the blue sea some hundred feet below them. Like every one in Ceylon, they had a vivid remembrance of Miss Gordon Cumming, who nearly walked the limp parson to death.[91] I got quite tired of her name, and heard far more about her than about the beautiful country with its orchids and elephants. A good old Cingalese waited on me in the hotel; he had been thirteen years there, and wanted to go with me all over the world, he said, "because he liked me." I wondered what "Elizabeth" would have said to my bringing home a very languid old native, with a round comb on the top of his gray hair, which was fastened in a most feminine knot beneath it, and who wore a jacket and petticoat! The women wore loose muslin shirts, very much starched, with enormous sleeves.

After eight days of slow stewing, I started in an open carriage (the coach) for Colombo with two young Oxford men for companions, thoroughly nice fellows, just come from China and Japan. We sent a boy up a cocoa-nut tree at the first post we stopped at, to ascertain if it were a fact that the leaves they tied on the trunks could rustle loudly enough to alarm the owners, when thieves climbed over them. I still doubt it, but the very reputation of their doing so may help to keep off thieves. The road was most interesting all the way, near the beautiful shore or through swamps full of pandanus and other strange plants, with perpetual villages. I much missed the neat mat and bamboo houses of Java. In Ceylon they were mere mud-hovels, and everything was less neat, the people lazier, but the little bullock-carts were very pretty; every "gentleman" kept one, instead of the "gig" of England. Two wheels and a thatched "ugly" overhead made a cart, and I have often seen four people all sitting like frogs on their own heels under one ugly, with a noble little hump-backed beast to draw it, a mere miniature of a bullock, not bigger than the smallest pony, but going at a regular trot, his pretty gray skin generally cruelly tattooed in some elaborate pattern. The people, in all sorts of bright-coloured drapery, wore tall mitres on their heads. The priests of Buddha all wore yellow drapery and uncovered heads. There were plenty of flowers, many of those I remembered having seen in Jamaica.

Colombo is most unattractive, but cooler than Galle. All its houses seemed in process of being either blown up or pulled down. My hotel had "temporary" actually printed on its bills. I sent in my letters to the Governor, and he wrote me a kind note asking me to breakfast, and offering me all kinds of hospitality, but I was anxious to get up to Kandy, and Colombo did not attract me; so he gave me some more letters and sent me off in his own "garry" to the station, ordering a carriage to be reserved for me. Sir William Gregory had a mongoose brought up to show me, which ate buttered toast and snakes, killing the latter in the most clever way, springing on the backs of their necks, pinning them down, strangling them and never getting bitten itself. I have never heard any confirmation of the curious story of mongeese combining, one to amuse the snake while another killed it.

The railroad journey was a most beautiful one, mounting slowly up 1700 feet to the top of the pass, with superb views, reminding me of the Petropolis Sierra in Brazil, only less fine. Everything in Ceylon is on a small scale, but some of the granite or limestone tops are fantastic in shape, like castles or towers, and the trees are loaded with creepers. The cultivation is very scattered and poor after Java, but none the worse for that in picturesqueness. The tall Taliput-palm stands out grandly with its head of yellow flower-feathers, which only comes to it after thirty years of growth, and is succeeded by the equally magnificent but leafless head of fruit; then the whole falls by its great weight and dies. The great wild palm too with the maiden-hair-fern-like leaves, called the sugar-palm by the natives, and the ironwood tree, made a great show in the landscape, the latter with its brilliant pink leaves and shoots.

It was dark and raining hard when I reached Kandy, and I scrambled into one of the clumsy covered Irish cars of the country, beside a native in a red turban. He turned me out at the hotel, where a tribe of more idle natives looked on at me as I tumbled out over the muddy wheels. Nobody offered a hand to help or to lift my things; they never even thought of finding me a room till I got myself into a rage and scolded them. After a deal of hunting a key was found, which opened a long slip of a room with three beds and nothing else. When I declared that would not do for me the man said, "That very good, that double-bedded room." I said it was more, it was triple-bedded, but I must have a table and more room, on which at last they got civiller, found a good room, brought me some tea and a plate of half-cold hard salt beef and carrots. Such a lot of men to do it and no head! in the hotel of Ceylon's old capital, Kandy! What a contrast to the inn at Buitenzorg!

The Governor had told me Mr. Thwaites was going to Colombo to stay with him the next day, so I ordered a carriage at six, and drove over to the Botanic Gardens to catch him before he went.[92] I found the dear old gentleman delighted to see me; and, in spite of the drizzling rain, we had a charming walk round the gardens for two hours. He had planted half the trees himself, and had seldom been out of it for forty years, steadily refusing to cut vistas, or make riband-borders and other inventions of the modern gardener. The trees were massed together most picturesquely, with creepers growing over them in a natural and enchanting tangle. The bamboos were the finest I ever saw, particularly those of Abyssinia, a tall green variety 60 or 100 feet high. The river wound all round the garden, making it one of the choicest spots on earth. Mr. Thwaites showed me also his exquisite collection of butterflies, and promised to give me some of his spare ones. He kept that promise most generously; he never said anything he didn't mean, and detested everything false. He was one of the most perfect gentlemen I have ever known, and I longed to be able to stay a while to rest and paint near him and his beautiful garden. As I was taking leave, I pulled a letter from my pocket and asked if he knew Mr. L., to whom it was written, and if it was worth my while to give it? He said, Oh yes, he was his best and nearest neighbour, whom he always called the "Good Samaritan"; that I had better go and see him at once, as he was sure to be at home on Sunday morning.

So I turned down a pretty lane, and in five minutes found myself in the garden of Judge L., where his Worship was hard at work, digging in his shirt-sleeves, far too grimy to shake hands, but intensely hospitable. He made me promise at once to move my things and take up my quarters in his spare rooms, in the most perfect peace and quiet, close to the gardens and their good old director, and three miles from the gossip and "Kleinstädterei" of Kandy: it was the very nest I had been longing for. Mr. L. drove off to his work after breakfast, never returning until dinner-time, every day, and I worked all day long in undisturbed quiet. My kind host gave quarters to all the waifs and strays besides myself. My nephew Stuart K. S. and other young men used to come in for a night or two at a time; the house had seldom any empty rooms. Mr. L. started for a holiday trip to India shortly after my arrival, and persuaded me that I should be doing him quite a kindness by keeping the house going during his absence and employing his servants. I desired nothing better than to stay quiet. It was funny feeling so entirely at home in that little bungalow, and having it all to myself. Everything was left about as if the master were there—plate on the sideboard, doors and windows always open, the butler seldom at home, yet nothing was ever stolen. There was a deliciously sweet garden round me, and two dogs, two monkeys, and some other pets to keep me lively. The carriage was also to be at my orders, but I did not want it.

Kandy is a cockney sort of place, full of croquet, lawn-tennis, fashions, and scandal, but very pretty with its little artificial lake and its monastery and palace half-hidden among delicious gardens and groves of palms. The Governor-Agent and Mrs. P. lived in the old palace, and the rooms were full of quaint figures, half raised on the wall, picked out with white on a blue ground. One day Mr. P. took me to the temple and into its most holy chamber, in which the gold Dagoba or bell which covers Buddha's famous tooth is kept, always surrounded by piles of yellow and white flowers and ever-burning lamps. The original tooth was carried off from Kandy to Burmah many years ago, and is there still; but as both teeth were taken from an elephant, it did not matter much which elephant they came from. The Prince of Wales alone had been allowed to see it uncovered. The biggest emerald in the world is said to be also kept under that bell. Miss G. C. (my Mrs. 'Arris) had nearly caused a rebellion by looking through her opera-glass at the thing, when it was uncovered once during some national ceremony. I spent one day making a painting in the semi-darkness of the holy place, trying to give some idea of the yellow flowers, yellow gold, yellow priest, and yellow light of the thing, which in the hands of a great artist might have made a rich picture; but the want of air, the smell of burning tallow, of flowers, and of general Buddhism, was almost too much for even my endurance, and after an hour or so I was glad to get back to my happy garden home and to quiet.

My host was the most hospitable of men. Before he left for India he had friends to dinner most days; but it required a good deal of diplomacy out there to arrange pleasant parties, as many of the nearest neighbours were not on speaking terms with one another. I have often driven round the lake of Kandy (the Rotten Row of Ceylon) with friends in whose house I have been staying, who cut dead the people with whom I have been lunching; it was too foolish, and all for some perfectly trivial little offence. The so-called "religion" caused many quarrels, the young bishop having excommunicated all the old missionary set, and started all sorts of ritualistic fashions, which must have delighted those they both called heathens. So I saw as little as I could help of all these charming people, and kept quietly at Peradeniya, working either in my own garden or the Botanical close by, never going out without finding fresh beauties and curiosities of nature. Mr. Thwaites also never went into Kandy, and every evening towards dusk I either walked up to have a stroll with him, or he came down to me. He was a most charming companion, giving me always some new flower or fruit to paint, and having always a cover laid for me at breakfast, if it suited me to take it in his house instead of my own.

From my window, I could see a thick-stemmed bush (almost a tree) of the golden-leaved croton, and many pink dracænas, while under them were white roses as lovely as any at home. Then came the lawn, and a great Jack-tree with its huge fruit (two feet long when ripe) hanging directly from the trunk, and branches with shining leaves like those of the magnolia. A cocoa-nut was beside it—a delicious contrast, with its feathery head, masses of gold-brown fruit, and ivory flowers, like gigantic egret-plumes. A thick-leaved Gourka-tree stood also on the lawn, loaded with golden apples, but all hidden away under the leaves out of sight. The lawn was bordered by a hedge entirely covered with the blue thunbergia, hiding the road, along which great bullock-carts were constantly passing, drawn by splendid beasts with humps conveniently placed for supporting the cross-poles by which they dragged their loads. Many of them were milk-white, with long straight horns, almost parallel with their necks. Some of them had their horns curved like ancient lyres. The drivers were quite in character with their beasts.

On the other side of the road was an untidy bit of nearly level ground covered with mandioca (from which tapioca and cassava are made), looking very much like our hemp-plant; bananas, daturas, sunflowers, gorgeous weeds which much offended the tidy eyes of my absent host, but delighted me; a lovely white passion-flower ran all over it, as well as many kinds of lantana, a plant originally introduced from the Mauritius, now all over the tropics, and of every possible colour. Pretty hills of about 800 feet surrounded the wide valley covered with scrubby trees; but all looked on a small scale after Java. There was a noble avenue of india-rubber trees at the entrance to the great gardens, with their long tangled roots creeping over the outside of the ground, and huge supports growing down into it from their heavy branches. Every way I looked at those trees they were magnificent. Beyond them one came to groups of different sorts of palm-trees, with one giant "taliput" in full flower. I settled myself to make a study of it, and of the six men with loaded clubs who were grinding down the stones in the roadway while they sang a kind of monotonous chant, at the end of each verse lifting up their clubs and letting them fall with a thud. It was a slow process, but they like to work in their own way, and Mr. Thwaites said he knew by the noise they made they did work continuously, though slowly. If he had compelled them (as most English did) to work in another fashion, they would be sitting down and tired out for hours together. Their own tortoise-like way was well adapted to the climate, and amused them.

He had a clever Cingalese head-man, and employed another native to make paintings for him of all the moths and butterflies, with their caterpillars and larvæ, and the leaves they fed on, as well as of the fungi and flora of Ceylon. These paintings were done in water-colours, so exquisitely that one could see almost every hair in the insect's wings; they were all painted from the real thing, without any help from glasses.

I spent many a delightful hour looking over them and the beautiful collection of insects Mr. Thwaites had made, hearing all their habits and histories. He also took me to the different trees in the garden where they lived and fed, and showed me their nests.

Sir William Gregory was the only person the old director ever went to stay with. We both went to dine and sleep at the Pavilion the only night the Governor was in Kandy. The gardens were fine, but the house, from long disuse, looked very comfortless, as its master had not cared to live there since his wife's death. I was put into the huge state-rooms the Prince of Wales had occupied last. His Excellency showed me in, and looked himself to see if they had put my sheets on the bed, for nobody was there to be responsible but the gardener. I felt like a sparrow who had by a mistake got into an eagle's nest, it was such a monstrous place, with one of those odd bunches of flowers gardeners make all over the world, on the table—a dahlia in the middle surrounded by gardenias, then marigolds, geraniums, roses, and heliotrope. Government Houses too all over the world are nearly as incongruous. This one had a staircase only a yard wide leading to all these grand rooms. We were only a party of five, and after dinner we walked through the dark shrubberies to spend an hour with the Colonial Secretary and Mrs. B., who were also there for one night only.

The next morning at six I was at work on my sketch of the outside of the temple, and breakfasted in the old palace, when a party of Indian pedlars came and spread out their gorgeous shawls and other goods on the verandah. They made a fine foreground to the flowers and palm-trees beyond. When I got home I found at last a ripe Jack-fruit to finish my painting from. Denis, the butler, had been constantly looking up at the tree and promising me one "the day after to-morrow" ever since I came, and that one always disappeared and another was looked at with the same answer. Mr. L.'s fruit was always going to be ripe "the day after to-morrow." Though the natives did not steal spoons, fruit was considered common property. Denis was a nice quiet fellow, and took a real pleasure in making me comfortable. He brought me an extra nice tea when I came back that day, on a huge brass tray, with a figure of Buddha in the middle of it, a dancing elephant on each side, and a circle of peacocks all round the rim.

I spent three days up at Ramboddy, half-way up Nuwara Eliya, about 5000 feet above the sea, getting there partly by rail, partly by car. The situation is fine, surrounded by mountains with grand waterfalls all round it; but the country had been denuded of trees in order to grow coffee, which was treated in the usual English fashion—the bushes cut down like pollards to the height of ordinary gooseberry bushes, and the leaves pulled off to let in every ray of sun, and force them on unnaturally. The wind was very high, and I was nearly blown over when trying to paint the distant view from the head of one of the falls, with a foreground of datura-bushes loaded with their big white bells, on each side of the torrent with its huge boulders. I was still weak, though much better, but found myself unable to bear any rough journeys, was glad to get back to my quiet home again, and refused all invitations to leave it or pay visits.

On Christmas Day I found myself in the midst of small troubles. Denis was ill with fever. The two dogs were *said* to have been bitten by a mad dog, and were locked up in a wire chicken-cage (from which by the least push they might easily have escaped). On Thursday I was called out to see the housekeeper's baby girl, a month old, which had been found dead in the morning. I went into the cottage, and found the mother rolling on the ground, throwing dust on her head, howling, and writhing like a serpent. The man cried, and put the little cold stiff body into my hands, while the elder children demanded "sugar-plums," as they always did at the sight of me. Soon after it was buried down by the river, and a lot of crackers were let off to drive away the evil one. (Like the Chinese, the Tamils thought more of him than of the good spirit.) A hill clergyman came in to breakfast with me. He took a forty miles' ride every month, and always rested like other waifs and strays at Mr. L.'s. He said very possibly those people had let the child die as it was only a girl, and they had three already, and would have to give them marriage portions. It was quite a usual custom to starve them to death slowly, or poison them. That Mr. M'L. was a sensible man. He went about the hill-country to the different chapels once a month, sometimes finding three white men for a congregation, sometimes none at all. He had no idea of converting the natives, but said that near the large towns they had found out "it paid" to pretend to be Christians, so they did pretend. I did not believe anything about the mad-dog story, but thought the "boys" wanted to get them out of their way for some reason. Perhaps, as Denis was not about, they were going to have their friends on the premises, to whom the dogs would object, so I had poor old Dick out of his cage, and he walked up to the gardens with me as usual, carrying a stick and as grave as a judge (far more so than his master); but old Mr. Thwaites said I did wrong, and made me tie him up again, as he said if any one were bitten I should be blamed. I submitted, but could not bear to see the poor old dog so unhappy and a prisoner. The sight of him made me regret leaving home less, and I went to stay a few days with an eccentric old gentleman, called generally "the Baron," and his dear old wife at Kandy, who lived in a little house full of curiosities, quite under the shade of the Temple Garden, and close to its pretty lake with its gimcracky balustrade.

The Governor was coming to open the new waterworks, and a great fête was to be given in his honour. Thirteen elephants were collected to make a show. Some of these had been employed on the work itself, and, I was told, carried the great squared stones with their trunks and pushed them into their places, then made two steps back, took a good look to see if they were straight, came and gave a few more pushes, took another look, and were not satisfied unless the work were done with the greatest neatness. I had been asked to make a drawing of Lady G.'s grave, and went on two mornings to the cemetery, picking my way carefully through the long grass, so as to avoid the clever leeches which were clinging to the leaves, on the watch for the rare chance of human legs to fasten on and suck. They never got a chance with me, it only required care and short petticoats to avoid them. At the gate of that cemetery was a magnificent mass of white datura, with the small scarlet ipomœa all over it, making a most exquisite bit of colour. I had to wait nearly an hour for the key one day, and much enjoyed it.

The opening ceremony at the waterworks was most amusing. We all went down into the narrow valley, and half the white people would not speak to the other half. Mr. Thwaites, myself, and one or two other odds and ends, had to be used as buffers to keep them from touching one another. All the chiefs were collected in a lump, with such full starched petticoats they could neither sit nor stand. Pincushions with gold tassels at the corners and buttons on the tops were on their heads, and tremendous rings on their fingers. The elephants, which had had level places dug out of the hillsides for each of their four feet to stand on, were very nervous about slipping down the hill, and kept up perpetual moans. The Governor and other officials read long speeches in a high wind, which no one could hear or understand, and were very glad, like every one else, when it was over.

After this my old friend and I went down with the Governor in his special express to Colombo, where I again had the Prince of Wales's great empty room, and after a few days in that dreary grandeur I said good-bye to my kind friends, and went on to stay with Mrs. Cameron at Kalutara.[93] I had long known her glorious photographs, but had never met her. She had sent me many warm invitations to come when she heard I was in Ceylon. Her husband had filled a high office under Macaulay in India, but since then for ten years he had never moved from his room.[94] At last she made up her mind to go and live near her sons in Ceylon. Every one said it would be impossible; but when told of what she was going to do, he said that the one wish he had was to die in Ceylon! He got up and walked, and had been better ever since. He was eighty-four, perfectly upright, with long white hair over his shoulders. He read all day long, taking walks round and round the verandah at Kalutara with a long staff in his hand, perfectly happy, and ready to enjoy any joke or enter into any talk which went on around him. He would quote poetry and even read aloud to me while I was painting. His wife had a most fascinating and caressing manner, and was full of clever talk and originality. She took to me at once, and said it was delightful to meet any one who found pearls in every ugly oyster.

Her son Hardinge, with whom she lived, was most excellent, and had made himself liked and respected by all his neighbours. He spoke Tamil well. Just at that time he was much away, as the cholera had broken out and he had to go about taking precautionary measures against it, working very hard in very unhealthy localities, to the horror of his mother, who did not hide her feelings about it, quite the contrary; she was always in a fidget about something. Their house stood on a small hill, jutting out into the great river which ran into the sea a quarter of a mile below the house. It was surrounded by cocoa-nuts, casuarinas, mangoes, and bread-fruit trees; tame rabbits, squirrels, and mainah-birds ran in and out without the slightest fear, while a beautiful tame stag guarded the entrance; monkeys with gray whiskers, and all sorts of fowls, were outside.

The walls of the rooms were covered with magnificent photographs; others were tumbling about the tables, chairs, and floors, with quantities of damp books, all untidy and picturesque; the lady herself with a lace veil on her head and flowing draperies. Her oddities were most refreshing, after the "don't care" people I usually meet in tropical countries. She made up her mind at once she would photograph me, and for three days she kept herself in a fever of excitement about it, but the results have not been approved of at home since. She dressed me up in flowing draperies of cashmere wool, let down my hair, and made me stand with spiky cocoa-nut branches running into my head, the noonday sun's rays dodging my eyes between the leaves as the slight breeze moved them, and told me to look perfectly natural (with a thermometer standing at 96°)! Then she tried me with a background of breadfruit leaves and fruit, nailed flat against a window shutter, and told *them* to look natural, but both failed; and though she wasted twelve plates, and an enormous amount of trouble, it was all in vain, she could only get a perfectly uninteresting and commonplace person on her glasses, which refused to flatter.

She also made some studies of natives while I was there, and took such a fancy to the back of one of them (which she said was absolutely superb) that she insisted on her son retaining him as her gardener, though she had no garden and he did not know even the meaning of the word. As she could not flatter me by machinery, she did so by letter to one of her sisters, and read me a description of myself which might serve as an elegy when the subject has been put under the ground and out of the way of blushing. People often talk to me of the quickness with which young girls make friendships, but I never heard of any so quickly made as this with Mrs. Cameron; and when I admired a wonderful grass-green shawl on her shoulders, she said, "Yes, that would just suit you," took a pair of scissors, cut it in half from corner to corner, and gave one half to me (which I have on at this moment). She had brought out a treasure of a maid called "little E," who made herself quite happy, and helped her mistress unselfishly and devotedly.

While I was at Kalutara I saw the first live snake I had seen in Ceylon. I left my sketching chair under the trees when I went in to breakfast one morning, and on my return saw a beautiful bright-green thing on the back of it waving in the wind. My spectacles not being on, I thought some one had put down some new grass or plant for me, and put out my hand to take it, when it darted off and was lost, and "I did not remain!" It was a riband-snake from the branches of the trees, said to be poisonous. Since that day I have always worn spectacles, and have seen no more live snakes.

I left Kalutara in the midnight of the 21st of January 1877, the whole family, including little E, going down the hill to the Judge's house with me to wait till the coach came. I had tried in vain to find some means of going by day over that beautiful high-road to Galle, but could not even get a bullock-cart, so was packed into the public conveyance with four natives and lots of bundles. They all crammed themselves into the least possible room, and the sea and palms were so beautiful that I almost think the enjoyment balanced the discomforts. After a day's rest at Galle I went on board the *Scindh*, a splendid French steamer, on the 24th, which brought me to Aden by the last day of the month, and to Naples on the 11th of February. When I first went on board, my old friend Mrs. 'Arris pursued me: "Miss Gordon Cumming, I believe," said a tall Anglo-Indian, taking off his hat, and it took some argument to persuade him I was not that famous traveller.

It was all summer as far as Aden, but the morning before reaching that place a sudden wave rushed in at the port-windows of the saloon as we were sitting at breakfast, and before they could be closed another wave followed it. We all mounted on the seats to get our feet out of the water. It was a most absurd scene, with three inches of water over the floor draining into all the side-cabins, and soaking all the luggage which happened to be on the floors. These waves were said to be the remnant of a recent storm, and no more came, but it was bitterly cold in the Red Sea. I never dared to leave the cabin and its stove; but the French officers organised "regular marches round" three times a day, with the children and a band in regular tramp, to keep up the circulation, and had all sorts of active games—puss in the corner, blind man's buff, and dancing, in which all the French joined, while the English looked on. The poor piano was also much tortured. One old Dutchman went on "trying" things for hours, while another "tried" the violoncello most excruciatingly. It is strange how "a little music" is thought charming in a man, and how much a poor woman is required to do before she is even bearable. The party was not particularly interesting on board, but we had a large proportion of English gentlemen and ladies, and after all they are the most agreeable people one could meet in a ship. I seldom went on deck, but when I did I had all sorts of comfortable chairs pushed to me, shawls heaped on me, and other kind little attentions, and met with small unselfishnesses I did not find amongst other nations.

A good-natured young red-haired officer sat opposite me at dinner, with a pretty little sister-in-law sent home with two small babies, to save her life. She was a mere child herself, only twenty; and when she was ill, it was grand to see the young fellow cut up the children's food and carry it on deck. I only hoped they liked pepper, for he used to empty the castor over the plates. We had one horrid example of a Becky Sharpe—a pretty woman in a blue satin dressing-gown, who used to hang up her poor child in a hammock in the passage half a mile from her cabin at night, with scarcely any covering over it.[95] The poor little beauty was too frozen even to cry, and would certainly have died if a kind sailor on guard had not taken it to his own berth. The child was utterly neglected; such a lovely little creature, with golden hair and blue eyes, looking as if they saw right through this mortal world and out at the other side on things we know not of. Every creature in the ship petted it, except its father and mother. One evening I found it on one of the farthest tables in the saloon, all alone, to be out of the way, the silent tears running down its face from positive fear, for the ship was rolling and it had nothing to hold on by.

The good-natured Dutch mothers said its mother was not a woman but a devil! How ugly they were! I had not noticed it so much when living quite amongst them, but in that ship, among the English, French, and Italians, one saw it more; and they still practised their slipslop kind of dress, only putting on more slipslops one over the other, as the climate got cooler, till they looked like barrels.

The Anglo-Indians as a rule were innocent of foreign languages. I heard two of them grumbling over the wine carte: "So odd they put down neither Claret, nor Bordeaux, nor Burgundy! *Garçong! avez vous Burgundy? Macon?* no; I don't want any of those second-class wines, I want good pure Burgundy." They took an amazing quantity of stimulating drinks, and mixed champagne with porter, ale, claret, etc., both men and women, in a way which proved them to have wonderful constitutions left, to survive it, in spite of the much-blamed climate. The children were great fun. I heard a little boy telling a group of others his discoveries at the other end of the ship: "Then I saw a mamma poodle dog, a papa poodle dog, and their whole family of poodle dogs; and there was also a boy poodle, but the boy was very drunk."—"How did you know he was drunk?"—"Because whenever he was told to stand up or beg, he always tumbled down again."

The colours about Suez were more lovely than ever; every tint was pure, yet all harmonious, and all very faint, a mother-of-pearl shell the only thing like them; I wondered every time I passed why great artists did not go to paint there. Brett could do it as well as any one if he would. The wake of the ship, with the colours deepening as it came near out of the fairy-like distance, and a long perspective of sea-gulls following; the near ones were whiter than the white clouds, with bright green reflections on their lower side from the green sea below. The canal itself has an odd picturesqueness; sometimes it is a large ditch, then widening out into a broad lake, with the same delicate and pure colouring.

We arrived at Naples on the 11th of February. I had a twenty-franc piece and one or two Ceylon notes left in my pocket. It was Sunday, and I knew no bank would be open. The Carnival was going on, and tempted me to go and see it at Rome, rather than stay in Naples among strangers; so I asked the stewards if they would change my Ceylon notes—"Pardon, madame, but we do not take that money here." Then I asked them most humbly if they would accept a present of one as their fee—"Yes, madame, with many thanks" (showing the difference between taking and giving!) One of my fellow-travellers changed my only remaining one, as he was going back in a couple of months. I landed through all the noise and ragamuffins, and drove to the railway station, where I had to wait four hours for the train, having telegraphed to my old friend, Miss Raincock, to expect me, and to get me a room in Rome. I bought a time-list, and made up my mind I could not afford to go first-class or to have any luncheon. The Jew boy at the bookstall helped me, telling me my money would go further if I changed it into paper, which the railways were obliged to take. He went with me across the street, saw I got the right exchange, and said I gained two francs in the transaction; then made me go and put my luggage in the baggage-room and get a ticket for it, as things were not safe in the lobbies there: he gave me to understand that the railway porters in Naples were all thieves.

A nice Englishwoman who had an hotel near Naples came and talked to me, made me go and have a cup of coffee and put a glass of brandy in it before I started, when I found after all I could go first-class. It was very cold, and I was glad of the comfortable carriage and foot-warmer. The sunset was lovely over the beautiful Italian landscape. A pair of young Italians were kind to me, and insisted on my sharing their capital supper-basket; and I felt quite rested, and so delighted to see my old friend at the station in the same funny old bonnet I had last seen her in, and the same dear old face under it. She had found a nice room for me near her own quarters. Although the cold was intense, I could not resist wandering through the glorious galleries of the Vatican, with their marble floors and thorough-draughts. We went to see the poor horses goaded on through the Corso, and all the trumpery show and masquerading, and I wondered how reasonable men and women could make such fools of themselves. Rome was becoming more and more unattractive to me; the outside alone was glorious. That wide Campagna, with its broken aqueducts and distant views of mountains and city, with the domes golden in the sunset or sunrise, the views from the Pincio, with the fine foliage in the foreground, are still lovely; but its shops, palaces, dry old ruins, and conventional pictures, both old and new, attract me very little.

After three days' freezing, I went on for two days' delicious sunshine in General MacMurdo's lovely garden at Alassio, then to see Mr. Lear at San Remo with his cosmopolitan gallery of sketches, my brother-in-law, Sir J. K. S., and my niece at Cannes. He has always been a most kind friend to me, and had written to me regularly every month all the time I was away. It grieved me to see how suffering he was. I stayed three days with him, and never saw him again, as he only returned to die a few months afterwards. I had promised Mrs. Cameron to convey a large china tea-pot in a straw cover to her sister, Lady Somers, at Cannes. It had been no little trouble to me, all the custom-house people persisting it was a barrel of spirits! and when at last it was opened, I found to my horror the old lady had filled it full of Ceylon tea, which might have caused even more trouble if discovered by the officials. But it gained me the pleasure of seeing one of the most beautiful women who ever lived, and a beautiful face, like a beautiful flower, never fades from the memory. It is a pleasure for ever.[96]

I went straight through from Cannes to London in thirty-six hours, arriving at midnight on the 25th of February 1877. After which I enjoyed six months with my friends in London and in the country, the chief event being a visit the Emperor of Brazil paid to my flat at eleven o'clock on the 20th of June, when he looked at all my curiosities and paintings, and told me about my different friends in his country, forgetting nobody that he thought I was interested in, with his marvellous memory. (He took me, between two visits, to a prison and a museum!) Another event was, that same Kensington Museum sending The M'Leod and Mr. Thompson to look at my different paintings, asking me to lend them for exhibition in one of their galleries.[97] Of course I was only too happy that they thought them worth the trouble of framing and glazing. I was still more flattered when I heard afterwards that in the cab on the way to my flat, Mr. T. had said to the Laird, "We must get out of this civilly somehow. I know what all these amateur things always are!" but in the cab going back, he said, "We must have those things at any price."

I employed the last few weeks of my stay in England in making a catalogue as well as I could of the 500 studies I lent them, putting in as much general information about the plants as I had time to collect, as I found people in general wofully ignorant of natural history, nine out of ten of the people to whom I showed my drawings thinking that cocoa was made from the cocoa-nut.

# CHAPTER IX

## INDIA

### 1877-79

I LEFT Southampton once more by the *Tagus* on the 10th of September 1877, touched for a few days each at Lisbon, Gibraltar, and Malta, and landed at Galle, in beautiful Ceylon, on the 15th of November, took my passage to Tuticorin by the next steamer, and spent some of the intervening days in visiting old friends in the island.

Ceylon looked even more lovely than it did the year before. The cocoa-nuts, with their endless variety of curves, were always a marvel to me, how they kept their balance, with their heavy heads and slender trunks leaning over the golden sand, and within a few yards of the pure clear sea waves. The moon shone gloriously, silvering all the bananas and palm-trees, and the phosphorus glittered on the sea.

I took a carriage from Pantura and drove ten miles to Kalutara. The road was a series of beautiful pictures all the way. I found Mrs. Cameron much as I left her, the old man even younger and happier; and I had the greatest difficulty in escaping without the other half of that green shawl, its enthusiastic owner running down a short cut to overtake my carriage and fling it at me again, but she missed her mark. Poor thing! her life is over now. I wonder who wears that bit of green cashmere, and if it ever will meet my half again.

I went to Peredeniya the next day, and up to the Botanic

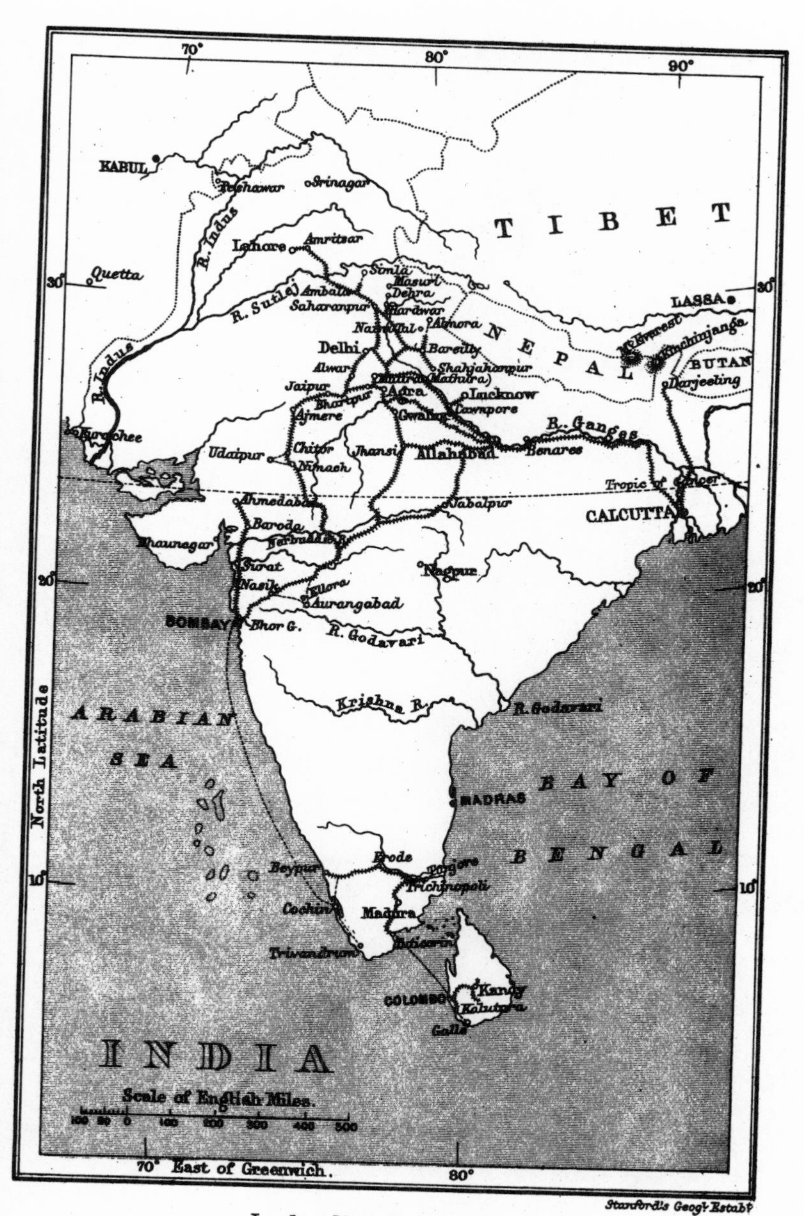


London: Macmillan & Co.

Garden to see my old friend the Director. I had a long stroll with the dear old man, who looked much aged, and so delicate that a touch might have knocked him down. The next morning I started before daylight to catch my steamer, but the heavy rains had broken down the roads, loosening great rocks which blocked the passage entirely. I had good friends who stuck by me all day. We met the Chief Justice, who shared the contents of a capital luncheon-basket with us, or we should have been starved, for there were no provisions on the sierra. We did not reach Colombo till 5 P.M., when I found my steamer had left hours before, so I had another week to wait on the island.

At last I embarked and had a good voyage to Tuticorin, in an excellent ship with pleasant passengers.[98] There were 500 coolies packed at the other end of the boat, returning to their homes, having acquired enough fortune by their labour to buy an umbrella and a woollen plaid apiece, and to pay their journey to see their friends, after which they would probably come back to earn more money. They were graceful figures of a beautiful dark bronze colour, very shiny, and loaded with bangles on arms and legs. They seemed always to live and to laugh on nothing.

The captain did not half like letting me risk the six-miles' row over the sand-breakers in a loaded boat, but at last we started and had a good sea. We were two hours and a half landing, but got in dry, then had to scramble for rooms at the rest-house as the train had gone. Friends were kind to me as usual, and the next morning they started me with a basket of provisions in the train for Madura.[99]

The first part of my Indian journey was over white sand covered with palmyra- and fan-palms, and cacti; then came cotton, quantities of millet, Indian corn, gram, and other grains. The richest crops were just ripening, and ought to be the best cure for famine. But I also saw many human wrecks, all skin and bone, quite enough to teach one what starving looked like. The stations were crammed with coolies, beautifully picturesque figures, loaded with bangles: some of these were as big as curling-hoops.

Mr. Thompson met me at the Madura station, and took me home to his comfortable bungalow.

The next day I was quite dumbfounded by the strangeness of the old Temple of Madura. It was full of darkness and uncanniness, with monkeys, elephants, bulls and cows, parrots, and every kind of strange person inside it. The god and goddess lived in dark central stalls to which no unbeliever is allowed entrance; but two small black elephants with illuminated faces, painted fresh in red and white every morning, wearing wreaths of flowers round their necks, were admitted into that "holy of holies," with a youth riding on the head of each, and carrying a silver vase of water. The dignity of that proceeding was tremendous. The elephants had attendants with tom-toms, great fans, and feather-flappers, to keep them cool and free from flies. At a particular spot near the centre the two separated and went off in different directions, one to the god, the other to the goddess, and were soon lost in the darkness. Each of the great beasts carried a heavy chain in its trunk, by which it was afterwards fastened up. They performed all sorts of tricks, refusing to pick up coppers; silver, though it were only a tiny threepenny-bit, they always handed over to the attendant priests, a very ill-looking race.

In these temples there is an endless variety of courts and columns, more grotesque than beautiful, with dragons, griffins, gods and goddesses, larger than life. In the entrance are money-changers, and all sorts of merchandise, a gorgeous variety of bright-coloured cloths with gold borders, which both men and women wrap round them like a petticoat all over India: these are made at Madura. In one part of the temple is a hall of a thousand pillars, all different. The judge took me to see all the temple jewels, which, if sold, might have helped to stop the famine in South India. The sapphires were as big as nutmegs, and there were other grand stones in abundance. One elephant-cloth was gorgeous, all embroidered with seed-pearls, gold, and silver. The priests offered us a dance, and the dancing-girls began wriggling to their wild music, seeming just the right things for the place; but my friends did not approve of them, and we did not stop.

I made a sketch of one of the small inner temples, in which the god and goddess were married every year. When we came away, wreaths of sweet trumpet-flowers were put round our necks, and a lime to smell in our hands (a very necessary luxury in such a locality). Dancing-women also came and saluted us. They attitudinised backwards and forwards slowly, making angular movements of hand and arm, and muttering compliments, while a band of musicians accompanied their movements with drum, fife, and voice, one old man jerking his head backward with every yell in a way that made one expect to see it roll off every moment.

I painted a sunset view of the grand tank outside the town, with its island-temple and palm-trees, grand old banyan-trees and other temples on its edge. The English people drive round and round it every evening, and make that drive their chief gossiping spot.

Starvation, floods, and fever were all round. The railway was washed away in nine places, and I could not have left it even if I had wished. The gardens were all flooded, the great river rising, tanks breaking on all sides. The pandanus pines alone seemed to enjoy it, being buried in water up to the tops of their odd stilted roots. Every one was taking opium, so I followed the fashion, prevention being better than cure. The punkahs were kept going day and night, to blow away the mosquitoes; and if the punkah-pullers went to sleep, Anglo-Indians had squirts by their bedsides, to squirt water at them through the Venetian shutters.[100] The natives had grown so much into the habit of expecting this, that they used to hold umbrellas over their heads (made of matting).. They got two rupees a month for half a night's work. I saw some terrible remains of starvation about the streets. The relief-camp was on the other side of the river. It was so flooded that people could not cross without difficulty, and were constantly washed away and drowned, and I have seen them climb up the stone posts put to mark the ferry, and cling on to them for hours, till an elephant was sent to bring them on shore.

The squirrels in the roof were a great amusement to me. I watched one of those tiny creatures playing close to me with its child. At last the young one got on its mother's back, was carried right up to the roof, and along the rafters to the corner where the nest was, when a whole tribe of little creatures with ridiculous tails came out to meet it, with much chattering. Then the old one came down to watch me again. They were very tame, and not longer than one's middle finger, with tails rather longer than themselves, and the most intelligent ears.

The rain came on again worse than ever. It was impossible to get away and dangerous to stay. The river was overflowing, and we were only a couple of yards above it, tanks bursting on all sides. No road out of Madura was practicable for more than a mile : all were lost in the floods.

At last I got away in a special carriage ordered for the judge and his wife, and we reached Dindigal at sunset, where Mr. M. met me, and drove me to his charming wife and home. They pressed me to stay and make an excursion to the Palani Hills, 8000 feet above them, which were most lovely ; but I could not linger, and at five o'clock I was in a long chair on the shoulders of four coolies, with eight more to relieve them, and two peons to drive them on. They carried me splendidly along the broken railroad, changing one by one under the bamboo-poles without stopping, and setting me down to walk over the gaps and planks where the road was yet unmended. At last we reached some waggons, full of coolies, which took me to the train, and so into Trichinopoli, where Colonel F. met me and drove me out to his home at the camp. The river here was broader than the Thames, but so shallow that people were walking through it in all directions, instead of over the noble bridge. The road beyond the bridge was arched over by the interlaced branches of banyan-trees, with their hanging roots tied up in knots to get them out of the way of the passengers' heads. The streets seemed very gay, after poor fever-stricken Madura.

On the 24th of December 1877 I reached Tanjore by the earliest train, asked a policeman I saw to show me the way to the Doctor's, and walked under his porch about nine o'clock, to his great surprise, as he was sitting among his books deep in work, having expected me by a later train. Living with him was like living with a live dictionary, and was a delightful change. He had another clever man spending his Christmas holidays with him, Dr. N., who had written a book about Indian snakes, and the two talked deliciously together. I had a delightful upper room full of windows, looking over some miles of country. My table was loaded with different valuable botanical books (including MSS.) An old ayah sat on the floor in a corner to wait on me and watch me, much to my discomfort, but as the doctor said it was absolutely necessary for her to do so, I submitted. He had all sorts of sacred Hindu plants ready for me to paint (he having undertaken to write their history at the same time, and to publish it some day with my illustrations). He made me feel quite at home, and in no hurry. He and his friend showed me the splendid temple, lingering over all its rare bits of carving and inscriptions till I felt at home there too. I know no building in its way nobler than that temple of Tanjore. The colour of its sandstone is particularly beautiful; its whole history is inscribed round the basement in characters as sharply cut as if they were done yesterday. I did one large painting of the outside, driving every afternoon to the point of view I had chosen, where the Princess of Tanjore had ordered a small tent to be put up for me, and a guard of honour to attend me!

It was comical, but most luxurious. After the men had saluted me, they retired to a distance, and did not worry me. Dr. Burnell used to bring his papers and work there also when he had time; he worked day and night, never resting.

One afternoon we went by invitation to the Palace, in our best clothes. We drove in to the court through the gates, and were saluted by six elephants, all throwing up their trunks and roaring at us, much to the disgust of our big English horses, who had an especial dislike to them and to their ways. A flourish of drums and trumpets also greeted us, and his Excellency, the husband of the Princess, came down to meet us. He gave a hand to each, and led us under three scarlet umbrellas into a large hall, and up to the centre of it, over gorgeous carpets, till we reached a screen of silk interwoven with gold thread, chairs placed in a semicircle in front of it. Here I was delivered over to Miss Wolff, her Highness's English teacher, and led behind the mysterious curtain, where I found the Princess seated. After giving me her hand, she stooped down, lifted the curtain a little, and put out her hand for Dr. Burnell to take, who said such pretty things to her, that she kept her attention fixed nearly all the time on his talk with her husband on the other side of the curtain, thereby saving me the trouble of amusing her. She was a very pretty woman of perhaps thirty, with clear olive skin. Her head and limbs, nose and all, were loaded with ornaments, her toes all covered with rings and enormous anklets. Her drapery was of silver and gold embroidery, and she had a very sweet smile and voice. Behind her was a square frame like a shower-bath, covered with gold kincob, under which she walked when in public; but, except on that one expedition to Delhi, she had never left home. She showed me the medal and ring given by the Queen-Empress.

I felt a real pity for the poor secluded woman. She showed me photographs of all our Royal family, and a full-sized portrait in oils of herself, done by a native artist, not from herself, but from one of her little nieces who was supposed to be like her. The jewels and dress alone were taken from the real thing. I was asked the rather embarrassing question: "Was it not a good likeness?" She gave me a photograph of herself, with her autograph; and when I asked how it was that a photographer was allowed to look at her or to do it, I was told he had his head in a bag, and was supposed not to be able to see her! The husband sat at the corner of his side of the curtain, and looked round it at the Princess and myself, talking intelligently in good English to me, while Dr. Burnell talked in Tamil through the curtain to the poor woman. The little nieces were very pretty, with shining bronze skins, and draperies of green and red shot with pure gold. One of them has since married the Maharajah of Baroda, and their wedding-festivities lasted forty days. When we took leave, garlands of flowers were hung on our necks and wrists, we were sprinkled with rose-water, then led back by the great man to the carriage, in the same way we had come, looking like prize beasts, the little doctor with his thin worn face and long hair quite smothered by his flowers, real pink oleander, while I had a still more absurd artificial wreath covered with birds cut out in pith.

The great man between us was gorgeous, though his beard looked as if it had been cut by himself in the dark with a blunt pair of scissors. Drums and trumpets and elephants all saluted us again, and a highly decorated camel performed some feats in our honour as we drove off. The next day the Princess sent us two great trays of sweetmeats and requested to have my photograph and address in England.

The real hot weather came, and to me was enjoyable. I was very sorry to leave Tanjore and its good talk, such as I was little likely to meet for months to come. The F.'s put me up most kindly again, and the next morning the Colonel took me to Seringham, the largest temple I ever saw: a perfect city in itself, but very dirty and rubbishy. The view from the roof was the most curious part of it, and gave one the best idea of its great size. The terminalia trees produced a strange effect, with their rectangular branches and deep-red leaves. We dined with the Judge that night, who said I must also see the Sira Temple near Seringham; he would send a peon to show me the way. So I went, and was glad, for in many ways it was more interesting and picturesque than the other.

I reached Erode at sunrise the next morning, getting into a carriage full of sleeping people at five. It was all dark, and no lantern, all a muddle, but one of the men struck lucifer matches one after the other, till I had cleared a corner to sit in, and I watched the gradual coming of daylight. At noon-day I reached Matapolium, and after breakfast was packed into a small open carriage, with two ponies to drag it, not much bigger than cats. They went at full gallop all the way to the hole in the wood where the "Tongas" were kept. Into one of these I was transferred, and hoisted on the shoulders of six natives, who carried me up the hill beautifully, but fought continually among themselves, and at last got so much in earnest that they set me down on a narrow shelf under the most magnificent bank of hanging ferns and creepers, and after abusing one another till their eyes seemed starting out of their heads, they went "to have it out," rolling over one another on the ground, and trying to throw each other over the precipice. One old man alone kept clear of the row, and signed to me to be quiet also; so we waited till the party of savages had fought enough, when they came back bleeding, picked me up again, and continued to jog on, and when they saw the end of their journey near, they got into the highest good-humour, and set up a monotonous chant, snatching flowers from the banks and flinging them on to my knees. Such flowers! Verbenas, ipomœas, scarlet rhododendrons! They were delighted to get their extra rupee, and deserved it, for they had taken good care of me, and the fighting was their own affair: if it amused them, it did me no harm.

Grey's hotel at Kunur consisted of a number of small bungalows, dotted about on lovely terraces and gardens, round the central boarding-house, where the master and mistress lived, and had their chief kitchens. I had a most luxurious little house all to myself; it was furnished with carpets, fireplaces, four-post bedsteads, and every kind of English luxury and absurdity. An old woman kept the fire going of an evening, and washed my clothes; and a grand man in a turban brought down four great covered dishes twice a day, with tea at seven and three. I interviewed the pompous old landlord, and went round his garden with the nice landlady. They had a heliotrope hedge six feet high, with a perfect mass of sweet flowers on it. The whole hillside was one sweet garden. The quiet was most delicious in my little house covered with lovely creepers, small scarlet and white passion-flowers, with exquisite tea-roses in abundance, and every other sweet flower, real wild rhododendron-trees all round. Every hill was tinged with red blossoms; they were scraggy, shabby trees, not bigger than English apple-trees, the flowers decidedly poor, with a white eye, but of the deepest red. I took some walks through the rain and clouds. Then, as I felt my limbs and ankles beginning to swell and stiffen, I decided to avoid rheumatism by coming down again to a warmer climate, resisting an invitation to go and stay at Utakamund.

A carriage brought me by a longer road, changing horses three times. The creepers were gorgeous: many kinds of passion-flower and ipomœa. One of the latter, a large white variety, with deep-purple tube and centre, was grand. It grew in great clusters lika cœrulea. There was also a miniature copy of that one, which quite covered the trees it crept over, often weighing down a great bamboo. But there were no high forest-trees, as in Brazil or Borneo. A very little more rain would have rendered the roads impassable, so that I was glad I came down when I did. Great rocks and landslips narrowed it in many places, and the constant traffic of the bullock-carts ploughed it up. Those great beasts had their horns painted blue, green, yellow, and red, never a pair alike. The lower I got the more beautiful was the jungle, but also the more feverish, so I took the first train on to Podanur Junction, and camped for the night in the ladies' waiting-room, then went on again westward, along the banks of a sandy river-bed, often full of beasts and of people enjoying the cool water, while on the other side were the Nilgherry hills piled one over the other.

At Beypur I found a large room over the station, a hundred yards from the sea, with a garden between me and it. Also a servant engaged for me by a friend of Dr. Burnell, who had come from Cochin on purpose to attend on me. I enjoyed being at Beypur close to the sea, with no dirty town. I could walk on the rocks and sands, watching the shrimps, crabs, and other queer creatures in their own home-circles. I found an old Scotchman, who gossiped on the little pier as he would have done at home, and was delighted to find a new listener. I made a long sketch of the river and distant mountains, with endless cocoa-nuts in the middle distance, ferry-boats, and picturesque people. It was very pleasant sitting on the clean sand, but it was hot. The jack-crows were the chief objection to my quarters at Beypur. They flew in at the window and stole every small thing they saw; I caught one just hopping off with a tube of my precious cobalt one day, and only came into the room just in time to make him drop it.

I had been waiting some days for the steamer, but suddenly determined, from what I was told, to go to Cochin by back-water instead; so went by rail to Shoranur, then took two bullock-carts, with noble sleek beasts in each, put my man Alex. and the trunks in one, myself in the other on a clean mat and pillows. The driver sat on the pole, pinching the bullock's back and tail with his fingers, first one, then the other, then both at once, and *da capo.* At dark I arrived at Trichur, where we found the only two big boats were out, so I decided to sleep at the travellers' bungalow, and to take a small one in the morning. I made myself some good tea in my machine, and started before daylight in a canoe, with a fair breeze to fill the sail, which was made of six mats sewn together. The canoe was about a yard wide, and had a seat between the two roofs of matting, on which it was too sunny to sit. But I kept near it, so as to see well out. Alex. was under the other roof, and two men and a boy made the whole crew. It was most enjoyable. Sometimes we were in a big lake, so shallow that people and storks paddled about amongst the red and white floating nymphæas, sometimes in narrow canals, where tall cocoa-nuts rested their heavy heads against one another from the opposite banks, with little huts under them full of amphibious, jolly-looking mortals, who passed half their lives in the water collecting a kind of bivalve shell-fish, squatting in the shallow streams and scraping them up in their fingers, to be deposited in floating baskets by their sides. They made curry of the fish, and the shells were made into lime.

We were continually passing strange birds perching on tree-stumps, or fishing among the flooded rice-beds. They were so tame that some of the long-legged species used to march along beside the boat as if they liked company. We also passed lovely lilies, floating or standing just out of the water, both pink and white. While the sunset was still gorgeous, Alex. wanted to shut the boat up, and to put a mat over the central opening between the two roofs. I resisted, and insisted in keeping the light and air, for there was also a glorious moon rising. After that my treasure of a servant went raving mad, sobbing, screaming, throwing his arms about like a maniac, finally he threw himself and the roof over him into the water, which quieted him for a few moments. Then he began again: "O God, God, Jesus, Maria, save me! save me! O my wicked soul, God, God!" etc. Always the same cries, louder and louder, and in English. I scolded him well, told him he was either mad or drunk, piled my two trunks on the top of the seat in the middle of the boat as a barricade, and went to the farthest end of my roof, leaving him scowling and muttering, "It displeaseth me that you think me mad or drunk!" It struck me he might perhaps avenge the insult to his feelings, so I kept my candle-box close to my hand, and got as near the steersman as I could, and did not sleep. He raved all night, but at daybreak we arrived in Cochin, and he was sober again, carried my things into the rest-house and left them there.

Cochin was full of Christians and beggars. I went out for a stroll past the old church and Frank settlements and through the Jews' quarter, saw the synagogue which Dr. Burnell said was built in the seventh century, Cochin having been a port to which the old Egyptians used to traffic; later still King Solomon himself sent his ships there.

I started at four in the afternoon in a big cabin boat, with thirteen men and a tame old Moslem as a servant, instead of poor Alex., whose infirmities were well known in Cochin. My crew made a frightful noise all night, singing and rowing furiously. We passed over huge inland seas, rivers, and narrow canals again, and reached Quilon about twelve the next day. I decided to rest the night there in the bungalow, which is a mile from the river, and deliciously airy, surrounded by cashew and mango trees. The former is quite the weed of the country, and the air was sweet with the scent of its abundant flowers. It was perfectly quiet up there too, though my boatmen did come up and camp round me. They only sang when their oars were going, and we walked back to the boat at four in the morning, through the bright moonlight. Thence on to Nevereya, where we left the boat and crossed the boundary in a bullock-cart. We went on in another canoe, hollowed out of one long tree, for twelve hours more, stopping to breakfast at a cocoa-nut farm within a stone's throw of the salt sea on one hand, and the backwater canal on the other.

Trivandrum is a model little capital, buried among tall trees. I stayed there with Dr. Houston, who got me rare plants, and told the men where to take me to see the prettiest views and to sketch. The little toy houses were something between Tyrolese and Arab, with tiny double-arched windows and slender marble shafts, so small that one could not get one's head through them. I met the Maharajah taking his walk one day. He shook his hand at me, and said, "I hope you are quite well." He always said that to all Europeans. There was a fine old temple, and a large holy tank. Twenty Europeans lived in Trivandrum, who came together on certain days when the band played in the gardens, to cut one another systematically, and to talk scandal.

I returned to Cochin very much done up, and hoped for a few days' rest, but heard that the steamer for Bombay would be in that afternoon, so had to be ready, and got well over the bar at the mouth of the harbour, and on board the very nicest little steamer I was ever in, the *Khandala* of the British India Line. It had but one objection to its comfort—the enormous armies of small red ants which swarmed in it. Bed, food, hair, portmanteaux, every place was full of them. Cans and bottles all had muslin tops tied on them, through which one had to pour the water. Those ingenious little creatures had a particular liking for a sponge; no matter how I hung mine, they were sure to find it out, and a long red living line was drawn up the wall or ceiling, and over the string into the sponge, which became alive with them, and it took me some time to weed them all out before I dare use it, for the tiny creatures stung if crushed or hurt.

The entrance to Bombay is very striking, with its numerous islands and abundant shipping. We got there at daylight. I was sorry to see the number of hideous factory chimneys and coal smoke, which are doing their worst to make Bombay as ugly as Liverpool. But the old town is most picturesque, with high houses and narrow streets full of life and colour, dirt and untidiness. Curiosities from the whole world are collected in those streets. I was not allowed to stay at the hotel. Sir Richard Temple's secretary, Major R., sent to ask me to move at once to Government House; then when I begged to be excused till the next day, on account of the ball that night, for which I had no proper dress, he sent down his brougham and ordered me to return at once; so resistance was of no use, I went as desired, and had a delightful set of rooms given me, opening on to the rocks, some hundred feet over the sea, at the extreme end of Malabar Point, with far sea-views on three sides of me, and delicious air. The ball-room and other state-rooms stood by themselves, surrounded by a wide verandah and garden, with bungalows for guests, and for the suite on the other side, each building detached from its neighbours.

That night the verandah and garden were illuminated with Japanese lanterns, the supper being under an open tent outside, gorgeous carpets laid down everywhere. There were some four hundred people gorgeously dressed, a few black-coated men amongst them, some Parsis with their wives, and other native men looking like fish out of water under the glittering chandeliers, mixed up with red coats and the low dresses of the English ladies. (My old turned black silk was perhaps still more incongruous.) The Governor, however, walked me about round and round all the rooms and gardens, as if I were a grandee, telling me about every one, and all their histories and belongings, with as much minuteness as any gossipy old lady. He had only just returned from a two months' famine tour, and was to start off in two or three days on another, but told me to make my headquarters always at his house, and to go and come as it suited me, while I stayed in India. He seemed never to rest. Whatever he did he did with all his might, putting his whole energy into it for the time being; but he could no more stretch out time enough for all than I could (and time has been my constant enemy all my life). He had untiring strength, and demanded more from those about him than people with less power of hard work could bear with impunity.

The sunrise every morning from the rocks behind my room was beautiful. It used to come up like a round red ball behind the purple hills and hanging smoke of the city some five miles off, and the red-coated servant used to bring my *chota hazra* or early breakfast out on that rock to me every morning. Around me were wild peepul-trees, full of berries; erythrina-trees, with their red flowers just opening, and wild cherries. Below all was the sea, and a perfect fleet of boats with bright sails going off after fish. The first morning Sir Richard Temple gave me a walk before breakfast, showed me all the odd trees, the beautiful stable of horses, and the great stretch of brown rocks, up which a perpetual stream of women came, carrying water-jars on their shoulders, full of salt water for the roads. Every afternoon the carriage drove to "the Mole," about six miles off, where every other carriage in Bombay and all the fashionable people also went, for no particular purpose but to look at one another. I went once or twice with Mrs. R. C., but did not find it entertaining. The Parsi ladies made a great show about all the public places of Bombay, as they dressed in the very brightest China silks, and seemed for ever walking about showing off their newly-acquired liberty with uncovered faces; but they spoiled their beauty by wearing a tight band just over their eyes like nuns, hiding all their hair. The rich natives all seemed to delight in driving about in English carriages drawn by fine horses.

Mrs. C. very kindly found me a servant, a Madras man,

who called himself John. He had a gorgeous turban, bright black eyes, and most limp long figure. He and I started off by rail for Neral, after a week of luxurious idleness at Government House; and there I mounted a pony and rode up to Matheran, which, Mr. Lear had written me, was "a highly Divine plateau." The views were certainly fine, having strangely shaped rocky hill-tops in the middle distance, with almost vertical strata of different coloured trap rising some 2200 feet from the great plain and distant sea and islands about Bombay. The colours were magnificent. The floating clouds blended the whole into exquisitely rich pictures. The air up there too was refreshing after that of the hot capital.

A few bored European soldiers were the only inmates of the hotel. All were depressed with having nothing to do; the monkeys alone seemed busy, and the trees swarmed with them. They were so tame that they hardly took the trouble to get out of my way as I rode up the hill, looking upon me decidedly as an intruder, and thinking me very rude not to go up the trees and get out of their way, who had the business of life to attend to, and were collecting nuts and fruit for their wives and families; and when I saw the depressed loafers around walking on two legs I did not wonder at the monkeys' contempt for them.

But a few days were enough at Matheran, and I rode down, took the rail on to Lanawali over a magnificent piece of engineering, up the Bhor Ghat, one of the finest bits of scenery I ever saw. No one going to India should miss that ascent, though, like myself, he went no farther than Lanawali, where there is an excellent small hotel on the top of the pass, on a flat broad valley or filled-up lake, surrounded by mountain-tops, all running into very horizontal strata, gold, pink, and brown in the sunlight, with shadows of the purest purple. The plain was covered with golden stubble, fine trees dotted about, and stacks of rice and corn were built up on stilts or in the middle of the spreading trees. Flocks and herds were grazing all about, with picturesque figures of men and women to look after them.

I had the house to myself, and thoroughly enjoyed it; but my chief object in coming there was to visit the Cave of Karli, and I had asked the station-master at Khandala to send me a pony. "Pony no got come," the man said, and after waiting an hour I set off to walk. My man misunderstood the directions, and did not believe in any one wanting to go to Karli; so I walked down nearly to Khandala, then met the pony coming up at a canter. At last we came to the final climb over the hard volcanic rocks, and first to a splendid tree of the *Jonesia Asoka*, full of orange flowers and delicate young lilac leaves. The priest of the temple found me one fine flower growing through a honeycomb full of honey, which had been built round its stem. Now this was a very curious thing. Did the buds push their way through the honey and wax, or was the thing built quickly round them? I never satisfied myself which was the first perfected. The cave itself, more interesting than beautiful, is accurately described by Mr. Fergusson, who also gives an engraving of it in his *History of Indian Architecture.* While I was sketching outside it a very sacred man came out, all painted and whitened, and produced a pot of red paint, which he daubed over several of the carvings, with a defiant look at me; I believe as a sort of precaution against the effects of my evil eye. I took his portrait, and as he did not take mine I think I had the best of it. He was dressed in yellow paint and a red mantle. . . .

The steam-launch had been ordered by the Governor himself, before he left Bombay, to take me to the Island of Elephanta, and on my return I spent a day among its strange old idols and semi-darkness, looking out from it on the dazzling blue sea and sky, lilac hills, and graceful fan-palms. Strange that such a dry spot should be so feverish! But the poor man who lived near it and acted as guardian, said he was never without fever, and every one who had tried to live there had suffered in the same way.

On the 26th of February I left Bombay at eight in the morning. A railway took me up another splendid pass to Nasik Road, where I transferred myself, my luggage, and John into two low dog-carts, drawn by ponies, with the driver sitting astride on the pole, his feet clasped under it. At the end of the pole was a crossbar, which rested against the ponies' shoulder, over a pad, and was simply kept in its place by a loop of rope. The little things went at a great pace, and took about an hour to drag us over seven miles. Sir George Campbell wanted to make Nasik the capital of India. It is certainly a most picturesque old town, with steep busy streets gaudily coloured and carved in wood and stone; the river banks lined with temples most beautifully ornamented, and paved with stone, having grand flights of steps, and many causeways across. It looked like a series of tanks rather than a river. To the Hindus it was very sacred. Their long series of temples led to its very source, which was said to be refilled by the Ganges itself once in every thirteen years, by underground and rather incomprehensible channels. There was a perpetual fair going on by the river banks, among all the temples; crowds of gaily-dressed people were always bathing, washing, and filling their chatties or water-jars, far too many for any comfort in sketching. I was overwhelmed with the amount of subjects to be painted, and could do nothing well. The streets were as attractive as the river, particularly the metal-bazaars, where every one seemed to be trying who could make most noise. Some of the old chatties of mixed copper and brass were tempting to buy, but too heavy to carry, and I resisted the temptation.

We drove out a few miles from Nasik to an isolated round hill, which had a complete circle of cave-temples round it called the Chenmar Luna, with three hundred steps leading up to them. Many were much ornamented and coloured inside, and the views over the plain of the Godavery were fine. It was cold in the night and early morning, and I had to wear gloves to keep my hands from being chapped and frozen. We had nine hours of dust and jolting on to Aurungabad, changing ponies every six miles. I found a nice bungalow there, and, after a bath, had time to drive round the town before dark. It belonged to the Nizam. Its people were a wild set, who seemed rather inclined to mob me when I tried to do some shopping. My change was given me in cowrie shells. The great sight was a tomb of King Aurangzib's beautiful daughter —an imitation of the Taj in white marble, looking very pure amongst the green trees and flowers which surrounded it.

The next morning we went on to Daulatabad, the famous Indian fortress. It has 120 feet of sheer perpendicular precipice all round it, and many subterranean passages, stairs, and halls, through which one must pass to get to its top. I saw a tiger-trap outside, which had lately caught its game, a sick kid being the bait. After leaving that horrible fortress, we drove on to another ruined city—Roza, which has some very beautiful mosques, and tombs of kings, with doors of silver and gold, and half-precious stones in the inlaid pavements. In one of these tombs, four large earthenware chatties were hung between each arch, for the doves to build their nests in. They would have made a pretty study of white and gray. It was a curious change to come suddenly among Moslem things and ways again. Just outside the walls of Roza was the edge of the high table-top and precipice. The whole was covered with tombs of all sorts; one large one had been fitted up as a mess-bungalow by the officers at Aurungabad. I had leave to stay there for two nights. It was an octagonal building, with a domed roof, and high arched recesses: a most delicious room, comfortably furnished, and supplied with shilling railway novels!

A mile farther down the steep road was Ellora, where I found twenty-four caves of every age and variety of design, but all insignificant compared to the great Kylas, which is a perfect cathedral, standing on the backs of some hundred elephants, nearly as large as life, all cut out of the solid rock. I never saw any building so impressive and so strange. The front of the temple is level with the face of the cliff, which surrounds it on the three other sides. They seem to have begun by cutting a deep moat round those three sides, so as to leave the square block in the centre detached, after which they excavated the different halls and staircases, covering the whole with the richest and most fantastic carving. Galleries are cut in the outer surface of the surrounding cliff, by which one could get the best views of the huge building at different levels. It happened to be a Hindu festival while I was there. People came to the caves from all parts, and camped outside, which made it still more exciting and picturesque. My attempts at painting were much hindered by the ants, which seemed to have an especial taste for oil-paints, and they ate a good deal of me up too on their way to my palette. I wished I could have had a tent and plenty of time in such an interesting place, for such buildings could not be sketched in a hurry. In the plain below were grand tanks, tombs, and trees, and a kind of fair was going on among them.

At 2 A.M. I was rolling on again, in one of the comfortable carriages of the G. I. P. Railway, and slept nearly all the way.

The old town of Jabalpur is full of picturesque bits; one tank especially, surrounded by lime-trees, white temples, and palaces. Masses of picturesque boulder-stones are scattered about in the neighbourhood of the town, with houses and tombs under and over them. Some seven miles off are the famous Marble Rocks, so cold and hard, dipping their perpendicular scarps into deep green water, with the most exact reflections under them. A good bungalow is built on a hill above—quite an ideal place for an artist to stop at; and on the banyan-tree hanging over the house was a group of half-tame monkeys, who were accustomed to being fed by the people who made picnics there, and were most impertinent. One of them nearly ran off with a paint-brush I dropped, but found me too quick for him, and retreated chattering.

I started at night again for Agra, which I reached on the morning of the 14th of March. The ground all round the city was pure dust,—one ate it, breathed it, drank it, slept in it,—but the place was so glorious that one forgot the dust entirely. I went that same afternoon to the Taj, and found it bigger and grander even than I had imagined; its marble so pure and polished that no amount of dust could defile it; the building is so cleverly raised on its high terrace, half-hidden by gardens on one side, and washed on the other by the great river Jumna. The garden was a dream of beauty; the bougainvillea there far finer than I ever saw it in its native Brazil. The great lilac masses of colour often ran up into the cypress-trees, and the dark shade of the latter made the flowers shine out all the more brightly The petræa also was dazzling in its masses of blue. Sugar-palms and cocoa-nuts added their graceful feathers and fans, relieving the general roundness of the other trees. The Taj itself was too solid and square a mass of dazzling white to please me (as a picture), except when half hidden in this wonderful garden, though on the river side it was relieved by wings and foundations of red sandstone. The gates, which are chiefly of that beautiful material, would in themselves be worth a journey to see, so graceful and exquisitely finished are they. It was some days before I mounted the terrace and went inside. Like a great snow-mountain, I felt I wanted to know it well from a distance before I dared approach nearer; but the more I studied it, the more I appreciated its marvellous detail and general breadth of design. The interior is most elaborately inlaid with jasper, serpentine, amethyst, and other half-precious stones—many of which have been ruthlessly picked out by barbarians—of different tints. The old palace-rooms in the Fort were even more lovely in their way; and I used to go there every afternoon, and to the Taj in the morning. Some of the balconies hanging over the old walls of the fortress seemed too fine for human beings to live in. I feared to break them with my weight as I wandered about them, with their windows of marble lace-work, all so pure.

My friends drove me over to Sikandra, the tomb of Akbar, a wonderful building with magnificent gates, standing in a large walled park full of fine trees. The patterns on the marble tomb at the top of all were exquisite, and I spent a morning in trying to trace them. The tomb now has nothing but sky over it. Fergusson said it was intended to have a marble dome; the judge said no, only an awning. I think both: a dome in the centre merely to cover the tomb, and an awning stretched from it to the side to shelter the pilgrims who came to visit it. One night I went with Mrs L. to see the Taj by moonlight. "All the world" was there, with a band, ices, tea, and scandal. I preferred it by daylight, with its setting of trees and coloured flowers, and perfect quiet.

I was not home till one o'clock, and the next morning off to Sikandra at six; and after a long day's work on the roof, I dressed at the bungalow, had a bath, and tea, and drove to dine at Judge R.'s on my way home. The next day I had a touch of fever, and went to Fatehpur for three days' change, to try and cure it. My lodging was a two-storeyed building, ornamented inside and out with a perfect sampler of marvellous stone embroideries, every panel being of a different pattern, in red sandstone, and all too perfect to be picturesque. There were many other exquisite houses and palaces, with great paved courts between them. Even the beams in the roof were made of stone; they were fifteen feet long, supported by deep niches and hanging pomegranates. A lovely marble mosque and tomb were also on that flat hill-top, with walls and gates all round, all too perfect, though entirely deserted, Akbar having left the place and moved his capital to Agra.

It was a melancholy scene of desolation, without any growing things to humanise the dry stones. I was too ill to do much else than doze, and trace a few of the patterns. I began to hate all architecture, however beautiful, and to long for green growing things again.

I saw odd things on the road back to Agra: six camels carrying six haystacks; camels pulling double-storeyed human cages on wheels at a trot; and the common cab or *ekkeh*, with a high shower-bath awning over the passengers, who sat on their own heels under it, with a wheel on each side of them, the driver on the pole between two bullocks, pinching their backs with his finger-ends continually to keep them going.

But I was very ill, and found it of no use fighting longer with the dry heat of Agra, so started by the rail at nine at night for Bareilly. At five in the evening of the next day we started again, in a great box on wheels, with a board to put up between the seats to make a bed of, changing horses every six miles. So we jolted through the darkness, stopping in the middle of the night to feed. I drank five bottles of soda-water that night, and still felt thirsty, and wondered I survived it. But at daylight we were close under the mountains, and we turned in among them, and stopped at Kanibagh, a place with large mango-trees all loaded with bloom, and a running river below, which did one's eyes good to see after the months of dust. I felt better, and could eat again, and after a night's rest was carried up the hills in a dandy (by two bearers at a time, changing places every five minutes), in four and a half hours, to Naini Tal. In spite of the earliness and dryness of the season, I saw many beautiful flowers: great masses of blue plumbago, pink gentian, white and pink bauhinia, and the judas tree, a perfect mass of pink without any leaves. Other bushes were covered with small red or blue flowers, looking like almond or peach blossom. The hills were marvellously blue, piled one over the other beyond them. I never saw such abundance of pure colour; but they said in a few days all that blossom would shake off, and I found it was so. When I returned a few days after to sketch, it was already gone.

At the top of the pass, I came suddenly on bazaars and a bustling native village, then descended to the lovely green lake, with a road on each side, high hills all round, dotted with English villas, and a little town at the farther end—all very unpicturesque. The Anglo-Indian, having apparently been tired, like myself, with the amount of exquisite buildings on the plain below, had determined to make those he built here as prosy and ugly as possible, and had succeeded admirably. I was the second arrival of the season, and had the best rooms in the hotel, on the ground-floor, with a delicious balcony in front, in which I sat and rested in a "long chair" all day, and looked out at the lovely lake and wooded hillside opposite. People were always going and coming, giving me entertainment unknowingly. I was shaded by pretty ilex trees, with white buds and white backs to the leaves, pinkish young shoots, and feathery flowers. Scarlet rhododendrons and white clematis were almost the only flowers out. The air and sun were delightful, though it was really cold at night. The lake gave a certain moisture to the air, which I had long been wanting.

One young officer, with a retriever-dog, shared the big house with me, and was kind. My servant was kind also in his way; he delighted in bringing me food, watching every mouthful greedily I did *not* swallow, and which he could stow away afterwards in his greasy bundle or brass pot. But I had a letter of introduction to Dr. Cleghorn; sent for him, and asked him to doctor me, which he did in the kindest way, and most efficiently.

In less than a fortnight I was a new creature, and able to walk over the hills, which were just then showing their spring foliage. My one officer had increased to eight before I left. Each of them had one or two dogs, and they were all very good to me, the one lady. But the snow also came down into the valley, so I determined to go down again, to seek my sacred plants in the Saharanpur gardens.

The last day in Naini Tal I spent shivering over an unwilling fire in a room too dark to work in, from the clouds. I found it difficult to get men to carry me down. The natives have a perfect horror, like cats, of getting wet. John got ten together before breakfast, and showed them to me, all huddled up in a circle, looking like bundles of damp dark blankets. After breakfast all but three had decamped, though it was not actually raining. We could not wonder at their dread of getting that one garment wet, poor things. My friends, the nine young officers, all offered me waterproof sheets and cloaks. I took one to cover my knees, and started at last, greatly delighted at the prospect of getting warm again. All the blossoms which had made the way up so lovely were gone, but the lower valley was lined with the "sâl" (*Shorea robustea*) in full bloom, a perfect cascade of yellowish-white flowers scenting the air, like our own lime-flowers in May. It is one of the best of Indian timber-trees.

At Ranibagh, at the foot of the hills, troops of helpless women and children were trooping up in different stages of fever and limpness. The bungalow was overflowing, and the one strong practical Englishman was wanted everywhere, cuffing and kicking obstinate coolies, making bargains, holding babies, cording up extraordinary packages, helping every one. Not "his" work, he said, for he had come in for rest, but he could not bear to see those poor English women and children left to the mercy of native servants alone. Between whiles he sat and talked to me, and told me of a lovely lake near, quite covered with pink lotus flowers. I took one of the returning carriages to Bareilly, then on by Aligarh to Saharanpur.[101] The next morning I drove to the gardens soon after daylight, and called on Mr. Duthie, almost before he was dressed; but he soon came down, and walked about the gardens with me. He found out the trees I wanted, few of which were yet in flower, but he said he would let me know in time to come down and paint them, even if I were up in the hills. He had expected me for some time, and had arranged with Doctor and Mrs. J. to take me in when I came. I very soon knew every one in Saharanpur, but I did not stay long, as my plants were still some way from flowering.

On the 24th of April I started in a dawk carriage (a heavy wooden close fly) for Rurki, where we crossed a fine bridge, guarded by two splendid stone lions, over the Ganges Canal, which is one of the grandest pieces of engineering in the whole world—not a sluggish ditch, but a rushing snow-fed river. There were massive locks and gates, looking solid as the work of old Egypt. I was suddenly asked where I was going, and had not the least idea, beyond remembering that it was to the house of the Head of the College for Engineers. We soon found Major B. in the road, looking out for me, and his wife gave me a most kind welcome and a comfortable room.

The next morning Major T., the head of the canal works, called on me in his dog-cart, and drove me up the edge of the canal to Hardwar. It was one hundred and seventy feet wide in some parts, and ten feet deep. The rivers (then dry) which rushed across it in the rains were taken, some over, some under it, in a very ingenious way. Major J. drove the most beautiful thoroughbred horse, and had sent on two others to different stations, so that we went very fast, and arrived at his bungalow at the junction of the river and canal in time for breakfast; after which he took me to my own quarters—a delightfully airy little house at the very upper end of the fork, with rushing water on three sides of me, and Hardwar about a mile above me. I could just see a few of its domes and towers through the trees, with lilac hills beyond, and the snowy mountains over them (when the clouds allowed them to be visible). It was not picturesque, but decidedly wholesome quarters.

The colonel commanding the district sent up a peon with John and the luggage, to stay and take care of me, and an elephant was put at my command; but one ride was enough. I did not enjoy his slow, slouching walk and high-over-everybodyishness. I had a quilt to sit on, with only his driver's turban to catch hold of if I got nervous; and when he went quite to the edge of the bank I felt my feet hanging over the rushing waters beneath. It did seem a risky seat; but there was really no danger, his back was so broad. He would have carried me down the steep bathing-steps to the water, where the holy fish eat the holy Brahmins, if I had not cried out, "Stop!"

The town was a perfect museum of rare old buildings, marvellously carved and painted. On one wall was represented the taking of some city by the English, who fired off cannon like pistols, and had a brandy-bottle under the other arm. We went on beyond the town and through the river, till we came to the last canal-dam. There we descended from the top of the big beast, and trusted ourselves to a small bamboo trustle bedstead fastened to the inflated skins of four huge beasts, something between buffalo and deer, which floated on the water, with their legs in the air, looking most helpless and imbecile. Head and all were shaped naturally, and all the flesh and bones extracted in some ingenious way through one leg, which was tied up and tightened with a stick. The Major and I sat on the frail bit of basket-work, and two men rested their bodies on the outer skins, paddling with their feet, and brought us down the river very quickly, perfectly smoothly, and safely. We were landed close to my home. After this first experiment, I had these two men, carrying their deer-skins on their heads and shoulders, following me every morning when I went to sketch, after which I floated back in the same easy way to breakfast.

Hardwar is like Benares, all on one side of the river, and from the opposite bank one gets the finest view of its strange old buildings, built at different times and by different races of Hindus, in all kinds of fantastic architecture, backed by wooded cliffs and hills. I used to work hard all day, and then went over to the other bungalow to dine with the Major in the evening. One night a tremendous thunderstorm came on. It was dark as pitch; every lantern was blown out as soon as lighted; the ground flooded with water; but as there was every prospect of its going on all night, I started right ahead, and nearly ran into an aloe hedge. The Major and his servants all tumbled about in other directions among the big trees, and, strange to say, got home safely in time. The next morning the clear river had turned into a torrent of yellow mud, and only the flattest of rafts could pass under the bridge over the surplus water of the canal. Many were dashed to pieces every year at that point. The great snow-peaks were perfectly clear and white after the storm.

Hardwar is a most enjoyable place for an artist, full of picturesque bits of street views. I went one morning into a room overhanging the river, with three fantastic windows and many-coloured hangings. A tomb was in the centre of the room, covered with green satin drapery, and a real live Fakir, entirely naked, was there too, with a long white beard and some dabs of yellow paint on him, who stared like a wild beast (as he was). I longed to stay and paint the scene, but it would have taken long, and the holy man might possibly have got hungry and eaten me up.

The people at Hardwar seemed a fine race, and were most friendly. I was bored by the attentions of the police, Captain B., their chief, having sent directions to all the inspectors round, telling them to report themselves to me (literally translated, they were for the time being to consider themselves under my orders). Deputations were continually pursuing me, asking me to write my name on a big envelope and state that I was satisfied with them. I had a perfect troop of people guarding me everywhere. I started again in a *dooly*—a box with shutters carried on four men's shoulders, with four others to relieve them. They carried me very badly, stopping each time they changed shoulders or men. We went through fifteen miles of jungle; got to the half-way bungalow after six hours, finding all the European rooms full (and I did not fancy having a windowless cell cleared of the natives, who were already sleeping like bundles on the floor); so I said I would sleep in my own *dooly* in a corner under the sky, and the English people there gave me a share of their supper.

At Dehra, where I stayed four days, I painted some more of the sacred plants, which I caught in flower. Dehra is famous for its bamboos, which, however, had followed the fashion and flowered the year before, so looked at their worst when I was there, though they were throwing up fresh canes from the roots. The old ones were very shabby, dead, or dying. It was the same year that all the English bamboos flowered and died, as well as those in Spain and France. In India they only died down and started afresh.

At Dehra I also had the luck to get rid of John! I had hardly seen him for some days (a common occurrence), and said to my landlord I feared he was ill again. "Ill, ma'am! He's drunk, and he's been dead drunk ever since you came." I was too glad of an excuse to be quit of him, so paid him his enormous wages, and his journey back to Bombay, and he took it quietly and went. I felt free again, and drove off to the foot of the hills, then was carried up the zigzags to Masūri, feeling determined to be tied to no more idle, lying servants, but to pay local people whenever I was well, who would take far better care of me than if I had a go-between like John, to make them do all his work, and only get half the money given for them.



END OF VOL. I

# *Textual Notes*

1. Roger North (1653–1734), a successful lawyer, became attorney general to James II's queen. With the revolution of 1689 his public career was over. In 1689 he bought Rougham, which is still the North family estate. He became a country gentleman and wrote—but did not publish—his own versions of the history of the period. His *Autobiography* was first published in 1887.

2. Roger North wrote biographies of his three successful older brothers, none appearing in print until the 1740s. Francis, also a lawyer, whose success in politics made him attorney general to the king and then Keeper of the Royal Seal, is immortalized in *Life of Lord-Keeper North*. The *Life of Sir Dudley North* tells of Sir Dudley's practical business career as a merchant who spent more than twenty years living in Smyrna and Constantinople and returned to become an MP. The *Life of Sir John North* is the story of the academic brother, a professor of Greek and Master of Trinity College who died before he was forty.

3. Vater Schmidt was Bernard Schmidt, a famous seventeenth-century organ builder who came from Germany to England, where he was known as Father Smith. Organ maker to the crown, Schmidt built many of the most famous organs in England, including the ones at Whitehall, Westminster Abbey, and St. Paul's Cathedral.

4. Dudleya North's collection of Oriental manuscripts is an instance of what Marianne North frequently presents as her heritage of family interest in places, and things, outside the West.

5. John Boydell was a famous eighteenth-century engraver on copper plates, particularly of landscapes and cityscapes. He is even more famous for becoming a print publisher of engravings by other English artists. The important result was that European collectors and

artists for the first time were able to become familiar with the accomplishments of English engravers. Boydell went on to become the Lord Mayor of London.

6. William Kent, like Bernard Schmidt and John Boydell, is another artistic name familiar to nineteenth-century English gentry. Naming these people at the opening of her *Recollections* is one way Marianne North establishes her artistic, and aristocratic, credentials. Kent was an early eighteenth-century painter (of the walls and ceilings of the mansions of the rich), a sculptor, and a landscape designer. Many people hired him, even though he was considered a terrible painter. Hogarth called him a "contemptible dauber." He is probably best known for doing the statue of Shakespeare in Westminster Abbey.

7. Dr. George Butler, a well-known educator in the early nineteenth century, was headmaster of Harrow, the elite private school for the children (boys only, of course) of the British ruling class, for twenty-four years.

8. Lucie Austin, whose married name would be Lady Lucie Duff Gordon, went on to fame when her health forced her to live for several years in Egypt. Her account of her life there, *Letters from Egypt,* first published in 1865, was a highly popular Victorian travel book.

9. The figure of the Methodist woman preacher roaming the English countryside in the early nineteenth century had been immortalized in the character of Dinah in George Eliot's 1859 novel *Adam Bede.*

10. Marianne North was explicitly not religious, which is why she couldn't see the "benefit." Her god was science.

11. Davies Gilbert, another eminent scientist, was president of the Royal Society from 1827 to 1830.

12. Phrenology was the study of the shape of, and bumps on, the head, believed to be indicators of character.

13. Mrs. Thomas John Hussey published a book on British fungi, *Illustrations of British Mycology,* which appeared from 1847 to 1855.

14. In August 1847 Marianne North was almost seventeen years old.

15. There were revolts against monarchies throughout Europe in 1848 and 1849, beginning in France.

16. The infamous Lola Montez, born in 1818 in Limerick as Marie Gilbert, was divorced for adultery and began dancing on the London stage in 1843. She toured Europe, and in 1847 the elderly King Ludwig of Bavaria saw her performance in Munich, fell in love, made her a citizen, then a baronne, then a comtesse, and gave her an income and a mansion. She effectively ruled Bavaria. After an insurrection the king was forced to abdicate in 1848 and Lola was banished. She then danced in England and America. In 1859 Lola renounced wicked ways, settled in New York to do good works, soon became ill, and died in 1861 at only forty-three.

17. In 1848 Vienna was the center of the Austrian revolutionary effort to free itself from the imperial authority of the king's rule.

18. Thomas and James Coutts founded the bank Coutts and Co. in the late eighteenth century. Bank to many of the British aristocracy, it supplied them with letters of credit which could be presented for cash at many European banks.

19. Michael Dahl (1656–1743) was a Swede who became a famous court portrait painter in London. Dahl was known for his accuracy, his rendering of his subject in a realistic rather than an idealized way. Like the other artists North mentions, Dahl created the kind of art North admired.

20. Mrs. Norton is probably Caroline Norton, granddaughter of Richard Brinsley Sheridan. Another infamous Victorian woman, Caroline Norton was involved in two famous court cases. First, her husband, who by all reports was very mean-spirited, accused her of adultery with Lord Melbourne. Both were acquitted. Caroline Norton and the case became the sources for George Meredith's successful novel *Diana of the Crossways*. Caroline Norton lived apart from her husband and made an excellent living as a popular writer. Then in 1853 her husband took her to court for a second time, now

to claim the proceeds from her writing. Indeed, in England for most of the nineteenth century divorce was impossible (unless you were married to Byron) and the husband had legal control of the children and wife's income. Caroline Norton then published a famous pamphlet exposing the injustice of *English Laws for Women in the Nineteenth-Century.* The case, the pamphlet, and Caroline Norton's courage in publicly battling her husband's cruelty are partly what led to changes in the laws during the latter half of the century which would grant married women their financial and parental rights.

21. The Great Exhibition of 1851 was the idea of Prince Albert, husband of Queen Victoria. In many ways a tribute to the new technology, it was a triumphant moment in Victorian culture. It consisted of exhibits of new and marvelous things sent from various "civilized" countries which the prince had invited to contribute, the whole housed in a gigantic building of iron and glass called the Crystal Palace. The United States sent, among other things, an artificial leg, false teeth, and a Colt repeating pistol.

22. William Henry Hunt (1790–1864) was primarily a watercolor painter, known for his faithful renderings of everyday people, familiar landscapes, fruit, and common objects. Hunt's "true to Nature" realism was an example of the kind of exactitude in art that North admired.

23. Edward Lear (1812–1888) began his career drawing illustrations of birds. Lear was a well-known visual artist who also wrote a whimsical book for the children of some friends, *A Book of Nonsense.* After 1836 Lear lived abroad, sketching and writing about his travels.

24. Francis Galton (1822–1911) was Charles Darwin's cousin and an explorer, travel writer, and amateur scientist in his own right. Very well known in the Victorian period, Galton coined the term *eugenics,* the study of race improvement. He wrote a book on *Hereditary Genius* in 1869. The idea of eugenics, or controlled evolution, was the basis of the Nazi concept of the master race.

25. Sir Thomas Brassey (1836–1918), later lord of the admiralty, was the son of Thomas Brassey (1805–1870, married to Maria), a famous

railroad contractor who made a huge fortune building railways all over the world. He started in England, then built them in France, Italy, Canada, Australia, and India. His daughter-in-law Annie Brassey also became a global traveler, primarily on the yacht the Brassey fortune made possible.

26. John Addington Symonds (1840–1893) was a famous Victorian intellectual, a bisexual, what we would now call an independent scholar, and Marianne North's brother-in-law. Bad health required that Symonds and his growing family live primarily in the warmer, drier climates of Europe. Symonds wrote extensively, including books on Dante, the Greek poets, and the history of Renaissance Italy. He wrote his own poems, translated Latin poems, and produced biographies of such well-known writers as Percy Shelley, Sir Philip Sidney, and Ben Johnson.

27. Elizabeth Gaskell (1810–1865) was a famous Victorian novelist. Her books include *Mary Barton, North and South, Ruth, Cranford, Sylvia's Lovers,* and *Wives and Daughters*. Gaskell was a contributor to the journal edited by Charles Dickens, *Household Words,* and wrote a biography of her friend Charlotte Brontë.

28. Mrs. S. was Marianne's American friend Mrs. Skinner. Laura Ponsonby has pointed out in *Marianne North at Kew Gardens* that the travelers parted company because Mrs. Skinner had terrible tantrums, the references to which were "taken out of Marianne's memoirs when they were published after her death" (p. 18).

29. Charles Kingsley (1819–1875) was a well-known minister and author of such popular novels as *Alton Locke* (1850), *Westward Ho* (1854), and *Water Babies* (1863). Kingsley had been to Brazil and the West Indies, and his book *At Last* (1870) is an account of that journey. Arranging for letters of introduction from friends, acquaintances, acquaintances of acquaintances, relatives, friends and acquaintances of relatives, etc., was to be Marianne North's constant method for ensuring hospitality and practical help from resident Europeans in many remote parts of the world. The method was startlingly successful.

30. Mrs. Skinner's son came to remove her and her luggage when she and North had finally had enough of each other during their trip to Canada (*Marianne North at Kew Gardens,* p. 18).

31. James Thomas Fields and his second wife, Annie Fields, were well-known supporters of the arts. Annie Fields, herself a published poet, was the friend of many famous poets, including Longfellow. She was a strong supporter of Anna Leonowens, whose memoirs of being a governess in Siam were the basis of *The King and I.* Fields had been a partner in Ticknor and Fields, the famous Boston publishing house. He was the editor of the *Atlantic Monthly* and an influential patron of the arts in nineteenth-century New England. Throughout the *Recollections* North tends to record meeting those figures who were the most visible or famous in the places she was visiting. Clearly, a key aspect of what she did on her travels was to meet the "important" people in the region.

32. Louis Agassiz was a naturalist and professor of natural history at Harvard. Although Agassiz actually opposed Darwin's ideas about evolution, he did much to promote the natural sciences as a major field of study in the United States.

33. Elizabeth Cabot Agassiz was Professor Agassiz's second wife. Her "famous Amazon expedition" was a trip with her husband to Brazil resulting in her book *A Journey to Brazil,* 1868. Clearly, North's conversations with Mrs. Agassiz are part of the reason why North's next trip from England would be to Brazil.

34. Dr. Emily Blackwell (1826–1910), another famous personage whom one visited as one visited Fifth Avenue, was a New York City doctor who wrote with her sister Dr. Elizabeth Blackwell *Medicine as a Profession for Women,* 1860.

35. Frederick Edwin Church (1826–1900), "the first of living landscape painters," was part of what would be known as the famous Hudson Valley School. He had paintings of tropical landscapes as well as those of the Hudson Valley.

36. The *Alabama,* built in a British shipyard, was a Confederate ship that caused great damage to United States merchant ships during the Civil War. After the war the United States insisted on British responsibility for allowing the *Alabama* and other ships to take to sea against it and demanded recompense for losses suffered by United States citizens. In 1871 the "*Alabama* Claims" were referred to independent arbitration in Geneva, establishing what would become a historic precedent for peaceably settling international disputes.

37. Frederick, Lord North, the "ex-Prime Minister of England," died in 1792. If Marianne North had been his youngest daughter, she would have been 101 in 1871.

38. Marianne North's hosts were Dr. and Mrs. Campbell.

39. Jamaica had been a center for the slave trade since the British developed its plantation economy in the late seventeenth century.

40. Sir John Peter Grant was the British governor of Jamaica. Craigton was a house of his with a fabulous view. Jamaica had been controlled by the British since 1655, when they captured it from the Spanish.

41. The "rebellion" was probably the riots of 1865. A few years after the riots, British rule in Jamaica was "modernized." Instead of the old-fashioned colonial structure according to which an appointed governor, attorney general, and a few others were sent out from England to rule with very few immediate restraints on their methods, Jamaica was put under direct control of the Colonial Office. This meant a more bureaucratic and regularized form of rule but not necessarily more justice for the Jamaicans. Jamaica did not achieve internal self-government until the 1950s. In 1962 it became an independent member of the British Commonwealth.

42. Di Vernon was probably Diana Vernon, the heroine of George Meredith's novel *Diana of the Crossways.* This admirable fictional character was modeled after the real heroine Caroline Norton. See note 20.

43. A.D.C.: an aide-de-camp, an assistant.

44. A medlar is a small tree with a fruit tasting somewhat like a crab apple.

45. "Obeah" people are those practicing a form of belief involving sorcery. Obeah is popular in parts of the West Indies.

46. The coolie couple from India who worked in sugar plantations as slaves are reminders that the British used slaves not only from Africa but from their other colonial possessions as well.

47. A marmoset is a common kind of monkey with a squirrel-like tail.

48. Slavery in Brazil in 1872 was still legal, though many prominent Brazilians, including the royal family, were abolitionists. Indian slavery in Brazil was declared illegal in 1831; the African slave trade was abolished in 1831 and again in 1850, the second time with ways to enforce it; and children born to slave mothers were declared to be born free in 1871. In practice, the plantation owners did much as they pleased.

49. The Botanic Gardens at Botofogo, just outside Rio, had an Austrian director who was very helpful to North.

50. The "halls of Karnac" are the gigantic temples of Karnak, the religious capital of the ancient Egyptian Empire. This would be a familiar reference to upper-class readers in Victorian England.

51. *Infra dig* is an abbreviated Latin phrase meaning beneath one's dignity.

52. The "Imperial Princess" was Princess Isabel, daughter of Pedro II of Brazil. Acting as regent of Brazil during her father's absence in Portugal in the spring of 1888, Isabel sanctioned the General Assembly bill, the Golden Law (Lei Aurea), that called for immediate and unconditional abolition of slavery in Brazil. Both the princess and her father were abolitionists. Powerful and bitter ex-slaveholders then oversaw a military coup d'état which eliminated

the monarchy in 1889 and set up a conservative republic. Princess Isabel had "redeemed a race but lost a crown" (Robert Brent Toplin, *The Abolition of Slavery in Brazil* [New York: Atheneum, 1972], p. 256).

53. Mr. Gordon was the manager of what was probably a gold mine.

54. "The Company" might have been British or Portuguese. The British had extensive business holdings in Portuguese Brazil, in part because they had long had extensive influence in Portugal.

55. "An English colony which possessed slaves" was not so uncommon up through the second half of the nineteenth century, in practice if not by law. There was legal slavery all through the Americas, North, Central, and South. There was slavery in Siam. There were systems of servitude in India and Asia (and North America after 1865) which were not slavery in name only. In all these places the masters were Europeans, including British. Perhaps what was uncommon was a slaveholding colony which was entirely British. Certainly, in most of the places North traveled, she stayed with European masters served by indigenous or imported peoples in virtual, if not legal, conditions of slavery.

56. This sketch of happy slave boys being fatted up for market makes it clear that if many Brazilians of Portuguese descent were strongly antislavery, as were many people in England, including the British Parliament, Marianne North was not.

57. The time of "political troubles" probably refers to the civil wars from about 1835 to 1845 between liberals and conservatives after the liberals forced the abdication of Pedro I in 1831 in favor of his young son. Dom Pedro reached his majority and become Pedro II in 1840, and after that the revolts slowly ended.

58. Sir Edward Sabine (1788–1883) was a British general who served in Canada and pursued his interest in ornithology and magnetism. He made two arctic expeditions and served as president of the Royal Society from 1861 to 1871.

59. The emperor of Brazil was Pedro II, the liberal ruler who, with his wife Thereza, visited Marianne North in her flat to see her paintings during their trip to London in 1877.

60. In the mid-nineteenth century Brazil tried to dominate Uruguay. The dictator of Paraguay opposed this and invaded Brazil in 1864. The result was a long, bitter, wasteful war from 1865 to 1870, with Brazil, Uruguay, and Argentina allied against Paraguay.

61. Alexander von Humboldt, perhaps the best German scientist of his generation, was noted for cowriting with Aimé Bonpland *Essai sur la geographie des plantes* (1807). The book prefigured ideas about evolution in its attempt to formulate the relations between classifications of plants and their geographic distribution. Humboldt's ideas would have been well-known at Kew.

62. Ashantee, where North's cousin Dudley was wounded, is Ashanti, a region of Ghana, and now part of the Gold Coast in Africa. The events there constituted a classic moment of British jingoism and invoked the British public's imperialistic pride. North's cousin was probably wounded in 1874, not in 1872; these memoirs were written years after North's actual journeys. The British in the 1870s occupied territories, many taken over by treaty from the Dutch, that bordered on what was then the kingdom of Ashantee. There were border problems, and in 1874 the British army invaded Ashantee to demonstrate their hugely superior military capabilities and to force the king to sign a treaty which allowed them to do as they liked in the region. By 1902 the British had taken over Ashantee completely. Many British soldiers came home from the 1874 invasion seriously ill with fever. Very few were actually wounded in action, because there was virtually no fighting.

63. John Tyndall (1820–1893), a major British natural scientist and famed lecturer, gave a controversial open address to the British Association at Belfast in 1874 on the relations between science and theology. North was always attentive to any public event where science seemed to triumph over religion.

64. M. E. is North's friend Mary Ewert. She stayed only two weeks in Teneriffe.

65. Sir Charles Grandison, the epitome of the perfect gentleman, is the lead character in Samuel Richardson's 1753–54 novel *Sir Charles Grandison*.

66. Brigham Young, who helped found the Mormons and led them to Utah, was a "horrid old wretch" probably because of his practice of marrying many wives.

67. "Man, the civiliser," is European man. North was an early environmentalist, but her preservationist sympathies were offered only to indigenous plants, not to indigenous peoples.

68. Sir Henry Parkes was the British minister to Japan from 1865 for the next eighteen years and "had a hand in most of the moves toward westernization" (Ian Nish, *Britain and Japan, 1600–1975*, vol. 1, *Historical Perspective* [London: Information Centre, Embassy of Japan, 1975]).

69. Sir Rutherford Alcock was Sir Harry Parkes's predecessor as Britain's representative to Japan, beginning in 1859. Alcock's appointment came after the 1858 Treaty of Edo (now Tokyo), which ended Japan's seclusion by granting trading rights with Britain.

70. The "new state of things" in Japan in the 1870s probably refers to an important moment in Japanese history, the resignation of the shogun and the restoration of the monarchy in 1868. This event signified the abandonment of feudal institutions in favor of a centralized government. From a Western perspective it signified that the Japanese were now more open to modernization and to international relations.

71. The "genuine Gamp" was Mrs. Sarah Gamp in Charles Dickens's novel *Martin Chuzzlewit*.

72. Durian is an indescribably fabulous fruit, perhaps the most delicious tasting, and repulsive smelling, in the world.

73. The "Straits" were the settlements of Singapore, Penang, and Malacca, called the Straits Settlements from the mid-1820s. In 1867 they were grouped together as a British crown colony. Labuan was added in 1907. The Straits, as they were called, remained a colony until 1946.

74. Sarawak was a section of the island of Borneo. In the nineteenth century Sarawak was ruled by the Brooke family, known as the white rajahs. Sarawak became a British crown colony after World War II, and in 1963 it became one of the thirteen member states of Malaysia (other sections of Borneo are now part of Indonesia).

75. Sir James Brooke, the first white rajah, was given the lands making up Sarawak by its then ruler, the sultan of Brunei, for helping him to end a long and vicious civil war and to drive out the pirates who terrorized the island.

76. The ranee was a young English girl who had married Charles Brooke, son of Sir James and the second white rajah, when she was only twenty and he, in his forties, had come to England to look for a wife to provide heirs for his beloved Sarawak. Margaret lost her first three children but then had three more, all boys. Having performed her wifely duty, she was sent back to England.

77. Alfred Russel Wallace (1823–1913) was one of the most famous Victorian naturalists, working both around the Amazon and then in the East Indies. His best-known book is the classic *The Malay Archipelago,* which appeared in 1869. In 1858 Wallace, down with fever in the Malay jungle, wrote a scientific paper on his theory of evolution by natural selection and mailed it to Darwin. Rather disturbed, Darwin, along with Joseph Hooker, had the paper presented publicly along with some of his own work, showing that he had also had this idea (which he had, for many years). Then, in a flurry, Darwin finally finished *The Origin of the Species,* published in 1859.

78. Labuan Island off Borneo is now part of Malaysia. Labuan would be the fourth Straits Settlement.

79. Sir Charles Lyell (1797–1875) was probably the most distinguished British geologist of the nineteenth century. His 1830–33 *Principles of Geology* described the divisions of the earth's strata as gradually formed over an unimaginably long time. Geology challenged many Victorians' faith in the biblical account of creation. The missing link was the hypothetical extinct creature, ape-man or man-ape, believed in the nineteenth century to link human beings to their ape ancestors.

    Collectors were people sent out from Kew, from other botanic gardens, or by private businesses to collect plants around the world.

80. Dr. Arthur Burnell, the Sanskrit scholar, was a friend of Edward Lear's whom North had also met by chance on a boat to Java. Dr. Burnell became one of her dearest friends. She visited him in India and for years she carried on a wittily frank correspondence with him. He died in his forties.

81. Batavia, the capital of Dutch-held Java or the Netherlands East Indies, is now Jakarta, the capital of Indonesia.

82. The Dutch residents in Java, unlike the British in their colonies, ruled by the indirect means. They functioned more as judges or policemen. While they certainly had final authority, direct control of provinces often was put in the hands of native authorities.

83. "Obliged by law" refers to Dutch law, of course.

84. Buitenzorg is now called Bogor. The biggest Dutch colonial garden was there.

85. Soerabaja is now called Surabaja.

86. Cinchona is the tree from South America which the British, through the botanists at Kew, imported to Southeast Asia to grow in plantations because its bark is the base for quinine, the drug that treated malaria.

87. Djocia, the "biggest town in Java, the residence of its native Sultan," is now Jogjakarta.

88. "Boro-Bodo, or Buddoer," is the monastery at Borobudur.

89. Ramadan, the holy month of fasting from dawn until dusk, the ninth month in the Muslim year, ends in a great feast.

90. Galle is a town on Ceylon, an island just south of India which became a British colony in 1798. The island finally achieved independence in 1948 as a member of the British Commonwealth and is now Sri Lanka.

91. Miss Gordon Cumming in Ceylon was a relative of the Victorian lion hunter R. G. Gordon Cumming. His 1850 book on being a lion hunter in South Africa made him extraordinarily famous.

92. North wrote to Joseph Hooker in January 1882 that "your friend Mr. Thwaites was my best friend in Java" (Royal Botanic Gardens, Kew: The Director's Correspondence, English Letters, 1859–1900, 161), but George Thwaites superintended the botanical gardens in Ceylon. He was part of the Kew connection.

93. Julia Margaret Cameron (1815–1879), an international award winning photographer, was born in Calcutta and married, then moved to England with her family in 1848. In 1875 she returned east, and she and her husband settled down in Ceylon. Mrs. Cameron took up photography when she was fifty. See the frontispiece for her wonderful picture of Marianne North.

94. Thomas Babington Macaulay, a well-known Victorian historian, spent 1834–38 in India, from 1834 to the end of 1837 as a member of the Supreme Council of India. He is credited (or discredited) with founding the educational system in India with a focus on English rather than Eastern courses of study and with establishing the Indian penal code.

95. Becky Sharpe is the self-centered, conniving, and charming heroine of Thackeray's *Vanity Fair.*

96. Julia Cameron died in 1879, a few years before North wrote the bulk of her *Recollections*.

97. The exhibition of North's paintings in one of the galleries of the Kensington Museum in 1877 before she went off to India required a great deal of cataloging work. This was the beginning of North's sense that she needed a gallery to display the paintings permanently.

98. North was to travel in India for over a year. She had left home in September 1877 and was back in England by March 1879.

99. Madura is now Madurai.

100. Anglo-Indians were the British who lived in India, many of whom were born there.

101. The gardens at Saharanpur were one of the many satellite gardens the British maintained in their colonies to cultivate commercial crops.

102. The second volume of the *Recollections* begins with an account of the continuation of North's trip to India, then goes on to recount the rest of her journeys.

# *Selected Bibliography*


Allan, Mea. *The Hookers of Kew, 1785–1911*. London: Michael Joseph, 1967.

Amherst, The Hon. Alicia. *A History of Gardening in England*. London: Bernard Quaritch, 1895.

Arac, Jonathan, and Harriet Ritvo. *Macropolitics of Nineteenth-Century Literature: Nationalism, Exoticism, Imperialism*. Philadelphia: Univ. of Pennsylvania Press, 1991.

*Athenaeum*. (London), no. 3280 (Sept. 6, 1890): 319, no. 3357 (Feb. 27, 1892): 269–70, no. 3425 (June 17, 1893): 755–56.

Atkins, Anna. *Sun Gardens: Victorian Photographs*. Text by Larry J. Schaaf. New York: Aperture, 1985.

Bean, William Jackson. *The Royal Botanic Gardens, Kew: Historical and Descriptive*. London: Cassell, 1908.

Bhabha, Homi K., ed. *Nation and Narration*. London: Routledge, 1990.

Bingham, Madeleine, Baroness Clanmorris. *The Making of Kew*. London: Michael Joseph, 1975.

Birkett, Dea. *Spinsters Abroad: Victorian Lady Travellers*. London: Basil Blackwell, 1988.

——. "A Victorian Painter of Exotic Flora." *New York Times*, Nov. 22, 1992, p. 30.

Bivona, Daniel. *Desire and Contradiction: Imperial Visions and Domestic Debates in Victorian Literature*. Manchester: Manchester Univ. Press, 1990.

Blunt, Wilfred. *In for a Penny: A Prospect of Kew Gardens, Their Flora, Fauna, and Falballas*. London: Hamish Hamilton, 1978.

Bonta, Marcia Myers. *Women in the Field: America's Pioneering Women Naturalists.* College Station: Texas A&M Univ. Press, 1991.

Brantlinger, Patrick. *Crusoe's Footprints: Cultural Studies in Britain and America.* New York: Routledge, 1990.

——. *Rule of Darkness: British Literature and Imperialism, 1830–1914.* Ithaca, N.Y.: Cornell Univ. Press, 1988.

Britten, James, and George S. Boulger. *A Biographical Index of Deceased British and Irish Botanists.* London: Taylor and Francis, 1931.

Brockway, Lucile H. *Science and Colonial Expansion: The Role of the British Royal Botanic Gardens.* New York: Academic Press, 1979.

Brooke, Margaret, the Ranee of Sarawak. *Good Morning and Good Night.* 1934. Rpt. London: Century Publishing, 1984.

——. *My Life in Sarawak.* 1913. Rpt. Oxford: Oxford Univ. Press, 1986.

*Catalogue of the Books, Manuscripts, Maps, and Drawings in the British Museum (Natural History).* London, 1933.

Chapple, J. A. V. *Science and Literature in the Nineteenth Century.* London: Macmillan Education, 1986.

Chaudhuri, Nupur, and Margaret Strobel, eds. *Western Women and Imperialism: Complicity and Resistance.* Bloomington: Indiana Univ. Press, 1992.

Conrad, Robert Edgar. *Children of God's Fire: A Documentary History of Black Slavery in Brazil.* Princeton, N.J.: Princeton Univ. Press, 1983.

——. *The Destruction of Brazilian Slavery, 1850–1888.* Berkeley: Univ. of California Press, 1972.

——. *World of Sorrow: The African Slave Trade to Brazil.* Baton Rouge: Louisiana State Univ. Press, 1986.

Cortazzi, Sir Hugh, and Gordon Daniels, eds. *Britain and Japan, 1859–1991: Themes and Personalities.* London: Routledge, 1991.

Crisswell, Colin C. *Rajah Charles Brooke, Monarch of All He Surveyed.* Oxford: Oxford Univ. Press, 1978.

*The Critic, a Weekly Review of Literature and the Arts* (New York: The Critic Company), 17 (n.s. 14), no. 352 (Sept. 27, 1890): 160, 23 (n.s. 20) (May 29, 1893): 203–4.

Darwin, Charles. *The Correspondence of Charles Darwin.* 6 vols. Cambridge: Cambridge Univ. Press, 1985–90.

Desmond, Ray. *Dictionary of British and Irish Botanists and Horticulturists, Including Plant Collectors and Botanical Artists.* London: Taylor & Francis, 1977.

*Dial* (Chicago), 13 (May 1892): 15, 15 (Aug. 1893): 64–66.

Dickens, Molly. "Marianne North (Illustrated)." *Cornhill Magazine* (London: John Murray), no. 1031 (Spring 1962): 319–29.

Duff Gordon, Lady Lucie. *Lady Duff Gordon's Letters from Egypt.* 1865. Rpt. New York: Praeger, 1969.

Ellegard, Alvar. *Darwin and the General Reader: The Reception of Darwin's Theory of Evolution in the British Periodical Press, 1859–1872.* 1958. Rpt. Chicago: Univ. of Chicago Press, 1990.

*The Garden, an Illustrated Weekly Journal of Horticulture in All Its Branches,* ed. Miss Jekyll and T. E. Cook (London: Hudson and Kearns), 58 (Christmas 1990): 300–301.

*Gardeners' Chronicle* (London: W. Richards), 23 (Jan. 17, 1885): 77–78, 24 (Sept. 5, 1885): 296.

Geikie, Sir Archibald. *Annals of the Royal Society Club: The Record of a London Dining-Club in the Eighteenth and Nineteenth Centuries.* London: Macmillan 1917.

Green, J. Reynolds. *A History of Botany, 1860–1900.* 1909. Rpt. New York: Russell & Russell, 1967.

Grosskurth, Phyllis. *John Addington Symonds: A Biography.* London: Longmans Green and Co., 1964.

Hemsley, W. B. "In Memory of Marianne North." *The Journal of Botany, British and Foreign,* ed. James Britten (London: West, Newman & Co.) 28 (1890): 329–35.

Hooker, Sir Joseph Dalton, ed. *Hooker's Icones Plantarum; or, Figures, with Descriptive Characters and Remarks, of New and Rare Plants, Selected from the Kew Herbarium.* 3d ser., vol. 5. London: Williams and Norgate, 1883–85, p. 57.

Hopkinson, Amanda. *Julia Margaret Cameron.* London: Virago Press, 1986.

Jameson, Fredric. "Third World Literature in the Era of Multinational Capitalism." *Pretexts: Studies in Writing and Culture,* 3, nos. 1–2 (1991): 82–104.

*The Journal of Horticulture, Cottage Gardener, and Home Farmer: A Chronicle of Country Pursuits and Country Life* (London: The Proprietor), 3d ser., 21 (July–Dec. 1890): 271–72, 51 (July–Dec. 1905): 364–65.

Lees-Milne, Alvilde. "Marianne North." *Journal of the Royal Horticultural Society* (London: Vincent Square), 89, pt. 5 (May 1964): 231–51.

Leonowens, Anna. *The Romance of the Harem.* 1872. Rpt. Charlesville: Univ. Press of Virginia, 1991.

Loudon, Jane Webb. *The Ladies Companion to the Flower Garden.* 3d ed. London: W. Smith, 1844.

——. *The Ladies Flower-Garden of Ornamental Annuals.* London: W. Smith, 1840.

——. *The Young Naturalist's Journey; or, The Travels of Agnes Merton and Her Mama.* London: William Smith, 1840.

McCracken, Donal. *Natal, the Garden Colony: Victorian Natal and the Royal Botanical Gardens, Kew.* Sandton, South Africa: Frandsen Publishers, 1990.

Meredith, George. *The Works of George Meredith.* 29 vols. New York: Charles Scribner's Sons, 1910.

Middleton, Dorothy. "Flowers in a Landscape." *Geographical Magazine* (London: Times Publishing Co.), 35, no. 8 (Dec. 1962): 445–62.

——. *Victorian Lady Travellers.* London: Routledge and Kegan Paul, 1965.

Mills, Sara. *Discourses of Difference: An Analysis of Women's Travel Writing and Colonialism.* London: Routledge and Kegan Paul, 1991.

Moon, Brenda E. "Marianne North's *Recollections of a Happy Life:* How They Came to Be Written and Published." *Journal of the Society for the Bibliography of Natural History* 8, no. 4 (1978): 497–505.

Morton, A. G. *History of Botanical Science: An Account of the Development of Botany from Ancient Times to the Present Day.* London: Academic Press, 1981.

*Nation* (New York), 54, no. 1405 (June 2, 1892): 417–18, 57, no. 1470 (Aug. 31, 1893): 162.

*Nature: A Weekly Illustrated Journal of Science* (London: Macmillan and Co.), 26 (June 15, 1882): 145–47, 155–56, 45 (April 28, 1892): 602–3, 48 (July 27, 1893): 291–92.

Nish, Ian. *Britain and Japan, 1600–1975,* vol. 1, *Historical Perspective.* London: Information Centre, Embassy of Japan, 1975.

North, Marianne, *Recollections of a Happy Life: Being the Autobiography of Marianne North.* Ed. by her sister, Mrs. John Addington Symonds. 2 vols. London: Macmillan, 1892.

———. *Recollections of a Happy Life: Being the Autobiography of Marianne North.* Ed. by her sister, Mrs. John Addington Symonds. 2 vols. London and New York: Macmillan, 1893.

———. *Recollections of a Happy Life: Being the Autobiography of Marianne North.* Ed. by her sister, Mrs. John Addington Symonds. 2 vols. London and New York: Macmillan, 1894.

———. *Some Further Recollections of a Happy Life: Selected from the Journals of Marianne North Chiefly between the Years 1859 and 1869.* Ed. by her sister, Mrs John Addington Symonds. London and New York: Macmillan, 1893.

———. *Some Further Recollections of a Happy Life: Selected from the Journals of Marianne North Chiefly between the Years 1859 and 1869.* Ed. by her sister, Mrs. John Addington Symonds. New York: Macmillan, 1894.

———. *A Vision of Eden: The Life and Work of Marianne North,* ed. Graham Bateman. Exeter: Webb and Bower, 1980.

Ponsonby, Laura. *Marianne North at Kew Gardens*. Exeter: Webb & Bower, 1990.

Pratt, Anne. *The Flowering Plants, Grasses, Sedges, and Ferns of Great Britain, and Their Allies, the Club Mosses, Pepperworts, and Horsetails*. London: F. Warne and Co., 1873.

———. *The Poisonous, Noxious, and Suspected Plants of Our Fields and Woods*. London: Society for Promoting Christian Knowledge, 1857.

———. *Wild Flowers*. London: Society for Promoting Christian Knowledge, 1852.

Pratt, Mary Louise. *Imperial Eyes: Travel Writing and Transculturation*. London: Routledge and Kegan Paul, 1992.

*Royal Botanic Gardens, Kew: Catalogue of the Library*. London: Darling & Son, 1899.

Royal Botanic Gardens, Kew:
The Director's Correspondence, Chinese and Japanese Letters, 1865–1900.
The Director's Correspondence, English Letters, 1854, 1857, 1859–1900.
The Director's Correspondence, Mascarene Islands Letters, 1866–1900.
Kew: North Gallery, 1879–96.
Letters to J. D. Duthie, vol. 2.
Letters to W. B. Hemsley, vol. 2.
Letters to J. D. Hooker, vol. 16.
Letters to the Shaen Family, 1875–84.
Marianne North Letters to Dr. Burnell, 1877–80.
Marianne North's Letters to A. R. Wallace.

Said, Edward. *Orientalism*. 1978. Rpt. New York: Random House, 1979.

Sangster, Ian. *Sugar and Jamaica*. London: Nelson, 1973.

*Saturday Review* (London), 73 (March 5, 1892): 282–83.

Schuler, Monica, *"Alas, Alas, Kongo": A Social History of Indentured African Immigration into Jamaica, 1841–1865*. Baltimore: Johns Hopkins Univ. Press, 1980.

*Scientific American* 63 (Oct. 18, 1890): 245.

Shanley, Mary Lyndon. *Feminism, Marriage, and the Law in Victorian England, 1850–1895*. Princeton, N.J.: Princeton Univ. Press, 1989.

*Spectator* (London), 68 (Feb. 27, 1892): 306.

Spivak, Gayatri Chakravorty. "Three Women's Texts and a Critique of Imperialism," in *"Race," Writing, and Difference,* ed. Henry Louis Gates, Jr. Chicago: Univ. of Chicago Press, 1986, pp. 262–80.

Stimson, Dorothy. *Scientist and Amateurs: A History of the Royal Society.* New York: Harry Schuman, 1948.

Strobel, Margaret. *European Women and the Second British Empire.* Bloomington: Indiana Univ. Press, 1991.

Symonds, Margaret. *Out of the Past.* London: John Murray, 1925.

Toplin, Robert Brent. *The Abolition of Slavery in Brazil.* New York: Atheneum, 1972.

Turrill, William Bertram. *The Royal Botanic Gardens Kew, Past and Present.* London: Herbert Jenkins, 1959.

Twining, Elizabeth. *Illustrations of the Natural Orders of Plants with Groups and Descriptions.* London: S. Low, Son, and Marston, 1868.

——. *The Plant World.* London: T. Nelson and Sons, 1866.

Weld, Charles Richard. *A History of the Royal Society, with Memoirs of the Presidents.* 2 vols. London: John W. Parker, 1848.

*Westminster Review* (London), 140 (July 1893): 93.